Helmut Günzler, Hans-Ulrich Gremlich

IR Spectroscopy

Further Titles of Interest

Helmut Günzler
Hans-Ulrich Gremlich

IR Spectroscopy

An Introduction

Translated by
Mary-Joan Blümich

Prof. Dr. Helmut Günzler
Bismarckstraße 4
69469 Weinheim
Germany

Dr. Hans-Ulrich Gremlich
Novartis Pharma AG
Analytics
WSJ-503.1001
4002 Basel
Switzerland

Translator: Mary-Joan Blümich

Library of Congress Card No.: applied for

A catalogue record for this book is available from the British Library.

Die Deutsche Bibliothek - CIP Cataloguing-in-Publication-Data
A catalogue record for this publication is available from Die Deutsche Bibliothek

ISBN 978-3-527-28896-0

Preface

This book is primarily intended for the beginning student as an introduction in using infrared (IR) spectroscopy for characterizing, identifying or determining a substance. The primary focus will be on spectroscopic techniques in the area of molecular vibrations. Besides basic knowledge in chemistry and physics, no special knowledge is presupposed. The reader will be first instructed about the design of various spectrometers as well as about the diverse methods for sample preparation and measurement. Finally, he or she will be introduced to the art of qualitatively interpreting the spectrum by suitable examples, whereby the latest developments using computers will be taken up. Sections about quantitative determinations, special application areas and about related methods within vibrational spectroscopy such as e.g. Raman spectroscopy are helpful in rounding off basic knowledge. References made to further literature sources will help orient the reader concerning special questions.

The individual topics are treated separately, so that the chapters can be read independently from one another. Corresponding side notes indicate when previous knowledge from prior sections is necessary. This quickly places the beginner in the position of operating a routine IR spectrometer and aptly preparing and measuring a sample. However, a systematic reading of the entire subject matter is recommended for the best use of the extensive possibilities offered by IR spectroscopy.

Theoretical considerations are inherently treated briefly in this introduction, although IR spectroscopy is considerably efficient in the field of determining molecular structures and dynamics, e.g. with using high-resolution Fourier transform (FT) and laser spectrometers. For more detailed coverage on this, reference is made to other works, because practical analytical questions predominate here.

Since the second edition has appeared, IR spectroscopy has again undergone further changes. For instance, today's laboratories are mainly equipped with FT spectrometers showing broadened possibilities regarding their attainable signal-to-noise ratio, spectral resolution and wider applicable spectral range from the visible to the far-infrared range. As the result of better instruments and techniques as well as the use of computers, the analytical laboratory is equipped with a versatile physical method for quickly identifying and quantifying substances having detection limits down into the picogram range. However, one must note the considerable expense of high-performance spectrometers needed to do the job, including accessories and their equipment with effective software.

The rapid development of IR spectroscopy has also opened up new application fields. Besides basic research in molecular spectroscopy and typical routine analysis, IR spectroscopy is used e.g. for process control in the analysis of animal feed and foodstuffs and in the clinical-chemical laboratory, to mention just a few new areas.

The variety of new techniques and instruments made it necessary to broaden the team of authors: H.M. Heise, scientist at the Institute of Spectrochemistry and Applied Spectroscopy in Dortmund assumed the tasks of presenting the Fourier transform techniques and special sample techniques. He also was in charge of revising the chapters about quantification, the techniques related to and bordering on mid-IR as well as about comparative spectra and expert systems. The second edition has been revised with new aspects incorporated in different places of the text. The proven concept of the introduction into the qualitative spectrum interpretation by Harald Böck, however, has been taken over from the second and first edition without considerable changes, whereby some parts as well as the literature have been updated. Hans-Ulrich Gremlich, head of the Optical Spectroscopy Laboratory at Novartis Pharma AG, Basel, Switzerland, has edited the present English translation.

We are pleased by the appreciative readers of our book and hope that the third revised edition will also be so kindly received. We welcome any suggestions or comments and would like to hereby thank those readers who have contributed towards improving this book through their constructive criticism.

Prof. Dr. Zlatko Meić (Zagreb) is commended for contributing the results of his normal coordinate analyses for benzene and the phenyl group. For constant support, we especially thank the Institute for Spectrochemistry and Applied Spectroscopy. Many colleagues have given valuable help in compiling this book, and we are particularly grateful to Dipl.-Phys. Andreas Bittner and Dipl.-Ing. Rüdiger Kuckuk (GC/FT-IR-coupling and spectral library search).

Dr. Helgard Staat was so kind to supply us with the violin-varnish spectrum, and Ms. Silke Sikorzynski transferred considerable revisions of the second edition into a text processing system. Dr. Wieland Hill gave us useful suggestions for improving Chapter 8. Special thanks are extended to Ms. Marianne Hillig who actively supported us with all of the additional figures in the new edition and who displayed remarkable perseverance and graphic competence.

Weinheim, Basel,
March, 2002

Helmut Günzler
Hans-Ulrich Gremlich

Contents

1 Introduction

1.1 Development of the Infrared Technique

Within the framework of his studies on the energy distribution in the spectrum of light, *Sir William Herschel* [1] performed an experiment, which would be fundamentally significant for the development of infrared spectroscopy:

Sunlight, entering the laboratory and passing through a prism, fell onto the surface of a table and was divided up into its spectral colors. Several mercury thermometers with blackened reservoirs were set up there for studying the thermal distribution. Surprisingly, the peak temperature was not found in the brightest observed range, i.e. in the yellow-green zone, but rather beyond red in the invisible radiation range. In further studies, Herschel proved that this did not concern a new type of radiation but rather obeyed the laws of optics like visible light. He named this range *infrared (IR)*.

Conducting studies that are more detailed was hardly possible as long as only the thermometer and, since 1830, also the thermoelement and the thermocolumn were available for detecting infrared radiation. More precise wavelength measurements could take place only after 1880 after *Langley* introduced the bolometer – a resistance thermometer. Measurements by *Rubens*, who penetrated the wavelength range around 300 μm by using the Reststrahlen-method named after him, inferred a continuous transition from the visible range of the spectrum to the infrared radiation range reaching to the Hertz waves.

The development of sufficiently sensitive detectors, the discovery of suitable prism materials and, finally, the introduction of the Echelette grating at the beginning of the 20th century were fundamentally important for the possibility of measuring well-resolved spectra.

But also then, infrared radiation was essentially a physical phenomenon and its thorough investigations were a domain of the physicists, insofar as the recording of a spectrum was very time consuming and tedious, demanding working nights in darkened and overheated basement rooms in order to maintain constant environmental conditions. For purposes of spectral analy-

sis, the infrared spectral range first gained importance, once
fully automated spectral photometers could be constructed. Such
an instrument was first developed in 1937 by *Lehrer* at BASF
AG in Luwigshafen on the Rhine [2]. Rapid development of instrumental techniques began after 1940, especially in the USA.
Consequently, it has been possible since 1950 to record well-resolved IR spectra within the course of a few minutes in standard
laboratory rooms. The first comprehensive and systematic cata-
log of infrared absorption spectra for analytical purposes was
published already in 1946 [3].

Whereas the accessible spectral region first began at 2–
15 µm and gradually advanced in stages to 50 µm only with the
help of various prism materials by using complicated and expensive conversion techniques, the range of between 2 and
50 µm can be continuously and fully automatically accessed by
using grating spectrometers. The latest development, beginning
in the sixties, allows analysis in the entire infrared range of between 780 nm and 2000 µm by applying the Fourier transform
(FT) technique.

The first use of an interferometer for spectroscopy dates back
to *Michelson*. A detailed account of the history of the interferometer has appeared on the 100th anniversary of his discovery
[4]. The advantages of the interferometric technique over the
dispersive technique were already recognized in the fifties
(among others, *Fellgett*). However, the first commercial FT
spectrometer appeared only after 1960. One problem was the
Fourier transform of the interferograms that was extremely complicated with the computers of the time. The rediscovery of the
FFT (Fast Fourier transform) algorithm in 1965 had enormous
consequences regarding broadened use of this interferometric
technique. As the result of the computer revolution in the eighties, Fourier transform has become so dominant that nowadays
only the term FT-IR is mentioned.

The interested reader may find a comprehensive and actual
historical survey of the development of the entire field of infrared technology in the account by *Jones* [5].

1.2   Applications
of Infrared Spectroscopy

IR spectroscopy is particularly important because of the high
information content of a spectrum and because of its variety of
possibilities for sample measurement and substance preparation.
Therefore, IR spectroscopy has become one of the most impor-

tant analytical methods for preparative as well as analytical che-
mists. It stands on the same level as nuclear magnetic resonance
(NMR) spectroscopy, mass spectrometry (MS) and ultraviolet
(UV) spectroscopy and, depending on the problem, can be used
alone or in a suitable combination there of in reaching the desired result. Knowledge of its possibilities and limitations is decisive for optimum application of IR technology [6]. Consequently, the following should give an introductory overview of
which information can be especially gained from the IR spectrum, to allow the reader to be quickly informed about the basic
applicability of the method for a special case. The reader is
kindly requested to excuse the unavoidable use of a few terms,
which will be defined in later chapters.

1.2.1　Direct Evidence on Constitution

By direct evidence, we mean inherent information in a spectrum
of an unknown sample that can be derived alone, without the
aid of comparative substances, through theoretical or empirical
correlations. In IR spectroscopy, such correlations exist between
the position of absorption bands within certain abscissa ranges
of the spectrum and particular structural groups. For instance,
the presence or absence of carbonyl functions, hydroxy groups,
amino groups, nitriles but also double bonds, aromatics and many
other structural elements can most likely be recognized upon first
glance. By considering other areas of the spectrum and, if need be,
by enlisting empirical correlation tables from the literature [7, 8],
a closer examination of the position and intensity of these bands
allows, in most cases, a more precise classification of the recognized structural group: ketone, acid or ester; primary, secondary
or tertiary alcohol; substitution type of aromatics etc.

The possibility of directly stating the structural groups, which
is often very more difficult or cannot be derived by other methods, constitutes the essence of IR spectroscopy and substantiates
it as one of the most important methods of instrumental analysis. The focus of the empirical application undoubtedly lies in
the clarification of the constitution of organic molecules. Inorganic atomic groups, too, show characteristic absorption spectra.

1.2.2　Substance Identification
by Spectral Comparison

The position and intensity of absorption bands of a substance
are extremely specific to that substance. Like a fingerprint of a
person, the IR spectrum is highly characteristic for a substance

and can be used for identifying it. The high specificity is based
on the good reproducibility with which the coordinates of the
absorption maxima (generally, wavenumber and transmittance)
can be measured. Of course, the best-suited bands for the identification are those, which can be classified to the carbon back-
bone of the molecule. These kinds of absorption bands are frequently found in the easily accessible wavenumber range of
about 1500 and 650 cm^{-1}, often called the "fingerprint region"
(see Sect. 6.3.1).

In addition, two factors are decisive for successful substance
identification through the IR spectrum:

1. The generally high number of absorption maxima occurring.
   Aside from molecules with high symmetry and a low number
   of atoms, at least 5, but mostly 10–30 and more bands appear in this range.
2. The large number of comparative IR spectra previously measured. The number of spectra categorized in different collections and published in the literature currently exceeds
   150 000 [9].

With such a large amount of comparative material, the spectral
search can hardly be handled manually anymore. Various systems were developed to make locating the spectrum of an unknown compound possible or, by using the empirical formula
(from the elementary analysis), to alleviate the search for a
comparative spectrum in corresponding ordered databases. The
use of electronic data processing technology has decisively advanced this area, so that today it is possible to compare within a
few minutes an unknown spectrum with the entire number of
catalogued and published IR spectra.

1.2.3 Quantitative Analysis

If I_0 is the intensity, or radiant power, of monochromatic radiation entering a sample, and I is the intensity transmitted by the
sample, then the ratio I/I_0 is called the *transmittance* of the sample. It is given the symbol T and labeled on the y-axis of a spectrum. The percent transmittance ($\%T$) is equal to $100T$. If the
sample cell has the thickness b, and the absorbing component
has a concentration c, then the fundamental equation governing
the absorption of radiation as a function of transmittance is

$$T = I/I_0 = 10^{-abc} .$$

The constant a is called the *absorptivity* and is characteristic for
a specific sample at a specific wavelength. This equation using

transmittance is usually transformed by taking the logarithm to the base 10 of both sides of the equation and replacing I/I_0 with I_0/I to eliminate the negative sign

$$\log_{10} I_0/I = abc \ .$$

The term $\log_{10} I_0/I$ is given the symbol A and is called the *absorbance*. The absorption law is called Bouguer-Lambert-Beer law or commonly *Beer's law* and is usually written as function of absorbance as

$$A = abc \ .$$

The product of the concentration c and the thickness b is a measure of the relative number of the absorbing molecules in the infrared beam. The unit of the absorptivity a varies with the units used for b and c. The absorbance A is alternately given by

$$A = \log_{10} 1/T \quad \text{or} \quad A = \log_{10} 100/\%T \ .$$

Beer's law is considered to be additive. In a mixture, the absorbance at a given wavelength is equal to the sum of the abc values for each component

$$A = \sum_i a_i b c_i$$

where the summation is over all the i components present. This implies that the radiation absorption by one component will not be affected by the presence of other components.

Two conditions are implied in the derivation of Beer's law. The first is that the resolution element being measured is monochromatic, i.e. the intensity of the region of the spectrum that is actually measured must have a small spread of wavelengths. The second condition is, that the absorptivity a should not change with concentration, e.g. by aggregation effects. If cell thickness and wavelength are held constant, a plot of concentration versus absorbance for a single component will be a straight line, if Beer's law holds. If Beer's law does not hold exactly, the plot will be slightly non-linear but can still be used for analyses. In general, calibrations are necessary which include the entire range of interest.

Based on these principles, each component of a mixture can be quantitatively determined from the IR spectrum, if a sufficiently intensive absorption band can be found that is not disturbed or is disturbed to a known extent by the other components of the mixture or by the solvent. Standard deviations of $s \geq 1\%$ can be attained under optimal instrumental conditions and if absorption bands are unaffected (see Section 7.3).

Because the evaluation is most often supported by a computer, more complicated multi-component analyses are also possible given a wide spectral range and a corresponding sample standard (so-called multivariate calibration; also see Section 7.5).

The concentration ranges and substance quantities that may be determined by using IR spectroscopy have changed drastically as opposed to earlier times when it was assumed that IR spectroscopy is unsuited for trace analyses. Quantitative analyses were mainly conducted for the concentration range of between 1 and 100%, whereby the absolute masses were in the microgram range. By means of the new techniques, concentrations of below 0.01% are determinable, whereby volume fractions in the ppb-range are measurable in the IR spectrometric gas analysis without enrichment steps e.g. by using special techniques. By using so-called cryotechniques, the minimum amount of substance needed for identifying many substances can be reduced to below 1 ng.

Time-dependent processes can also be followed by quantitative IR spectroscopic measurements, such as e.g. chemical reactions and order changes by stretching of a polymer film. Reaction rate constants and their order were often determined as well as their equilibrium positions and activation parameters. Another example is time-resolved studies of biomembranes during which structural changes could be immediately followed.

1.2.4   Other Uses

Aside from the aforementioned more practical uses of IR spectroscopy, important data about molecular design can be derived from the IR spectrum. This is obvious when one considers that an IR spectrum results from the interaction of electromagnetic radiation with the vibrations and rotations of the molecule ("rotational-vibrational spectrum"). Accordingly, data about the moments of inertia, bond lengths, energy constants and symmetry properties can also be calculated from the gaseous phase spectrum. In general, the spectra of the substance are used in various phase states.

The IR spectrum is particularly important for calculating thermodynamic constants, because a certain fraction of the specific heat of a substance is established in the vibrational energy of the atoms. This vibrational fraction of the specific heat can thus be calculated from the sum of the molecular vibrations, which can often be determined by using the Raman spectrum.

References

[1] Herschel, W., *Philos. Trans.* **1800** MDCCC, 284.
[2] Lehrer, E.Z., *Techn. Phys.* **1942**, 23, 169.
[3] American Petroleum Research Program 44.
[4] Giacomo, P., *Mikrochim. Acta III*, **1987**, 19.
[5] Jones, R.N., *Analytical Applications of Vibrational Spectroscopy-A Historical Review*, in Durig, J.R. (ed.), Chemical, Biological and Industrial Applications of Infrared Spectroscopy, Chichester: John Wiley & Sons, **1985**, p. 1.
[6] Gremlich, H.-U., *Infrared and Raman Spectroscopy*, in Ullmann's Encyclopedia of Industrial Chemistry. Vol. 5, 6th Edition, Weinheim: WILEY-VCH Verlag GmbH, **2000**.
[7] Colthup, N.B., Daly, L.H., Wiberley, S.E., *Introduction to Infrared and Raman Spectroscopy*, 3rd Edition, San Diego: Academic Press, **1990**.
[8] Lin-Vien, D., Colthup, N.B., Fately, W.G., Grasselli, J.G., *The Handbook of Infrared and Raman Characteristic Frequencies of Organic Molecules*, Boston: Academic Press, **1991**.
[9] Davies, A.N., *Nachr. Chem. Techn. Lab.* **1989**, 37, 263.

2 Absorption and Molecular Design

2.1 Fundamentals

2.1.1 Electromagnetic Radiation

2.1.1.1 Nature of Electromagnetic Radiation

From electricity studies, it is known that moving electrical charges induce magnetic fields and, inversely, that changes of the magnetic field create an electric field. Vibrating electrical charges therefore cause a periodic change of electromagnetic fields, which propagate as electromagnetic waves linearly into space with the speed of light. Depending on their appearance or their effect on material and on human senses, one speaks of various types of radiation (e.g. light, heat, X-rays) which differ from one another only with regard to wavelength or frequency but which are identical physically. The wavelength range of the electromagnetic spectrum comprises a wide scale ranging from radio-waves to γ-rays (Fig. 2-1).

2.1.1.2 Parameters and Units

Wavelength: Electromagnetic radiation is characterized by its wavelength λ with the base unit m. The wavelength is typically

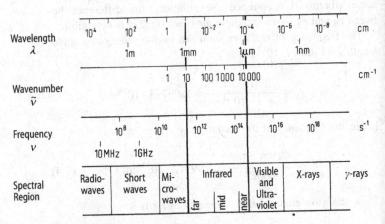

Fig. 2-1. The electromagnetic spectrum.

expressed in the units μm and nm. The unit μm is most practical for metric wavelength measurements in the mid-infrared (commonly just infrared (IR)) region. In the near-infrared (NIR) region, however, the unit nm is typically applied as in the visible (VIS) to the ultraviolet (UV) spectral region. These units of length as well as other useful units are given as follows:

$$1\mu m = 10^{-6}\ m = 10^{-4}\ cm = 10^{-3}\ mm = 10^3\ nm \qquad (2.1)$$

Frequency: Another typically applied parameter for characterizing wave motion is the frequency v, defined as the number of oscillations of the electric (or magnetic) radiation vector per unit of time (Fig. 2-1). The frequency unit is s^{-1} (oscillations per second), often specified in Hertz (Hz).

The proportionality constant of the correlation between frequency v and wavelength λ of radiation is its rate of propagation, namely the speed of light c having the unit cm s^{-1}:

$$v = \frac{c}{\lambda} \qquad (2.2)$$

In a vacuum, the speed of light is $c_0 = 2.99793 \cdot 10^{10}$ cm $s^{-1} \approx 3 \cdot 10^{10}$ cm s^{-1}. However, in a room filled with material, its value is smaller and can be calculated according to the equation

$$c_n = \frac{c_0}{n} \qquad (2.3)$$

whereby c_n denotes the speed of light in the corresponding medium and n the refraction index of the medium at the respective wavelength. Because the frequency v is a constant characteristic parameter for a vibration, but c_n depends on the refraction index of the medium, it follows according to Equation (2.3) that, in various materials, various wavelengths are measured for the same vibrational occurrence. Nevertheless, the difference between c_0 and c_n for the medium air ($n = 1.00027$) is negligible:

The frequency of radiation with the vacuum-measured wavelength $\lambda = 1$ μm $= 1 \cdot 10^{-4}$ cm is

$$v_0 = \frac{c_0}{\lambda} = \frac{2.99793 \cdot 10^{10}}{1 \cdot 10^{-4}} = 2.99793 \cdot 10^{14}\ s^{-1}$$

In air, the frequency of this radiation is

$$c_n = \frac{c_0}{n} = \frac{2.99793 \cdot 10^{10}}{1.00027} = 2.99712 \cdot 10^{10}\ cm\ s^{-1} \qquad (2.4)$$

Thus, the wavelength of this frequency in air is

$$\lambda = \frac{c_n}{v_0} = \frac{2.99712 \cdot 10^{10}}{2.99793 \cdot 10^{14}} = 0.99973 \cdot 10^{-4} \text{ cm}$$

The importance of the frequency v of the electromagnetic radiation lies in its direct relationship to the elementary phenomenon of IR spectroscopy: the interaction between the electromagnetic alternating field and the vibration of the atoms in the molecule. Because of the large numerical values ($\approx 10^{13} \text{ s}^{-1}$), this parameter is seldom used in practice.

Wavenumber: A third parameter that has gained acceptance especially in IR spectroscopy and that has become more important than frequency and wavelength is the wavenumber $\tilde{v}$, the number of waves per unit length (also see Fig. 2-1). The wavenumber expressed in cm^{-1} is the number of waves in a 1 cm-long wavetrain. The wavenumber $\tilde{v}$ is related to the wavelength λ by

$$\tilde{v} = \frac{1}{\lambda \text{ [cm]}} = \frac{1 \cdot 10^4}{\lambda \text{ [µm]}} \tag{2.5}$$

An important argument for using the wavenumber as parameter is its proportionality to the frequency v and thus to the energy of the electromagnetic alternating field described by

$$\tilde{v} = \frac{v}{c} \text{ cm}^{-1} \tag{2.6}$$

$$Wavenumber = \frac{Frequency}{Speed \ of \ light}$$

In IR spectroscopy, both wavenumber and wavelength are used, whereas in Raman spectroscopy (see Section 8.3) only wavenumber is employed.

2.1.1.3 Radiation Energy

For producing electromagnetic radiation, some type of energy has to be expended. Best known is the emission of radiation in the visible region by heated material such as e.g. the radiation of a white-hot tungsten wire of a lamp. Then too, energy released during a chemical reaction can be emitted as radiation (chemical luminescence). Thus, electromagnetic radiation is an energy carrier. Energy (unit: Joule) and radiation frequency have the following relationship:

$$E = h \cdot v = h \cdot \frac{c}{\lambda} \tag{2.7}$$

The parameter $h = 6.626 \cdot 10^{-34}$ Js is the Planck constant and c, the speed of light.

Radiation frequency and energy are therefore directly proportional to one another. The position of various types of radiation in the total electromagnetic spectrum also reflects their energy: the short-wave or high-frequency γ rays possess high energy, while the long-wave radiowaves have low energy (Fig. 2-1).

As during the creation of radiation, an interaction of the electromagnetic wave with material can result in an energy transfer, whereby the radiation energy $E = h \cdot v$ is absorbed in the molecular system of the material and e.g. is converted into thermal energy. As will be seen in the following section for the infrared region, this energy transfer is bound to very particular conditions. Thus, every material is permeable within a wide spectral range that means it doesn't interact with the radiation, while in other regions, a substance is totally or partially impermeable.

While the release of radiation energy is called *emittance E* of a sample, for most common infrared applications (near-, mid- or far-infrared) data are presented in a light absorption format. In practice, this involves comparing light transmission through the sample with the background transmission of the instrument. The output, normally expressed as *transmittance T*, is I/I_0, where I is the intensity (or power) of light transmitted through the sample, and I_0 is the measured intensity of the source radiation (the *background*) illuminating the sample. Instead of transmittance T, also the *absorbance A* is used, especially for quantitative measurements (see Section 1.2.3). Absorbance and transmittance are linked by $A = \log_{10} 1/T$ or $A = \log 10\ I_0/I$.

The absorbance or transmittance is a very characteristic, frequency-dependent property of every chemical substance. Its measurement and interpretation are the purposes of optical spectroscopy.

2.1.2 Molecular Design

It is known that atoms and molecules cannot always be described in simple pictures because of the often-complicated models of applied quantum mechanics with which atomic and molecular processes can be calculated. However, the models that will be considered here originate in classical mechanics, which can help us understanding simple spectroscopic relationships.

2.1.2.1 The Atom

For fundamentally understanding the interaction between radiation and material, it doesn't matter if one considers the atomic model from the viewpoint of quantum mechanics (Bohr) or that of wave mechanics (Schrödinger). The classic, graphic Bohr atomic model is chosen here for reasons of brevity, easy comprehension and practicality. According to this model, an atom consists of a positively charged nucleus encircled by negatively charged electrons or, as Sommerfeld postulated to broaden the model, a positively charged nucleus encircled by negatively charged electrons following eccentric-elliptical orbits. Because the number of electrons equals the atomic number, i.e. the ordinal number of the atom in the periodic table, the atom overall is electrically neutral.

Each of these electron orbits has a particular energy content E and is characterized by several so-called *quantum numbers* which are half or whole numbers. In other words, the energy of the electrons is quantized. If an electron jumps from one orbit to another, this occurs in defined quantum leaps, which obey exact selection principles and involve the uptake or release of energy according to the energy difference existing between the orbits in question:

$$\Delta E = E_1 - E_2 \qquad (2.8)$$

Here, E_1 and E_2 denote the energy of the electron in orbit 1 and 2, respectively.

The energy uptake can take place in different ways, e.g. by impact with other atoms but mainly by interaction with radiation. In the latter case, the frequency condition (frequency principle)

$$E_1 - E_2 = h \cdot v \qquad (2.9)$$

must be fulfilled, i.e. the frequency of the incident light must exactly correspond to the energy difference between the orbits concerned. The return of an electron from an excited high-energy level to a lower energy level results in the emission of radiation, likewise according to Eq. (2.9).

In wave mechanics, electron orbitals are replaced by wave functions which are calculated from the *Schrödinger equation* [1] and which represent the probability of finding an electron in a particular region. Quantum numbers (see Section 2.1.2.2), selection rules (Sections 2.1.2.2, 2.2.1.1, 2.2.1.2) and transition probabilities (Section 2.2.1.2) between the energy states result from this mathematical equation and can be calculated. This will be exemplified in the next section.

2.1.2.2 The Molecule

Molecules are compounds consisting of two or more atoms that maintain certain distances from another through interaction between the electrons of the outermost orbits. This results in other energy forms: vibrational and rotational energy.

In a molecule, the atoms are situated in a defined spatial position to one another. The distances of the atomic nuclei are determined by the sum of all forces, which act between all atoms. Through energy uptake, the atoms can be excited to vibrate around their state of equilibrium. Like the energy of the electrons, the vibrational energy is also quantized, i.e. it can assume only certain values described by quantum numbers. With excitation of vibrations by electromagnetic radiation, the frequency principle (Eq. 2.9) must likewise be fulfilled. The frequencies, which come into question, lie in the IR radiation region with wavelengths of between 780 nm and 50 µm (corresponding to wavenumbers of $12\,800-200$ cm^{-1}). The transition of a molecular vibration of an excited state into the next lower or into the ground state can take place by radiation emission or, radiation-free, by energy released to the surrounding area (e.g. for gases, conversion into translational energy by collision).

With the input of energy quantities that are lower than those needed for vibrational excitation, the molecules can only be induced to rotate. Especially for small molecules, the frequencies suited for exciting the likewise quantized energy levels of the molecular rotation partially lie in the longwave region of the IR spectrum, i.e. the far-infrared region (over 50 µm = below 200 cm^{-1}), but generally in the microwave region (see Fig. 2-1).

Molecular Vibration

For describing molecular motions, one starts out from simple physical models in which the atoms are thought to be points of mass kept together by weightless elastic springs. Here, the electrons are totally disregarded. The simplest case is a diatomic molecule given as the model of a harmonic oscillator, which can be physically conceived as a vibrating mass (m) on an elastic spring (Hooke's law). The spring is characterized by the elastic force constant k.

The corresponding vibrational equation is advanced by classical mechanics, the solution of which is represented by a sine motion of both points of mass (see Fig. 2-2). The total energy depends on the maximum displacement Δr_{max}. During the vibration, potential and kinetic energy (formula $V = 1/2k\Delta r^2$ and $T = 1/2m(dr/dt)^2$ with dr/dt as the velocity) are constantly interconverted, like in a pendulum, whereby the sum of both forms of energy, however, remains constant.

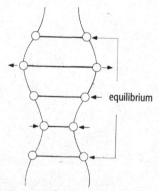

Fig. 2-2. Vibration of a diatomic homo-nuclear molecule around the equilibrium.

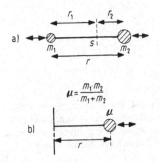

Fig. 2-3. Definition of reduced mass:
a) Weight model of a diatomic molecule.

m_1, m_2 atomic masses
r interatomic distance
s center of mass
r_1, r_2 distance of the atoms to the center of mass
b) Alternative model in which the reduced mass μ vibrates against a fixed wall or rotates around a fixed axis within the distance r.

Any given diatomic molecule can be mathematically described as a so-called one-dimensional harmonic oscillator; elucidated as follows: The origin of the molecular coordinate system is placed into the center of mass of the two-particle vibrator. By defining the center of mass coordinate s by $(m_1 + m_2)s = m_1 r_1 + m_2 r_2$ and given that $s = 0$, the new distances are denoted by r_1 and by $r_2 = -(m_1/m_2)r_1$. By additionally introducing the reduced mass $\mu = m_1 m_2/(m_1 + m_2)$ and the distance between the atoms, r, as internal coordinates (Fig. 2-3), one obtains a differential equation for the corresponding vibration

$$\mu \, d^2 r/dt^2 + k\Delta r = 0 \tag{2.10}$$

for which equation $r(t) = r_{max} \cos(2\pi vt)$, corresponding to preordained boundary conditions, can be found as a time-dependent solution. The frequency of the resulting vibration is calculated to be

$$v = 1/2\pi \sqrt{k/\mu} \tag{2.11}$$

Both atoms vibrate around the shared center of mass in opposing phase and, for various masses $m_1 \neq m_2$, with different amplitude.

Because classical mechanics do not suffice for describing the molecular motion, the description must be broadened to quantum mechanics models. For this transition, corresponding mechanical parameters are assigned in quantum mechanics to so-called operators, which act on the state function of the model system in question. The operator which corresponds to the total energy is called the Hamiltonian $H = T + V$. The stationary, i.e. time-independent energy states of a quantum mechanics system yield the solution of the so-called eigenvalue equation of this operator with $H\psi = E\psi$. This is the Schrödinger equation assigned to the system in which ψ represents the eigenfunction and E the corresponding energy value of the system. For a one-dimensional single particle system, the Schrödinger equation is given by

$$-\frac{\hbar}{2m}\frac{d^2\psi}{dx^2} + V(x)\psi = E\psi \tag{2.12}$$

where the Planck constant h appears in $\hbar = h/2\pi$. In general, the state function of a system is complex, so that the probability of finding the particle in the coordinate element dx can be specified by the product $\psi\psi^* dx = dW$, whereby ψ^* denotes the corresponding conjugated complex function. The functions are normalized so that for the probability summed over the entire area, the following applies: $\int dW = \psi\psi^* dx = 1$ (also see Fig. 2-4).

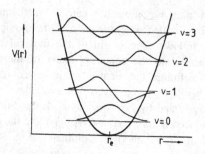

Fig. 2-4. Potential energy with the first discrete energy states and the corresponding eigenfunctions for the linear harmonic oscillator.

The following result is found for the discrete energy states produced, the energy eigenfunctions of the one-dimension harmonic oscillator:

$$E_v = hv(v + 1/2) \quad \text{with} \quad v = 0, 1, 2 \ldots \qquad (2.13)$$

The quantity v is the oscillator frequency from classical mechanics and the parameter v is the vibrational quantum number which may be all positive integers including zero, and which characterizes the ground state of the system. However, only certain transitions are allowed, namely $\Delta v = \pm 1$. These are the so-called selection rules, which can be derived from the time-dependent Schrödinger equation that describes the interaction of the oscillator with electromagnetic radiation.

As can be seen from Eq. (2.11) and assuming the same molecular geometry, the vibrational frequency is dependent on the atomic masses and the binding forces acting between them. In general, the constant k denotes the force constant in the molecular model. An example of possibilities for practical estimations is elaborated below.

Example: After isotoping an atom in the molecule, the changed vibrational frequency can be calculated (due to the proportionality, the corresponding wavenumber also), because for the same force constant, only the reduced mass is changed. Considered here is the well-localized stretching vibration of methanol, characterized by $v(OH)$ and of monodeuteromethanol $v(OD)$. The applicability of the equation $\tilde{v}_{OH}/\tilde{v}_{OD} = \sqrt{\mu_{OD}/\mu_{OH}}$ can be checked with the experimental wavenumbers. The reduced mass is $\mu_{OH} = 16/(16+1) = 0.941$ and $\mu_{OD} = 16 \cdot 2/(16+2) = 1.777$, respectively. The experimental values (wavenumber in cm^{-1}) are:

		Liquid	Gaseous
$CH_3 - OH$	$\tilde{v}_{OH}$	3328	3681
$CH_3 - OD$	$\tilde{v}_{OD}$	2467	2718

To become familiarized with vibrational frequencies and vibrational energies, the reader may figure the corresponding conversions for the vibration of carbon monoxide $v(CO)$:

Given is $\bar{v}=2143$ cm^{-1}, and the frequency is calculated at $v = \bar{v} \cdot c = 2143$ cm$^{-1} \cdot 3 \cdot 10^{10}$ cm s$^{-1} = 6.429 \cdot 10^{13}$ s^{-1}. By using Eq. (2.13), the energy of the relevant ground state is given to be

$$E_0 = 1/2hv = 0.5 \cdot 6.62 \cdot 10^{-34} \text{ J s} \cdot 6.429 \cdot 10^{13} \text{ s}^{-1}$$

$$= 0.213 \cdot 10^{-19} J$$

and the energy of the first excited state, with $v=1$, is three times larger than that of the ground state ($E_1 = 3/2 hv = 0.639 \cdot 10^{-19}$ J).

Molecular Rotation

According to classical mechanics, the *rotational energy* E_r of a molecule is defined by

$$E_r = 1/2I\omega^2 \tag{2.14}$$

with the moment of inertia I (see Fig. 2-3)

$$I = m_1 r_1^2 + m_2 r_2^2 = \frac{m_1 m_2}{m_1 + m_2} r^2 = \mu r^2 \tag{2.15}$$

and the angular velocity $\omega = 2\pi \, v_{rot}$.

The energy of the rotating system must also be calculated here according to the principles of quantum mechanics. The solution of the corresponding Schrödinger equation yields the discrete energy eigenvalues:

$$E_r = \frac{h^2}{8\pi^2 I} J(J+1) \quad J = 0, 1, 2 \ldots \tag{2.16}$$

The parameter J is the rotational quantum number. It may be all positive integers starting from zero.

Consequently, for the rotational energy, the following applies:
1. In contrast to the vibrational state, the molecule in the rotational ground state $J=0$ has no rotational energy.
2. The rotational energies are smaller the larger the moment of inertia $I = \mu \cdot r^2$ is.

2.1.3 Infrared Absorption and the Change in Dipole Moment

Based on Bohr's atomic model, it is obvious that electromagnetic radiation can interact with an electron, being a moving charge, and that energy can be released to it. Likewise, a vibrating or rotating atomic group can be associated with the motion of an electric charge, especially when the charges of the atoms in a molecule are asymmetrically distributed or when the charge distribution becomes asymmetric by the vibration of the atoms.

Consequently, electromagnetic radiation of the corresponding frequency can only then be absorbed by a molecule if a *change* in the dipole moment is associated with vibrational excitation of the atomic group concerned. Diatomic molecules with the same atoms, in principle, cannot be excited to vibrate, because they do not have any dipole moment. Molecules consisting of various types of atoms, however, can always interact with infrared radiation (Fig. 2-5a). When a dipole moment is not present from the very beginning, at least those vibrations are excited for which a dipole moment results from antisymmetric displacement of the center of charge (Fig. 2-5b). Vibrations showing no change of the dipole moment are characterized as infrared inactive (Fig. 2-5c).

The excitation of a molecule to rotate or the transition into a state of higher rotational energy can accordingly only then occur, when the molecule as a whole shows a dipole moment or a dipole moment is induced during a vibration.

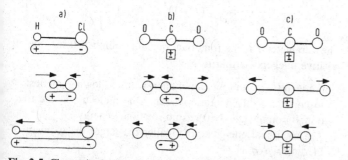

Fig. 2-5. Change in the dipole moment during molecular vibrations.
a) Hydrogen chloride: the dipole moment is changed during vibration.
b) Carbon dioxide, antisymmetric vibration: in contrast to the dipole-free equilibrium state, the negative center of charges of the O-atoms and the positive charge of the C-atom oscillate during vibration.
c) Carbon dioxide, symmetric vibration: the centers of negative (O) and positive (C) charges oscillate in phase: no change in the dipole moment.

2.2 Absorption of Infrared Radiation

2.2.1 Infrared Spectra of Diatomic Molecules

As was seen in Section 2.1.2.2, a molecule can absorb vibrational or rotational energy, whereby a transition occurs from energy state E'' of the particular atomic group to the energy state E' having a higher quantum number (Eq. 2.12 and 2.16). Because the frequency condition from Eq. (2.9) must be simultaneously fulfilled, the following applies by taking into account Eq. (2.6):

$$E' - E'' = h \cdot v = h \cdot \tilde{v} \cdot c \tag{2.17}$$

and

$$\tilde{v} = \frac{E'}{h \cdot c} - \frac{E''}{h \cdot c} \tag{2.18}$$

Each of both quotients in Eq. (2.18) consisting of energy, the Planck constant and the speed of light is defined as *term*. Thus, the wavenumber of the absorbed radiation directly results from the term difference of an energy transition.

2.2.1.1 Rotational Spectra

According to Eq. (2.16) and (2.18), a *rotational term F(J)* is defined as

$$F(J) = \frac{E_r}{h \cdot c} = \frac{h}{8\pi^2 c \cdot I} J(J+1) \tag{2.19}$$

To simplify this, one defines

$$B \equiv \frac{h}{8\pi^2 c \cdot I} = \frac{27.986}{I} \cdot 10^{-47} \text{ cm}^{-1} \tag{2.20}$$

as the *rotational constant*.

The difference of two terms with the rotational quantum numbers J' and $J''(J' > J'')$ is then

$$F(J' - F(J'') = BJ'(J'+1) - BJ''(J''+1) \tag{2.21}$$

Because the wavenumber of the absorbed (or emitted) radiation is given by the term difference according to Eq. (2.18) and (2.21), and because the selection rule

$$\Delta J = \pm 1 \qquad (2.22)$$

applies for the change of the rotational quantum number, the line positions in the absorption spectrum of the rigid rotator can be specified by the equation

$$\Delta F(J' = J + 1 \leftarrow J'' = J) = 2B(J + 1) \qquad (2.23)$$

Fig. 2-6a depicts the term scheme of the *rigid rotator* up to $J=5$. The terms are drawn as horizontal lines corresponding to their rotational quantum number. As shown by inserting $J=0,1,2...$ into Eq. (2.21), the first line lies at $\tilde{v}=2B$ and, as the reader can see, the distances of $2B$ are equidistant.

Because a molecule is not a rigid object, however, but rather the interatomic distance r and, therefore, the moment of inertia I increases with higher rotational energy E_r through centrifugal forces, according to Eq. (2.20) the rotational constant B decreases with increasing rotational quantum number J. The term differences ΔF thus increase to a lesser extent than that expected from the calculation for the rigid rotator, and the line distances become shorter with increasing quantum number (Fig. 2-6b).

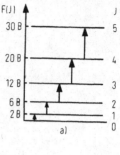

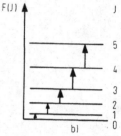

Fig. 2-6.
a) Term scheme of the rigid rotator.
b) Term scheme of the non-rigid rotator.

2.2.1.2 Vibrational Spectra

For the terms of an harmonic oscillator, a *vibrational term* $G(v)$

$$G(v) = \frac{E_v}{h \cdot c} = \tilde{v}\left(v + \frac{1}{2}\right) \qquad (2.24)$$

applies according to Eq. (2.18), (2.13) and (2.6).

The difference of two neighboring terms ($\Delta v=1$) is $\tilde{v}$, which infers an equidistant term scheme (see Fig. 2-7). For the selection principle applicable to the harmonic oscillator,

$$\Delta v = \pm 1 \qquad (2.25)$$

there, thus, should be only one single absorption band in the IR spectrum at the wavenumber $\tilde{v}$.

The harmonic oscillator model assumes that the attractive forces acting between the atoms of a molecule are proportional to their deviation from the equilibrium state. The attractive force between the atoms or their potential energy would accordingly increase infinitely with increasing distance (Fig. 2-8).

This, however, is empirically not the case, because the attractive forces acting between two atoms approach zero with sufficiently increasing distance, i.e. the potential energy reaches a

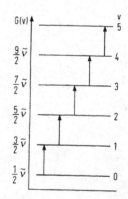

Fig. 2-7. Term scheme of the harmonic oscillator.

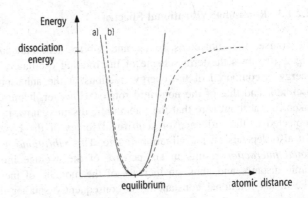

Fig. 2-8. Potential energy diagram of the a) harmonic and b) anharmonic oscillator (Morse potential).

boundary value. This boundary value is called the *dissociation energy*. If this energy quantity is applied to the molecule, the bond breaks. On the other hand, if the atoms near each other beyond the equilibrium ground state, repulsing forces start to act. Therefore, the potential energy increases exponentially on this side of the equilibrium state (Fig. 2-8).

Therefore, in reality an *anharmonic oscillator*, not a harmonic oscillator, is concerned here. Anharmonic oscillation occurs periodically but not in a sine wave and, consequently, the term differences decrease with increasing quantum number.

The most important difference between both oscillators is the selection rule, however. For the anharmonic oscillator, this selection rule is given by

$$\Delta v = \pm 1, 2, 3, \ldots \qquad (2.26)$$

This means that, for example, starting with the term $v = 0$ and aside from the transition to $v = 1$, transitions into higher terms are possible but with very greatly decreasing probabilities. This is elaborated in Fig. 2-9.

One refers here to overtones, because their frequency lies (although not exactly and generally somewhat lower) at the multiple value of the normal vibration ($v = 1 \leftarrow v = 0$), also called the fundamental vibration. In the IR spectrum, a fundamental vibrational band occurs at the wavenumber of the term difference $G(v = 1) - G(v = 0)$ as well as overtone bands, due to the anharmonicity of the vibration, at approximately the double ($\Delta v = 2$), triple ($\Delta v = 3$) etc. wavenumber but with greatly increasing intensity.

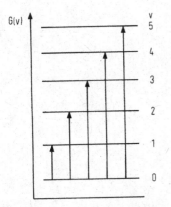

Fig. 2-9. Term scheme of the anharmonic oscillator with allowed transitions from the ground state.

2.2.1.3 Rotational Vibrational Spectra

Of course, the molecules rotate and vibrate simultaneously. Fig. 2-10 shows the term scheme of the rotating oscillator.[1] Its energy is composed of the energy fractions of the anharmonic oscillator and that of the non-rigid rotator. However, it must be taken into account here that the interatomic distance, now, is not subjected to the influence of centrifugal forces of the rotation but also depends on the vibrational state. This *vibrational-rotational interaction* results in an increase of the average interatomic distance and, thus, an increase of the moment of inertia. The actual rotational constant B_v is consequently smaller than the rotational constant which is denoted by B_e and applies for the equilibrium distance of the atoms.

$$B_v = B_e - a\left(v + \frac{1}{2}\right) \qquad (2.27)$$

In Eq. (2.27), a is a constant, which is smaller than B_e. However, one can recognize here that the difference between B_v and B_e always increases with increasing quantum number. The centrifugal distortion effect cited in Section 2.2.1.1 had been already mentioned as deviations from the rigid rotator, which depend on the rotational state of the molecule and diminish the term distance in contrast to the simple model.

The appearance of an IR spectrum in the absorption region of the rotational vibrations can be directly derived from the

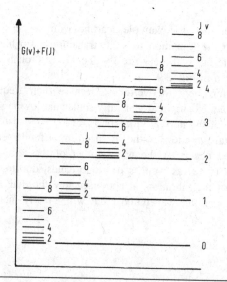

Fig. 2-10. Term scheme of the rotating oscillator (schematic! In a scaled presentation, the distances of the rotational energy levels are smaller relative to the vibrational energy levels).

[1] Because both forms of motion, i.e. vibration and rotation, are equal, the rotating oscillator can also be characterized as a vibrating rotator.

term scheme shown in Fig. 2-10, if the selection rules for the transitions are taken into account. These are already known from Eq. (2.22) and (2.26) for the rigid rotator and the anharmonic oscillator, respectively. For the rotating oscillator, $\Delta v = 0$ is also allowed now, so that the following selection rules apply:

$$\Delta J = \pm 1 \qquad (2.28\ a)$$

for the rotational transition, and in exceptional cases like paramagnetic molecules (NO), for example, also

$$\Delta J = \pm 0, 1 \qquad (2.28\ b)$$

and

$$\Delta v = \pm 0, 1, 2, 3 \ldots \qquad (2.29)$$

for the vibrational quantum number.

This means:
- The rotational transition, associated with a vibrational transition, takes place only into the next higher or into the next lower energy state.
- Vibrational transitions may occur between any given vibrational levels; the transition into higher terms corresponds to an energy uptake (absorption), a transition into lower terms corresponds to an energy release (emission).
- As the result of the selection rule $\Delta v = 0$, the rotating oscillator can also absorb corresponding energy quantities exclusively for rotational excitation. Besides the rotational vibration spectrum, there is therefore also a *rotational spectrum*, provided that a permanent dipole moment exists.

Fig. 2-11 shows the term scheme of a rotating oscillator for a given vibrational transition $\Delta v = 1$. Based on the selection rule cited above, the absorption spectrum of this vibrational transition consists of a larger number of individual lines, the wavenumber of which is produced, respectively, from the sum of the term differences of the vibrational transition and a rotational transition:

$$\tilde{\nu} = G(v') - G(v'') + B'_v J'(J' + 1) - B''_v J''(J'' + 1) \qquad (2.30)$$

$$\underbrace{\hspace{3.5cm}}_{\text{Vibrational transition}} \quad \underbrace{\hspace{4cm}}_{\text{Rotational transition}}$$

Here, B'_v and B''_v are the rotational constants affected by the atomic vibration in the upper and lower vibrational energy level, respectively, Eq. (2.27). The position of the individual lines in the spectrum is presented in Fig. 2-11b. One recognizes here, that $\Delta J = +1$ and $\Delta J = -1$ occurs for each line series. Both series

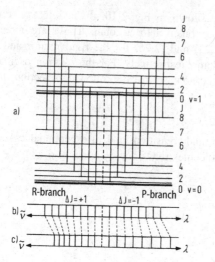

Fig. 2-11.
a) Term scheme of a rotating oscil-
   lator for the transition $G(v=1, J\pm 1) \leftarrow G(v=0, J)$;
b) Diagram of the rotational-vibra-
   tional band of this transition;
c) The same by considering the interaction between rotation and vibration.

are separated by a gap at the wavenumber corresponding to the prohibited, pure vibrational transition ($\Delta J=0$). The series with $\Delta J=1$ is called the *R-branch* and the series with $\Delta J=-1$ as the *P-branch*.

Due to the interaction between rotation and vibration (obtained in the difference between B'_v and B''_v, Eq. (2.30)), the line distances on both sides of the gap are unequal. With increasing distance from the center, the line distances decrease in the R-branch but increase in the P-branch (Fig. 2-11c).

Fig. 2-12 depicts the IR spectrum of hydrogen chloride as a typical spectrum of a diatomic rotating oscillator having a clearly recognizable gap in the center as well as decreasing and increasing line distance in the R- and P-branch, respectively.

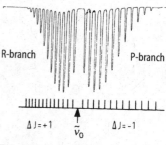

Fig. 2-12. Rotational-vibrational spectrum of hydrogen chloride (gaseous) for the fundamental vibration $v'=1 \leftarrow v''=0$ with $\tilde{v}_0 = 2884$ cm^{-1}.

2.2.2 Infrared Spectra of Multiatomic Molecules

2.2.2.1 Normal Modes of Vibration

By nature, diatomic molecules can perform only one single vibrational motion. The number of possible vibrational modes of multiatomic molecules can be calculated in the following simple way: each single atom can move in three spatial directions corresponding to $3N$ spatial coordinates for N number of atoms. Thus, a system of N points of mass has $3N$ degrees of freedom available.

In three of these movements, however, the atoms do not shift themselves relative to one another, but they all move in the same direction, thereby simultaneously changing the position of the center of mass. These are the *translational motions* of the molecule; in the gaseous aggregate state, these motions are subject to the laws of kinetic gas theory and do not belong in the field of molecular spectroscopy, because an interaction with electromagnetic radiation is impossible. Only the periodically repeating molecular translations in the crystalline association of a solid, the so-called lattice vibrations, are excitable in the long-wave IR spectral region and produce very intensive absorption bands. Another three movement combinations cause a rotation around the center of mass, so that the number of actual vibrational degrees of freedom is

$$Z = 3N - 6 \qquad (2.31)$$

Linear molecules, however, have only two rotational degrees of freedom, because the rotation around the molecular axis is not linked with any movement of the atoms or of the center of mass. Therefore, this molecular type has one more vibrational degree of freedom, namely

$$Z = 3N - 5 \qquad (2.32)$$

The number of vibrations of a molecule calculated in this way is called the molecule's *normal modes of vibration*, which can be excited independently from one another. Here, the atoms involved in the normal modes of vibration oscillate with the same frequency and move in phase relative to each other. A particular vibrational frequency is assigned to each normal vibration, . whereby the frequencies of different vibrations can nevertheless adopt the same value under certain prerequisite conditions, as will be seen in the next section.

2.2.2.2 Vibrational Modes

A triatomic linear molecule can accordingly carry out four vibrations (Fig. 2-13). In the vibrational mode v_1, both outer atoms of the molecule move symmetrically away either from or towards the valence direction of the central atom. If the masses of both these atoms are equal, the center of mass coincides with the center atom. Therefore, this type of vibration does not cause any movement. Because no dipole moment change is associated with this symmetric stretching vibration, it cannot be excited by electromagnetic radiation in the IR region. The vibration is infrared-inactive (but Raman-active, see Section 8.3).

For vibrational mode v_2, both outer atoms move in the same direction and thus antisymmetrically with respect to the central atom. The latter performs an opposing movement, whereby the center of mass is maintained. This *antisymmetric stretching vibration* is infrared-active.

In vibration v_3, the atoms move perpendicular to the valence direction, with the outer atoms moving in the same direction and the central atom in the opposite direction. This vibration results in a $180°$ change of the valence angle in equilibrium, which is why this vibrational mode is named *deformation vibration*. Because a dipole moment is also induced for this vibration, the vibration is infrared-active.

Inherently, vibration v_4 fully corresponds to vibration v_3; the direction of motion is turned only $90°$ out of the paper plane: whereas vibration v_3 occurs in the paper plane, the atoms in vibration v_4 move perpendicular to this plane. Because both deformation vibrations are described by various spatial coordinates, they are to be fundamentally seen as two different normal modes of vibration. However, if one considers the vibrational mode, it is easy to discern that the vibrational frequency in both cases must assume the same value. Such vibrations occurring with the same frequency are called *degenerate vibrations*, which are additionally characterized by *degrees of degeneracy*, i.e. the number of same-frequency normal modes of vibration. This case concerns a *two-fold degenerate vibration*.

Carbon dioxide is cited as a typical example for such a symmetric linear triatomic molecule. According to the following Table 2-1, both vibrational frequencies expected in IR, namely for the antisymmetric stretching vibration and the two-fold degenerate deformation vibrations, can be actually observed. The symmetric stretching vibration is only Raman-active.

In contrast, the distribution of mass and charge in carbon oxysulfide is asymmetrical. The molecule thus has a permanent dipole moment; both stretching vibrations are IR-active (Fig. 2-14 and Tab. 2-2).

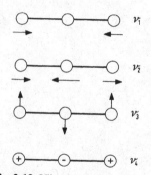

Fig. 2-13. Vibrational modes of a triatomic linear molecule.

Table 2-1. Normal modes of vibration of carbon dioxide [2]

Vibration	Wavenumber cm^{-1}	Vibrational mode	Remarks
1 v_1	(1285/1388)*	symmetric C=O stretching vibration	v_s IR-inactive
2 v_2	2349	antisymmetric C=O stretching vibration	v_{as} IR-active
3 4 $\}v_3$	667	deformation vibration	δ IR-active, two-fold degenerate

* Fermi resonance, see Section 2.2.2.4.

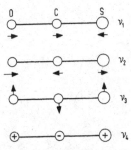

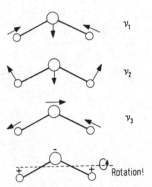

Fig. 2-14. Vibrational modes of a triatomic, linear molecule with asymmetric mass distribution.

Fig. 2-15. Vibrational modes of a triatomic non-linear molecule.

Table 2-2. Normal modes of vibration of carbon oxysulfide [2]

Vibration	Wavenumber cm^{-1}	Vibrational mode	Remarks
1 v_1	859	C=S stretching vibration	v (C=S) IR-inactive
2 v_2	2062	C=O stretching vibration	v (C=O) IR-active
3 } v_3 4	520	deformation vibration	δ IR-active, two-fold degenerate

A non-linear triatomic molecule is water. In this case, 3 N–6 = 3 normal modes of vibration are expected (Fig. 2-15 and Tab. 2-3). v_1 deals with a symmetric stretching vibration. Because of the angular structure, this vibration is connected with a change of the center of charges, so that the already existing dipole moment is altered. Therefore, the vibration is IR-active. Here, as also for the antisymmetric stretching vibration v_3, not only do the hydrogen atoms move but also the oxygen atom moves, to a necessarily lesser degree, for maintaining the center of mass. As one can easily see, only one single deformation vibration v_2 is conceivable, because a corresponding movement of the atoms perpendicular to the molecular plane would be the same as a rotation of the molecule. In agreement with Eq. (2.31) and in contrast to linear molecules, only 3 normal modes of vibration are possible here. The existing numbering of the normal modes of vibration is elucidated in Section 2.2.2.6.

Table 2-3. Normal modes of vibration of water (gaseous) [2]

Vibration	Wavenumber cm^{-1}	Vibrational mode	Remarks
1 v_1	3657	symmetric stretching vibration	v_s IR-inactive
2 v_2	1595	deformation vibration	δ IR-active
3 v_3	3756	antisymmetric stretching vibration	v_{as} IR-active

2.2.2.3 Overtones and Combination Vibrations

As already mentioned in Section 2.2.1.2, besides a vibrational transition into the neighboring term, the selection rule also allows quantum jumps into higher terms for the anharmonic oscillator as well as for the rotating oscillator. These transitions with

$\Delta v = +2, 3 \ldots$ are shown as absorption bands of the aforementioned overtones. Their intensities are much less than those of the fundamental vibration, but with accordingly high pathlength, particularly the overtones of CH- and OH-stretching vibrations that lie in the region of between 2800 and 3600 cm^{-1} can be observed very well in the so-called near-infrared (NIR) region (780–2500 nm, corresponding to 12 800–4000 cm^{-1}). Overtones of vibrations, the fundamental vibrations of which occur at smaller wavenumbers, still fall in the mid-infrared (IR) region, however, and thus are often not easily differentiable from fundamental bands. For example, the 1st overtone of the C=O stretching vibration of hexanal ($v(CO) = 1725$ cm^{-1}) is observed at 3433 cm^{-1} (Fig. 6-44).

Aside from the fundamental vibrations and overtones, still other mostly weaker bands occur, however, which approximately originate from a combination of a single or multiple frequency quantity of two or more normal modes of vibration, i.e. *combination vibrations*:

$$v_{kombi} = a \cdot v_1 \pm b \cdot v_2 \pm c \cdot v_3 \pm \ldots \qquad (2.33)$$

Binary combination bands, e.g. $v_i + v_j$, result from two different fundamental vibrations being simultaneously excited. Such combination vibrations are found regularly e.g. for benzene and derivatives in the region of between 1600 and 2000 cm^{-1}. The relationship between these combination vibrations and the corresponding fundamental vibrations is exemplified with polystyrene in Fig. 2-16 [3]. This hereby concerns the combination of respectively two fundamental vibrations in the region of between 800 and 1000 cm^{-1}.

As for overtones, the combination vibrations are not found exactly at the calculated wavenumber but rather somewhat lower, due to the anharmonicity.

2.2.2.4 Fermi Resonance

As previously discussed in Section 2.2.2.2, vibrations that inherently show the same frequency are called *degenerate*. Nevertheless, fundamental vibrations and overtones of various vibrational modes can accidentally possess the same energy and thus lie at the same frequency. These vibrations are named *accidentally degenerate*. The result of this accidental degeneracy is a repulsion of the energy level, a cleavage of the frequency of both vibrations and thus a splitting of the bands. This is connected with the fact that the actually weaker overtone borrows intensity from the fundamental vibration, so that two bands with similar intensity result. Such a resonance splitting is observed for the

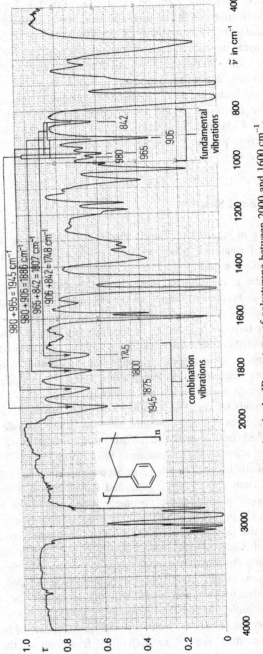

Fig. 2-16. Assignment of combination vibrations in the IR spectrum of polystyrene between 2000 and 1600 cm^{-1}.

symmetric stretching vibration ν_1 of CO_2 (Tab. 2-1, Fig. 2-13). The first overtone of the deformation vibration ν_3 coincides with the fundamental vibration ν_1 at ca. $2 \cdot 667 = 1334 \text{ cm}^{-1}$. These interact with one another, and two bands shifted by about $\pm 50 \text{ cm}^{-1}$ are observed at 1286 and 1389 cm^{-1} only in the Raman spectrum due to the symmetry.

The aldehydes are an often-observed example of this in practice: The frequency of the 1st overtone $2\delta(\text{OCH})^2$ coincides with the C–H stretching vibration $\nu(\text{C–H})$ of the aldehyde group; consequently, a double band occurs at ca. 2770 and 2830 cm^{-1} (see Fig. 6-44).

2.2.2.5 Band Shape

As shown in Figure 2-12, the IR absorption band having sharply delineated rotational transition lines (rotational fine structure) occurs only for gases.

In order to measure the line positions exactly, low pressures and a high spectral resolution of the spectrometer are preferentially applied. At these low pressures, one finds only a small, so-called *Doppler broadening* of the lines attributed to the distribution of the kinetic energies of the molecule to be measured. Increasing pressure causes a further line broadening resulting from more frequent molecule collisions. One calls this phenomenon *pressure broadening* of the absorption bands.

For liquids in which the rotation is strongly hindered, one no longer observes any single rotation lines but rather a single relatively broad, unstructured band. With the transition into a solid crystalline state where the rotation is fully frozen, a definite narrowing of the line half-width again occurs (also see Fig. 2-17).

Aside from the aggregate state and the effects of a limited spectroscopic resolution, the symmetry properties of the molecule strongly influence band structure, especially when the substances exist in gaseous state. The hereby-occurring rotational vibrational bands mostly show a fine structure, which can be theoretically described in a model. This is presented in detail in various monographs [4-7]. For the non-linear molecules, one differentiates between the asymmetric-top molecule, the symmetric-top molecule and the spherical-top molecule, depending on whether the principle moments of inertia (moments of inertia around the three axes of inertia) differ from one another or whether two or all three are equal in size. For each one of these molecular types distinguished from each other by symmetry properties, other quantum mechanical models and corresponding selection rules for the proper rotational quantum numbers apply.

---

[2] $\delta(\text{OCH})$: O=C–H deformation vibration.

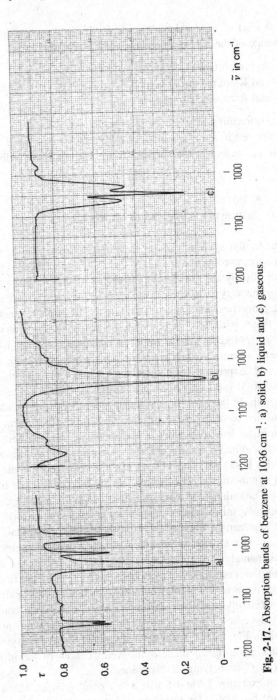

Fig. 2-17. Absorption bands of benzene at 1036 cm^{-1}: a) solid, b) liquid and c) gaseous.

Which kind of band exists depends on this and on the direction of change of the dipole moment relative to the top axes.

The most important types are:
- P- and R-branch, no Q-branch
- P- and R-branch, strong Q-branch.

The relationships between band type and symmetry are practically applied in the assignment of the vibration and the determination of molecular parameters from the rotational vibration spectrum.

2.2.2.6 Symmetry Properties of Molecules

The molecular symmetry plays such an important role in vibrational spectroscopy that a few of the basics must be discussed. As was already apparent in the vibrations of the triatomic molecules, symmetry properties exist, such as symmetric and asymmetric properties, which also determine the IR or Raman activity.

To begin with, lets us consider the so-called symmetry elements of molecules. These must be taken into account for the symmetry operations that transform a molecular structure again into itself. Expressed in mathematical terms, one says that the amount of all symmetry operations assigned to a molecule make up a group. In the case of vibrational motions of a molecule, let us consider so-called point groups; this means that during the vibrations, at least one point in the molecule, i.e. the center of symmetry, remains unaffected. To elucidate, we will take the molecules H_2O, NH_3 and CO_2 as examples. In general, the molecular symmetries can be covered by the elements listed in Table 2-4. Particularly for water, we find a C_2-axis, as well as two perpendicular mirror planes denoted by σ_v and σ_v' (Fig. 2-18).

The symmetry elements are similar for the pyramidal NH_3 molecule. The C_3-axis present here implies two symmetry operations, namely a $60°$ clockwise rotation and one in the other direction that corresponds to a double application of the first. In addition, three mirror planes exist in which the rotational axis and respectively one hydrogen atom lie. Much more complicated is the case for CO_2. Some of its symmetry elements are illustrated in Fig. 2-19. Shown here are the C_∞-axis, the symmetry center i, the mirror plane σ_h as well as three of the infinite number of C_2-axes lying in this plane.

As an example, Table 2-5 lists a few point groups, which are important for molecular vibration assignments. Likewise specified are the corresponding symmetry elements, whereby the numerals before them denote their respective number.

We can gain further insights into the group-theoretical view of molecular symmetry by using the symmetry operations on

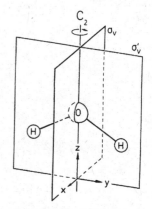

Fig. 2-18. Symmetry elements for H_2O (point group C_{2v}).

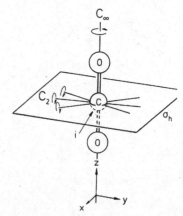

Fig. 2-19. Some symmetry elements for CO_2 (point group $D_{\infty h}$).

Table 2-4. Symmetry elements and operations

Symmetry elements		Symmetry operation
Symbol	Description	
E	identity	no change
i	center of symmetry	inversion at the center
σ	plane of symmetry	reflection at a plane
C_n	axis of symmetry	rotation around the axis by $360°/n$
S_n	rotation-inversion axis	rotation around the axis by $360°/n$ with subsequent reflection at a plane perpendicular to the rotational axis

Table 2-5. Important point groups in vibrational spectroscopy

Point group	Symmetry elements	Examples
C_1	E	CHFClBr
C_2	E, C_2	H_2O_2
C_s	E, σ	CH_3CH_2CN
C_{2v}	$E, C_2, 2\sigma_v$	H_2O
C_{3v}	$E, C_3, 3\sigma_v$	NH_3
$C_{\infty v}$	$E, C_\infty, \infty\sigma_v$	OCS
C_{2h}	E, C_2, σ_h, i	BrHC=CHBr(trans)
D_{2d}	$E, 3C_2, S_4, 2\sigma_d$	$H_2C=C=CH_2$
$D_{2d} \equiv V_h$	$E, 3C_2, 3\sigma, i$	$H_2C=CH_2$
D_{6h}	$E, C_6, 6C_2, 6\sigma_v, \sigma_h, C_2, C_3, S_6, i$	C_6H_6
$D_{\infty h}$	$E, C_\infty, \infty C_2, \infty\sigma_v, \sigma_h, i$	CO_2
T_d	$E, 3C_2, 4C_3, 6\sigma, 3S_4$	CH_4
O_h	$E, 3C_4, 4C_3, 3S_4, 3C_2, 6C_2, 6C_2, 9\sigma, 4S_6, i$	SF_6

molecular motions. To begin, let us consider simple translations in the atomic Cartesian coordinate system as was already illustrated in Fig. 2-18. A movement in the direction of the z-axis is given by T_z, while for example a rotation around the y-axis is labeled with R_y. In the case of the H_2O molecule, which does not show any degenerate vibrations, the result of the symmetry operations on the translation directions, represented by vectors or "arrows" on the atoms, can mean that either these remain unaffected or their direction becomes reversed. Both these possibilities are given mathematically by "+1" for the symmetric and with "−1" for the antisymmetric result. The types of vibrations occurring for water (see Fig. 2-15) can be transformed similarly, as can be easily seen. These transformation properties are typically presented in the so-called character tables, as exemplified below.

Table 2-6 depicts the corresponding information for the point group C_{2v}, which applies, among others, to H_2O. In the character

Table 2-6. Transformation behavior of a molecule with C_{2v}-symmetry (e.g. H_2O) with respect to the symmetry operations of its point group

C_{2v}	E	C_{2v}	σ_v	σ_v'		
A_1	1	1	1	1	T_z	a_{xx}, a_{yy}, a_{zz}
A_2	1	1	−1	−1	R_z	a_{xy}
B_1	1	−1	1	−1	R_y, T_x	a_{xz}
B_2	1	−1	−1	1	R_x, T_y	a_{yz}

table, the so-called symmetry species are listed in the first column to which the individual vibrations can be assigned. The first symbol (here A_1) is typically the total symmetric species. The next differentiation is based on the transformation properties of the symmetry operation, which includes the symmetry axis of greatest multiplicity (A symmetrical, B antisymmetrical). By agreement, A and B denote non-degenerate, E denotes two-fold degenerate and F, three-fold degenerate characters. Indices often used here are g and u that signify even and odd, respectively.

Besides the assignments of the molecular translations and rotations to the various species, the components of the polarizability tensor are listed which is fundamentally important for Raman spectroscopy, as we will still see in Section 8.3. What is important to know here is that the translations T_x, T_y and T_z transform themselves like the corresponding Cartesian coordinates x, y, z and the dipole moment components μ_x, μ_y and μ_z. Statements on the IR- and Raman activity of fundamental vibrations can be derived from the information in the character tables. Such a vibrational transition is allowed for the Raman spectrum, when corresponding components of the polarizability tensor are found for the fundamental vibration. These dependencies can be explained by quantum mechanics and group theory, but they should not be further elucidated here. However, important for molecules with a symmetry center is that IR-allowed transitions are forbidden in Raman spectroscopy and vice versa. This exclusion principle for vibrational bands can be usefully applied for the determination of structures with particular symmetry, e.g. whether they are linear or bent.

The following assignments can be made for the fundamental vibrations of H_2O (for their special denotation, lower case letters will be generally used for the symmetry characters, in contrast to that agreed upon for combination bands): a_1: v_1 and v_2, as well as b_2: v_3. The nomenclature and numbering of the fundamental vibrations of multiatomic molecules is typically oriented to the symmetry characters. First listed here are the total symmetrical vibrations with decreasing wavenumber and next, according to the same pattern, the subsequent characters in the character table.

Table 2-7. Correlation between the symmetry species of benzene- and phenyl vibrations

C_{2v}	D_{6h}	C_{2v}	D_{6h}
A_1	A_{1g}	B_1	A_{2u}
	B_{1u}		B_{2g}
	E_{1u}		E_{1g}
	E_{2g}		E_{2u}
A_2	A_{1u}	B_2	A_{2g}
	B_{1g}		B_{2u}
	E_{1g}		E_{1u}
	E_{2u}		E_{2g}

Table 2-8. Distribution of the symmetry species of fundamental vibrations of benzene and of monosubstituted benzenes (phenyl group)

Benzene, C_6H_6- (Point group D_{6h}):
in-plane vibrations (14): $2a_{1g}+a_{2g}+2b_{1u}+2b_{2u}+4e_{2g}+3e_{1u}$
out-of-plane vibrations (6): $a_{2u}+2b_{2g}+e_{1g}+2e_u$

Phenyl, C_6H_5X-(Point group C_{2v}):
in-plane vibrations (21): $11a_1+10b_2$
out-of –plane vibrations (9): $3a_2+6b_1$

Another example is presented here. As we can see in Table 2-5, D_{6h} is the point group for benzene. If one substitutes benzene, a reduction in symmetry generally results for the derivative. In the case of monosubstituted benzenes, a phenyl group, C_6H_5X is formed which, in turn, can be assigned to the point group C_{2v}. Thus, the selection rules for the activity of the fundamental vibrations are changed, so that all vibrations of a phenyl group become IR- and/or Raman-active. The correlation between the symmetry characters of benzene and phenyl vibrations is shown in Table 2-7. Upon substitution, the degeneracy is missing, and the number of the total active vibrations increases from 20 to 30 (see Table 2-8). Further details on these fundamental vibrations and their assignments are found in Section 6.8.

A detailed consideration of the group-theoretical possibilities in vibrational spectroscopy as well as data on all character tables would go beyond the scope of this introduction. Therefore, the interested reader may refer to other monographs [1, 4, 5] that give detailed information.

References

[1] Barrow, G.M., *Introduction to Molecular Spectroscopy.* New York: McGraw-Hill, **1962**.

[2] Schimanouchi, T., *Table of Molecular Vibrational Frequencies.* Consolidated Vol. 1, Nat. Stand. Ref. Data Ser., Nat. Bur. Stand. (U.S.), *39*, **1972**.

[3] Linag, C.Y., Krimm, S., *J. Polym Sci.* **1958**, *27*, 241.

[4] Hollas, J.M., *Modern Spectroscopy,* 2nd Ed. Chichester: John Wiley & Sons, **1992**.

[5] Herzberg, G., *Molecular Spectra and Molecular Structure,* Vol., 1: *Spectra of Diatomic Molecules,* 2nd Ed. New York: Van Nostrand Reinhold, **1950**. Vol. 2., *Infrared and Raman Spectra of Polyatomic Molecules,* 4th Ed. New York: **1949**.

3   Spectrometers

3.1   Design

First, instrumental foundations of absorption spectroscopy will be dealt. Although the infrared spectral region is the focus of interest, the same principles are also applicable to other regions. For observing the spectrum, one needs an instrument for measuring the electromagnetic radiation transmittance of a sample as a function of the wavelength (or wavenumber). The transmittance is defined as the ratio of transmitted to incident radiation power. The transmittance of a medium, if it can be measured independently of reflection losses on the cell windows, for example, is also characterized as the internal transmittance as opposed to the total transmittance.

Such an instrument for measuring spectra contains the so-called spectral apparatus as the most important element with which radiation of limited spectral regions may be isolated, as well as other different design elements that will be presented in further detail in the following sections. Whereas one previously used spectroscopes for recording spectra in the visible region, later on spectrographs were developed with which a spectrum could be photographically captured. With the help of specially sensitized photo plates, spectroscopy could take place e.g. in the short-wave near-infrared (NIR) region. Generally used nowadays are spectrometers that are equipped with one or several detectors for measuring the characteristic parameters of the spectrum. A spectrometer having the necessary basic elements is illustrated in Fig. 3-1.

Depending on the spectral apparatus used, one differentiates between the various types of spectrometers, subdivided into non-dispersive IR spectrometers in which no variable wavelength selection is possible, dispersive units and Fourier transform (FT) spectrometers. For variable wavelength selection, spectral apparatuses are used such as grating monochromators in dispersive instrument types. In addition, series with sequential filters are possible with which polychromatic light of all wavelengths is broken down into light beams of single wavelengths or, at least, narrow wavelength ranges. In FT spectrometers, the spectral splitting takes place via an interferometer allowing a wavelength-dependent radiation modulation. In most

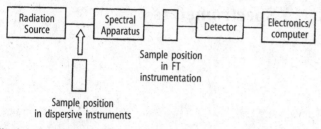

Fig. 3-1. Schematic design of an IR spectrometer.

cases, Michelson interferometers are available in which the two-beam interferences are converted by mathematical Fourier transform into spectral information. In the non-dispersive IR photometers, interference filters can be used which are based on a multi-beam interference. However, other effects such as optical absorption or reflection may be used in order to filter out more or less narrow-band spectral regions.

Another essential element for absorption spectroscopy is a radiation source which should show an as high as possible intensity in the wavelength area under investigation. Thermal radiators are mainly used which provide a very broad band, so-called continuum radiation. In contrast, the use of lasers for measuring absorption spectra presumes the use of monochromatic radiation, namely that having an extremely narrow band. This is typically referred to as the laser wavelengths.

The purpose of the optical system of a spectrometer is to transmit radiation from the radiation source to the detector, at best without loss. Lens systems of glass or quartz, as used in the visible and ultraviolet region, are useless in the infrared, because all radiation fractions above 2.0 or 3.8 μm (respectively, below 5000 or 2630 cm^{-1}) are absorbed. It is impractical to make lenses out of IR-transmitting material because of the high cost, possible sensitivity to air moisture and the energy loss near the material-related transmittance limits.

Thus, all IR spectrometers are equipped with mirror optics. The mirrors generally consist of glass, the surface of which is vapor-coated with aluminum or gold. A mechanical contact of the sensitive surface, for instance for cleaning purposes, must be avoided. Techniques such as the rinsing with organic solvents and distilled water or the application and removal of kollophonium are partially successful. Furthermore, the long-wave IR radiation is only slightly weakened by a slight clouding by dust particles, which are already clearly observable in visible light, yet is accompanied by an increasing fraction of scattered light.

The optical system used is equipped with a sample compartment in which the sample to be measured is placed in the path of the measurement radiation, in part in cells or in corresponding attachments with which measurement techniques other than transmission measurements may be applied. Advantageous for IR spectroscopy is that multiple techniques are available which are adapted to the different sample requirements. For dispersive spectrometers, as shown in Fig. 3-1, the sample compartment is situated before the monochromator for the following reasons: the scattered light caused by the sample can be removed by the spectral apparatus when this is placed in succession. The placing of the sample compartment before the monochromator, however, entails the disadvantage that the entire, still not split radiation acts upon the sample, whereby this can heat up considerably through absorption. With FT spectrometers in which the entire spectral range is investigated simultaneously, the sample compartment is placed behind the interferometer, so that the emitted sample radiation remains unmodulated.

The detector is used for converting the optical signal into easily measurable electrical signals, e.g. as measurement of voltage. This is aided by corresponding electronics for amplifying and digitizing the signals. While in earlier times spectra were analogously recorded with pens directly onto spectral paper, nowadays the spectrometer computer is an invaluable component offering multiple processing and storage possibilities for the spectra acquired. Plotters serve for presenting the spectra or the user is most often interested in a brief screening on the monitor. Archiving the spectral data can take place on digital data carriers.

Various principles exist for measuring the sample transmittance. For so-called two-beam instruments, the sample- and reference measurement occur almost simultaneously, so that the transmittance is available as the primary signal. Some dispersive instruments, but primarily FT spectrometers, are conceived as single beam instruments, despite them often containing two equivalent sample compartments. In this case, sample and reference signals must be measured consecutively, whereby the necessary quotient calculation is performed by the computer.

Because the technology of Fourier transform spectroscopy has made significant advances within the past years, routine laboratory instruments are mainly based on this technique. The quality of the optical and electronic components as well as the multiple application possibilities and degree of user comfort predominately determine the price of commercial IR spectrometers. Routine instruments for overview spectra in the range of between 2.5 and 25 μm (4000–400 cm^{-1}) currently cost between Euro 20 000 and Euro 40 000. However, sums of Euro 40 000 up to Euro 75 000 must be invested for dedicated instruments with higher spectral resolution and more flexible measurement possibilities. High-performance research instruments demand prices of Euro 125 000 and upwards.

3.2   Radiation Sources

In IR spectroscopy, Planck radiators are used as radiation sources for sample excitation. The intensity of the radiation emitted by a so-called black radiator is subject to the Planck radiation principle. According to this principle, the emitted spectral radiation power reaches its maximum (also see Fig. 3-2) depending on temperature (Wien's law). On the short-wave side of the maximum, the spectral radiation power curve falls steeply and, towards higher wavelengths, drops off flatter. Depending on the spectral region of interest, whether it be the far, middle or near-IR region, different sources are therefore used.

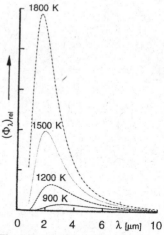

Fig. 3-2. Spectral radiation power curves for Planck radiators at various temperatures.

The thermal radiation sources in spectrometers typically show a somewhat lesser emissivity than the black radiator having the same temperature. The most frequently used radiation source in the mid-IR spectral region is the Globar, consisting of silicon carbide in the form of rods or helixes. As the result of its electrical conductivity in a cold state, a Globar can be directly ignited. At a burning temperature of about 1500 K, its power uptake is quite substantial, so that in most cases the source casing in the spectrometer is equipped with water-cooling. The Globar likewise has the advantage that its emissivity is relatively high to about 100 cm^{-1}, thereby allowing it also to be used as a source in the far-IR region.

In comparison, the Nernst rod, applicable as a radiation source in the mid-IR region, has a higher working temperature and consists of a few centimeters long and millimeters thick rods made of zirconium oxide with additives of yttrium oxide and oxides of other rare earths. The Nernst rod is relatively sensitive, mechanically speaking, and particularly upon improper mounting is prone to deformation possibly impairing the optical stability of the spectrometer. The oxide mixture has a negative temperature coefficient of the electrical resistance, i.e. its electrical conductivity increases with increasing temperature. This means that the Nernst rod is a non-conductor at room temperature and that an initial auxiliary heating is required for ignition. The normal operation temperature lies at about 1900 K, so that its emission maximum appears at between 1 and 2 µm. The intensity difference between the maximum and the region above 12 µm is considerable at about three orders of magnitude.

Some inexpensive spectrometers contain sources made of metallic helices, mainly of chromium-nickel alloys or tungsten with operating temperatures lying at about 1300 K, so that these are usually cooled with air-cooling. For the near-IR region, tungsten-halogen lamps are used exclusively, which allow a higher operating temperature and thus a higher radiation to be produced. Some instruments also use metallic conductors coated with ceramic. Thus, one finds heating coils made of platinum or a platinum alloy wound around ceramic rods that are covered by an additional sintered layer of aluminum oxide, thorium oxide, zirconium silicate or a similar material. One of their advantages is an unproblematic and extensively repair-free operation of these radiation sources.

For the far-IR region, particularly for measurements below 100 cm^{-1}, a mercury high-pressure lamp is suited, the plasma emission of which considerably surpasses the spectral radiation power of a black radiator of the same temperature. These radiation sources have their own voltage supply in the spectrometer and must be started by a high-voltage pulse. More demanding spectrometers often contain already two different radiation sources usable via corresponding electronics and optics.

Lasers are increasingly gaining importance nowadays. Only relatively narrow spectral regions can be scanned with gas lasers and this not always gapless (CO_2-laser 1100–900 cm^{-1}, CO-laser 2000–1800 cm^{-1}, NO_2-laser 900–910 cm^{-1}). In contrast, besides the known red line at 632.8 nm, the He-Ne laser, which we will discuss later regarding FT spectrometers, also delivers two lines in IR (1.152 and 3.391 µm) that cannot be easily detuned. Semiconductor-diode lasers show the advantage of larger tunability, whereby almost the whole mid-IR region can be covered with PbSnSe-type lasers. A disadvantage of these lead salt-diode lasers is the low operation temperature of between e.g. 15 and 90 K. In contrast, semiconductor-diode lasers such as those made of gallium arsenide, gallium aluminum arsenide and similar materials, which laser in the near-IR region, can be operated at around room temperature with Peltier cooling. Diode lasers are best suited for high-resolution IR spectroscopy, especially of gases at low pressures. Thus, these laser spectrometers, with a tunable radiation source, have a variety of applications (see Section 8.2 for more detailed information on this).

3.3   Infrared Detectors

Before the principles of spectral resolution can be discussed, the common radiation detectors for IR spectroscopy should be introduced. Their job is to convert the optical signals into further applicable electrical signals, whereby different physical material properties are used.

For one important group, i.e. thermal detectors, properties based on temperature changes resulting from absorption of radiation are decisive. Substantial for bolometers is the change of the electrical conductivity, while for thermoelements a change

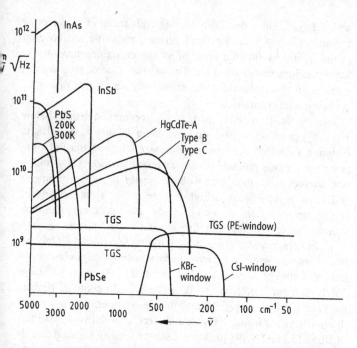

Fig. 3-3. Specific detectivity of
some common detectors.

of the thermotension is the decisive parameter. In pyroelectric
detectors, the temperature-dependent pyroelectric effect occurs.
Another further category of detectors is pneumatic detectors
consisting of various types. Accounting for a possibly low heat
capacity, these detectors have a mostly small and delicate design and therefore are mechanically quite sensitive.

The other group of radiation detectors is photoelectric detectors, called quantum detectors, in which the measurable effects
are based on a direct interaction of the photons with semicon-
ductive materials present here. Consequently, their sensitivities
depend very greatly on the wavelength. The basic principle behind detectors used in the infrared is the internal photoelectric
effect, whereby the phenomena of photoconductivity as well as
photovoltage are measured.

Besides their lifespan, other criteria for evaluating detectors
is the applicable wavelength range as well as the sensitivity, represented by the change of the measurement signal as a function of the change of the radiation power. Moreover important is
the signal-to-noise ratio (SNR) and the time constant of the response velocity. A characteristic parameter, the specific detectivity D^*, has been shown useful for comparing different detectors
with one another and is dependent on various measurable parameters:

$$D^* = F_D^{1/2}/NEP \text{ cm Hz}^{1/2} \text{ W}^{-1} \qquad (3.1)$$

whereby
F_D = detector area in cm^2
NEP = noise equivalent power in $W\,Hz^{-1/2}$.

The noise equivalent power NEP, here, is the incident radiation power on the detector that leads to a signal-to-noise ratio of one at the available electric bandwidth likewise of one within the subsequent electron amplification. This power can be defined and measured by the following equation:

$$NEP = \Phi/(\Delta f^{1/2} \cdot S/N) \tag{3.2}$$

whereby
Φ = radiation power in W
Δf = electric bandwidth in Hz
S/N = signal-to-noise ratio.

The sensitivity curves of a few detectors, important for IR spectroscopy, are compiled in Fig. 3-3.

3.3.1 Thermal Detectors

The thermal detectors use effects based on changes of the detector material resulting from temperature influences. A thermally produced photosignal depends on the incident flow of radiation and is thus dependent on wavelength. Temperature equilibrium is slower than for photon processes (μs); the time constants for the response rate are thus longer and lie in the ms-range.

Among the thermal radiation detectors, the *thermoelement* is most widely used. Its principle is based on the temperature-dependent change in the *thermotension* that forms on the contact site between two different metal or semiconductor alloys. The thermoelement itself is found in an evacuated capsule. The radiation enters through a window of infrared-permeable material (e.g. KBr or CsI, depending on the required wavelength range). Thermoelements are very sensitive mechanically. With proper handling, however, they can provide many years of service. The *bolometer*, much used earlier, has greatly lost importance compared to the thermoelement, but for special experiments, among others in the far-IR, Ge-bolometers e.g. are still a good choice. The bolometer is based on the *change in resistance* of an electrical conductor with temperature change resulting from radiation absorption. The absorption occurs e.g. on platinum mohr, on electrolytically precipitated nickel or on semiconductor layers (thermistor-bolometers). One of the disadvantages of this type of detector is that a special electricity source, maintained with high precision, is needed for its operation.

A so-called pyroelectric radiation detector is known for some time now, which is based on the polarization change of ferroelectric material below the Curie point. The pyroelectric effect for substances such as $LiNbO_3$ or triglycine sulfate (TGS, $(NH_2CH_2COOH)_3 \cdot H_2SO_4$), among others, is based on a spontaneous electrical polarization which can be measured as an electrical signal, when the material is used as a dielectric of a capacitor. With a rapid change of temperature, the internal dipole moment changes and, thus, the polarization, which can be detected as a voltage pulse.

The *Golay detector* belongs to a subgroup of pneumatic radiation detectors within the group of thermal detectors. The principle behind the Golay detector can be described as follows (see Figure 3-4):

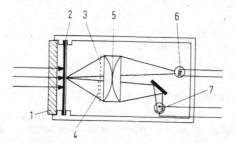

Fig. 3-4. Golay detector (principle).
1) IR-permeable window
2) blackened membrane with a reflective coating on the backside
3) grating
4) grating image
5) lens
6) light source
7) photocell.

Radiation enters through an IR-permeable window into a gas-filled cell and is absorbed there on a blackened film. The absorption heat causes an increase of the gas pressure to which the back wall of the cell, designed as flexible mirror, yields. This mirror is part of an optical system, in which the image of a line grating is superimposed on the grating. The radiation intensity changing in time with the modulation frequency causes a periodical bending of the mirror resulting from pressure deviations within the cell and thereby alters the coincidence of the grating and the image. A periodically changing photocurrent is produced which is proportional to the mirror deflection and thus to the primary radiation.

The advantages of the Golay detector are its large linear range and its applicability way into the far-infrared region. It is extremely sensitive to overexposure and, upon opening the monochromator, must be protected from foreign light by an automatic lid. Unfortunately, this kind of detector is no longer commercially available.

Photoacoustic Detector: Analogous to the Golay cell, the photoacoustic detection is principally based on the same physical conversion of radiation energy into an electrical signal. If the sample material, located on a solid foundation, is placed on the radiation-absorbing membrane in the Golay cell, and the spectrally separated light is modulated by a chopper, then only the wavelength range absorbed by the substance is converted into heat and therefore into pressure deviations during the radiation-free release of energy [1]. With a modulation frequency in the acoustic range (here, 20–1000 Hz), one can then use a microphone for the subsequent detection (also see Fig. 5.3).

3.3.2 Photodetectors

Electromagnetic radiation can interact in many ways with the material. Nowadays, most detectors used for IR radiation belong to the group of photoelectric detectors. Because of their higher sensitivity, they have widely replaced thermal detectors.

Photoconductivity: Incident radiation alters the electrical conductivity in the irradiated semiconductor material. The principle is simple here (Fig. 3-5): The photosignal is measured either as a change in voltage via the resistance R or as a change in current.

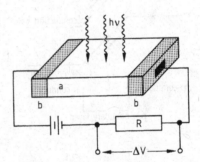

Fig. 3-5. Circuit diagram of a photo-semiconductor for measuring the photoresistance. a: semiconductor, b: contacts, R: resistance.

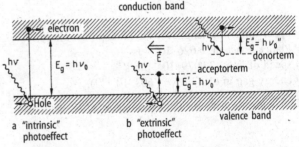

Fig. 3-6. Regarding the process of photoconductivity:
a) According to the electron-hole pair formation, both charge carriers migrate in the electric field. The energy gap E_g depends on the alloy.
b) The energy gap E_g can be diminished by spiking. One charge carrier stays bound to the foreign atom.

The elementary process of photoconduction is always the production of an electron-hole pair, whereby either both charge carriers are free to move in the electric field ("intrinsic process") or one of the two charge carriers is spatially bound ("extrinsic process") for the case where the electron-hole pair is formed on a spiking center (see Fig. 3-6).

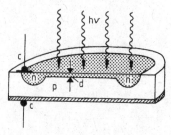

Phototension: Within or nearby an n-p-boundary layer, IR photons free charge carriers which produce a measurable tension that is proportional to the radiation intensity. The design of such a detector is illustrated in Fig. 3-7. The elementary process is again the formation of a charge carrier pair that is now separated by the n-p-boundary layer and thus creates an electrical tension (Fig. 3-8).

Fig. 3-7. Cross-section through a phototension cell. n: n-conducting material, p: p-conducting material, d: 0.5 μm of the n-layer, c: electrical contact.

The limit wavelength λ_0 from which the photoprocess begins, approaching shorter wavelengths (inferring higher energy photons), and a phototension can be measured, lies at

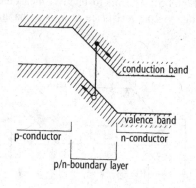

Fig. 3-8. Energy-band model of a p,n-boundary layer. The optically released electron migrates to the n-conductor, the hole in the p-material.

$$\lambda_0 = \frac{hc}{E_g} = \frac{1.24}{E_g} \qquad (3.3)$$

whereby

λ_0 = upper limit wavelength for the photoprocess in μm
h = Planck constant, $6.626 \cdot 10^{-34}$ Js
c = speed of light, $2.9979 \cdot 10^{14}$ μm s^{-1}
E_g = energy gap in eV (1 eV = $1.602 \cdot 10^{-19}$ J).

Table 3-1 gives values for the most important IR detectors.

Table 3-1. Energy gaps between the valence band and conductivity band and between the valence band/acceptor level and donor level/conductivity band (E_g) and application wavelength of the photoprocess (λ, $\tilde{v}_0$)

	Material	Operation Temperature (K)	E_g (eV)	λ_0 (μm)	$\tilde{v}_0$ (cm^{-1})
Photo-conductor	PbS	295	0.42	2.9	3500
	PbSe	195	0.23	5.4	1800
	Pb$_{0.2}$Sn$_{0.8}$Te } 77		0.1	12	850
	Hg$_{0.8}$Cd$_{0.2}$Te }				
Diode-semicon-ductor	Ge: Hg		0.09	14	700
	: Cd		0.06	21	500
	: B		0.0104	120	90
	Si: Ga		0.0723	17	600
	: As		0.0537	23	450
	: Sb		0.43	19	350

3.4 Spectral Resolution

By means of suitable optical mountings, a possibly narrow-band spectral region is isolated from the polychromatic radiation of the thermal radiation sources. Naturally, this succeeds only approximately with the spectral apparatuses at disposal. Monochromatic radiation, like the radiation of singly emitted spectral lines of excited metal vapors (e.g. mercury vapor lamp) or the radiation of a laser, can only be attained at considerable apparative expense, although the linewidth, also in the cited cases, is limited by various physical reasons. Therefore, aside from the natural linewidth caused by the lifespan of the excited molecules, the Doppler broadening is caused by the distribution of the kinetic energy of the atoms and molecules in a gas or the pressure broadening due to the collisions of the neighboring molecules.

Besides its resolution, the criteria for assessing a spectral apparatus are its applicable spectral range and the optical conductance, which is a measure for the light intensity of the optical arrangement used. The resolution of a spectral apparatus is defined by the distance between two neighboring absorption maxima having about the same height and which are separated by an absorption minimum, the transmittance of which is ca. 20% higher than the band maxima (see Fig. 3-9). The value is arbitrary and is not handled uniformly. In another practical procedure, a narrow-band absorption band, e.g. of a diatomic gas at low pressure, is recorded, and the resulting experimental half-width approximately gives the half-width of the spectral apparatus function, which is generally understood to be the resolution.

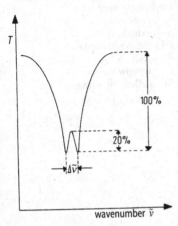

Fig. 3-9. Spectral resolution $\Delta\tilde{v}$.

With commonly used dispersive spectral apparatuses such as grating monochromators, resolutions up to 0.2 cm^{-1} are obtained, while even resolutions of 0.02 cm^{-1} are reached with specially designed spectrometers. In contrast, the resolution of prism monochromators can lie between 2 and 20 cm^{-1}. Resolutions up to 0.001 cm^{-1} can be attained with Michelson interferometers of commercial FT spectrometers.

3.4.1 Dispersive Spectrometers

The monochromator is the essence of a dispersive spectrometer. A monochromator consists of a splitting system, namely the optics and the dispersing element that splits the radiation: prisms or diffraction gratings. Aside from monochromators, there are also so-called polychromators with which several spectral regions can be selected simultaneously.

The splitting system, optically included in a monochromator, consists of two narrow slits, which limit the rays at the entry and exit of the monochromator. By moving the slit cheeks, the slit openings can be changed with high precision between ca. 10 μm and a few millimeters. Radiation power and resolution are thereby affected: with increasing slit width, the radiation power penetrating the monochromator increases and the spectral resolution becomes smaller.

The radiation diverging from the entry slit is collimated by a parabolic mirror, namely the entry collimator, and is deflected onto the dispersive element (grating or prism), the whole surface of which is illuminated. Depending on the type of optics, the collimated radiation travels back to a second parabolic mirror (Ebert mounting, Fig. 3-10) after reflection on the diffraction grating or after passing through the prism and reflection on a plane mirror, or goes back to the same parabolic mirror (Littrow mounting, Fig. 3-11) and is sharply imaged by this on the exit slit.

3.4.1.1 Diffraction Grating

While prism monochromators were used almost exclusively earlier, nowadays only diffraction gratings are being used as the radiation-splitting optical elements. By using diffraction gratings, one avoids the difficulties involved with prism materials (e.g. hygroscopic, reflection, absorption) and the frequent, complicated prism replacement that was necessary in using larger wavelength ranges. Moreover, gratings produce a much better resolution, constant over wide wavelength ranges (see Fig. 3.4).

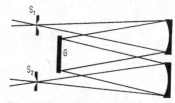

Fig. 3-10. Grating monochromator in the Ebert mounting.
S_1 entry slit
S_2 exit slit
PS_1; PS_2 parabolic mirrors
G   grating.

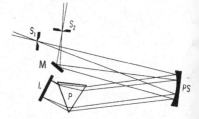

Fig. 3-11. Prism monochromator in the Littrow mounting.
S_1 entry slit
S_2 exit slit
PS   parabolic mirror
P   prism
L   Littrow mirror
M   deflection mirror.

The *basic form* of diffraction gratings can be obtained by a number of wires ordered parallel and equidistant to each other in one plane. With scratch gratings, slits scratched into a flat glass or metallic surface, on which incident light is diffracted, serves the same purpose. The main fraction of the diffracted radiation disappears by interference erasing. Light of a defined wavelength is observed only in a particular direction, and always when the pathlength of the light emitted from two neighboring grooves is a multiple of the wavelength (Fig. 3-12). The direction of the emitted light is given by

$$\sin a_n = \frac{n \cdot \lambda}{d} \tag{3.4}$$

whereby a_n is the angle measured from the grating perpendicular, λ the wavelength in cm, d the grating constant, i.e. the dis-

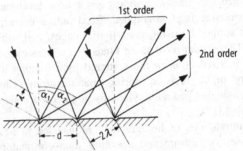

Fig. 3-12. Diffraction grating;
a_1, a_2: angle of the diffracted radiation in the first and second order;
D: distance between adjacent groves (grating constant) in the substrate;
$n \cdot \lambda$:   pathlength difference of the radiation with the same incidence angle diffracted from two adjacent grooves = wavelength of the n-tenth order (according to W. Brügel [2], p. 95).

tance between two adjacent wires (grooves) in cm and the index $n = 1, 2, 3 \ldots$ the corresponding ordinal number.

One disadvantage in using gratings earlier was their low light intensity. This disadvantage could be overcome by giving the grating grooves a very special form, which leads to an energy concentration of one or a few orders of magnitude in the spectra. Such gratings are called *Echellete gratings*. The optimum of the energy distribution depends on the blaze angle (see Fig. 3-13). The wavelength at which the maximum diffractive efficiency or the greatest relative radiation intensity is reached is called the blaze wavelength. The grating constant of the used reflection grating lies about in the range of magnitude of the wavelength of the incident radiation to be analyzed. The efficiency of the diffraction grating

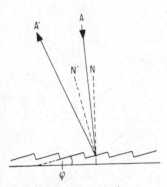

Fig. 3-13. Echelette grating.
N perpendicular to the grating plane
N' perpendicular to the fissure plane
A direction of radiation incidence
A' direction of the reflected radiation
φ Blaze angle.

is generally represented as a function of the ratio of the wavelength and grating constant (range of 0 to 2). This is understandable in that first-order gratings are often used.

A disadvantage of the grating is the overlapping of spectra having different orders. If one sets the grating monochromator to a certain wavelength λ, the wavelengths $\lambda/2$, $\lambda/3$, $\lambda/4$ etc. (i.e. of the second-, third-, fourth-order etc.) are also obtained in the same direction. These undesired wavelengths can be filtered out by mounting a prism monochromator before or afterwards, which allows a relatively wide wavelength range to pass through, but only in the desired range.

After the polychromatic radiation is split by the dispersion element, a dimensional separation of the individual wavelengths exists in the plane of the exit slit. By turning the grating or the Littrow mirror (Fig. 3-11), this wavelength interval passes by the slit opening and only a limited wavelength range is allowed to go through and be focused on the detector.

So-called *holographic gratings* are a new development. The grating structure is photographed, fixed onto a correspondingly prepared surface by using laser light sources, and subsequently etched. The advantage of these kinds of gratings over Echelette gratings is a reduction of the scattered light fraction due to the extremely high precision during manufacturing and, consequently, a higher radiation yield by a factor of one order of magnitude [3].

The introduction of holographic gratings led to new important classes of instruments: scanning monochromators having concave gratings as the only optical element as well as compact polychromators which allow the simultaneous recording of an entire spectrum, since the image plane of the concave grating can be optimally adapted to the photodetector array.

3.4.1.2   Interference Filters

Another already known dispersion element is recently being used again for very compactly designed instruments: the Fabry-Pérot interferometer [4] in form of the further developed interference filter. The principle of the interference filter [5] is based on the multiple reflection of a polychromatic light beam within a thin, dielectric layer between semipermeable silver layers. Due to interference, all wavelengths are eliminated except a narrow band section. If one designs the dielectric intermediate layer as a wedge and arranges it segmented in the circle of a disc and turns this disc past a slit, then a wavelength selection is obtained as a function of the disc setting. This type of monochromator is found in portable instruments for environmental analysis.

3.4.1.3   Acousto-Optical Tunable Filters (AOTFs)

A new type of dispersion element called the acousto-optical tunable filter (AOTF), having multiple special applications in the near-infrared region, has been introduced since the mid-eighties. It consists of a double-refracting optical material in which vibrations in the MHz-range are induced. The created acoustic wave runs perpendicular to the optical path of the filter, modulates the refraction index of the crystal and thus creates a type of diffraction grating. Its big advantage is the fact that no movable parts are necessary to tune the filter in a matter of microseconds over a wide spectral range.

3.4.1.4   Spectral Slit Width, Resolution

The dependency of the spectral radiation power of a narrow wavelength interval after passing through a slit shows a characteristic pattern. Depending on the instrument and on instrumental parameters, there is a mathematical probability function about that radiation fraction which finally reaches the detector. This is the so-called *slit function*. It can be described for identical slit widths for the entry and exit slit by a triangular function (Fig. 3-14), the peak of which lies at the wavelength set by the monochromator. The half-basewidth of this triangular function is defined as the *spectral slit width* $\Delta\lambda_s$ or $\Delta\tilde{v}_s$. It can be calculated according to Williams [6] and increases with increasing geometric width of the entry and exit slit. It decreases with increasing angular dispersion of the prism material, larger prism base selected as well as with longer focal distance of the collimator mirror. Good resolution corresponds to a small spectral slit width, whereby this is obtained with the given optics by a possibly narrow slit setting. The theoretically attainable *resolution* of a monochromator means the same as the spectral slit width with an infinitely small slit opening. Diffraction effects on the slits, however, limit this. Because the radiation source shows only a limited radiation power and the radiation detector a limited sensitivity, finite, in part, even relatively wide slit widths must be set in order to obtain a sufficient signal. The actually obtainable resolution is therefore always lower than the theoretical resolution of a monochromator (see Sect. 3.4).

The opening of the entry and exit slit is regulated in practice mechanically or electronically, so that about the same radiation power throughout the entire wavelength range reaches the detector. The function of a slit program thus accounts for the spectral energy distribution of the radiation source (Fig. 3-2). This is an essential difference compared to the later discussed Fourier-transform spectrometers in which the spectral signal-to-noise ratio is greatly dependent on the wavelength.

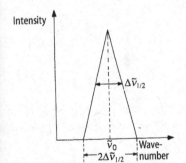

Fig. 3-14. Slit function.
$\tilde{v}_0$ wavenumber in the intensity maximum
$\Delta\tilde{v}_{1/2}$ spectral slit width (measured at half the height of the slit function).

3.4.1.5   Scattered Radiation

Besides the wavelength interval characterized by the set wavelengths and spectral slit width, a certain fraction of other wavelengths goes through the exit slit and reaches the detector. This is attributed to the imperfect mirror- and prism surfaces, dust or other impurities, inhomogeneities in the prism material etc. resulting in a diffuse scattering of unsplit radiation. Fractions of this scattered radiation can reach the exit slit via diverse routes and thereby impair the purity of the monochromatic beam. As the result of the Planck radiation energy distribution, especially short-wave scattered fractions near the energy maximum are dangerous in the range of higher-set wavelengths because of the wide slit widths necessary there.

3.4.2   Fourier Transform Spectrometers

All commercial mid-infrared (IR) instruments, sold for analytical applications, are based on an interferometric measurement – that is they are classified as Fourier transform infrared (FT-IR) instruments. Today, there are probably no more than 5% of dispersive instruments still in regular service for mid-infrared analyses.

In contrast to this, for near-infrared having experienced an impressive renaissance in analytics dispersive spectrometers are still being widely used [9]. A historical account on the introduction of the FT technique was already given in Sect. 1.1. The interested reader may refer to the monograph by Johnston [10] for further details.

3.4.2.1   Michelson Interferometer

As the monochromator is the heart of a dispersive spectrometer, the same applies for the interferometer in FT instruments. Figure 3-15 shows the schematic diagram of a classical Michelson interferometer. The beam of radiation from the source is split by a semi-permeable beamsplitter into two partial beams that are reflected on a fixed and on a movable mirror back to the beamsplitter where they recombine and are brought to interfere. The fraction directed back to the source is not used and gets lost. A shifting of the movable mirror changes the optical pathlength in this interferometer arm, whereby a phase difference between both partial beams results and, hence, a change of the interference amplitude. The intensity signal from the detector, as a function of the change of the optical pathlength corrected by a constant component, is called an interferogram.

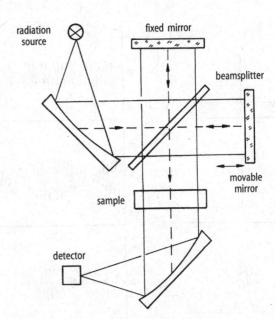

Fig. 3-15. Schematic diagram of a Fourier transform spectrometer with a classical Michelson interferometer.

For a monochromatic radiation source having a wavelength λ or a wavenumber $\tilde{v}$, as shown by a laser, a cosine signal (I_0 partial beam intensity) is obtained at the detector as a function of the optical path difference x, which is called *retardation*:

$$I(x) = I_0\{1 + \cos(2\pi\tilde{v}x)\} \tag{3.5}$$

With the same optical path in both arms of the interferometer, there is no phase difference for the two partial beams, which may then interfere constructively, i.e. both electric wave peaks are added. If the movable mirror shifts by $\lambda/4$, then the retardation between both partial beams is exactly $\lambda/2$ and a quenching interference results at the detector. For a radiation source with several wavelengths, the interference pattern corresponds to the sum of the cosine signals of all individual frequencies. The interferograms of various spectra are illustrated in Fig. 3-16 to acquaint the reader with the Fourier domain. For broadband spectra, the maximum of the interferogram, for which in the ideal case the same phase exists for all wavelengths (ZPD, Zero Path Difference), drops rapidly. The symmetrical interferograms obtained can be converted into the spectrum by Fourier transform:

$$S(\tilde{v}) = \int\limits_{-\infty}^{\infty} I(x)\cos(2\pi\tilde{v}x)\,dx \tag{3.6}$$

The result is likewise a symmetrical function for which the spectral data are doubled, mirrored at the zero point. A retrotransfor-

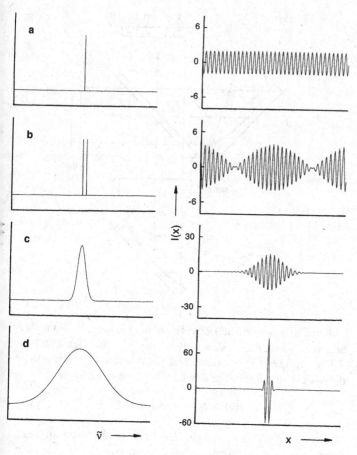

Fig. 3-16. Spectra (on the left) with corresponding double-sided interferograms (on the right).
a) monochromatic radiation
b) two narrow-band emission lines
c) band-pass filtered thermal radiation
d) broadband thermal radiation.

mation is also possible without information loss. In most cases, the experimentally obtained interferogram is asymmetric through phase shifting at the various wavenumbers due to optical and electronic effects, making a complex Fourier transform necessary (sine terms must also be considered besides cosine terms).

An analogy is given, when in a concert the human ear is hearing the complex sound wave produced by the orchestra. The auditory nerve transmits this complex wave to the brain. This in turn gives the information what instruments are involved with what intensity in the orchestra, i.e. the Fourier transform of the complex sound waves to frequency and intensity. Because a more detailed handling of this goes beyond the scope of this book, the reader is referred to other literature sources [11–13].

For technical reasons, the interferogram is not stored continuously but point-by-point. Regarding this, the interference pattern of monochromatic light, e.g. a He-Ne laser ($\lambda_{\text{He-Ne}} = 633$ nm or $\tilde{\nu}_{\text{He-Ne}} = 15\,800$ cm^{-1}) is simultaneously detected with a detector diode. The zero crossings of this reference laser sine signal or a multiple of these intervals define the retardation points, at which the IR interferogram is digitized. The frequencies of normal He-Ne lasers vary somewhat (<0.01 cm^{-1}) because of their mode structure, so that only stabilized single mode lasers are used for the highest demands as in the high-resolution IR spectroscopy.

Earlier FT spectrometers for the far-IR region contained simple Moiré systems with two transmission gratings somewhat turned back-to-back, whereby the one having the movable mirror was shifted. By means of white light as well as two diodes mounted behind the gratings, interference patterns, having an accuracy sufficient for the long-wave far-IR, can be detected as a function of the grating shift.

The distance Δx between two points in the interferogram determines, for signal-theoretical reasons, the maximum frequency or wavenumber, which can be clearly detected. On the other hand, it can be shown that any random signal can be digitized without loss of information when a sampling frequency exists that is greater or at least equal to twice the bandwidth of the signal to be measured. This is the so-called Nyquist sampling criterion:

$$\tilde{\nu}_{max} - \tilde{\nu}_{min} = 1/(2\Delta x) \qquad (3.7)$$

As was already mentioned, the Fourier transform of the interferogram delivers the spectrum and its mirror image. The discrete Fourier transform of the scanned function now leads to the situation than infinite replica of this are produced along the wavenumber axis. Due to this, other spectral regions are also calculable, which respectively must lie between an upper and lower convolution limit. Indications on the practically used spectral intervals are found in Tab. 3-2.

In order to satisfy the Nyquist criterion, the computer must be given the correct spectral interval limits, within which the spectral radiation power to be measured is not zero. For example, if the interval from 3950 to 0 cm^{-1} is prescribed but, due to

Table 3-2. Typical digitization point distances (in μm) realized with a He-Ne laser with respective spectral convolution limits (in cm^{-1})

Δx	$\tilde{\nu}_{max}$	$\tilde{\nu}_{min}$
0.316	15 800	0
0.633	7 900	0
	15 800	7900
1.266	3 950	0
	7 900	3950
	11 850	7900
	⋮	⋮

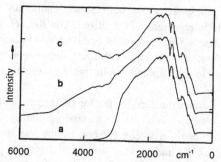

Fig. 3-17. Effects of incorrectly selected sampling frequencies in the interferogram. a) single beam spectrum a of deep-pass filtered interferogram, for which the double point distance in contrast to the standard setting is possible (Δx corresponding to $\tilde{v}_{max} = 3950\ cm^{-1}$); b) spectrum from an interferogram with adequate point distance ($\tilde{v} = 7900\ cm^{-1}$); c) faulty spectrum from the same interferogram with double point distance; the latter two spectra were shifted in the ordinate direction for clearer representation.

insufficient electronic deep-pass filtering (also see Sect. 3.4.2.2), spectral elements with corresponding intensity above the given limit are still determinable (e.g. with correctly selected double sampling frequency in the interferogram), these spectral data in the interval below the Nyquist limit are reflected (so-called aliasing) and falsify the spectrum (see Fig. 3-17).

For slow scan instruments, the mirror is moved at a constant speed or also in discrete steps (step-scan). The beam is modulated with a mechanical chopper and detected phase-sensitively. A more efficient interferogram recording for the mid- and near-IR region, however, can be performed with fast-scanning interferometers doing without a chopper. In this case, the radiation is used better, because the detector is not exposed to any light-dark phase. A prerequisite, however, is that fast detectors are available. Depending on the detector used, the applied mirror speeds lie between about 0.05 and 5.0 cm s^{-1} and are frequently selectable in steps. For typical Michelson interferometers, the optical frequencies or wavenumbers can be transformed into the acoustic frequency range by using this technique. With c_s denoting the mirror speed, the equation for this is given by:

$$v_s = 2c_s\tilde{v} \tag{3.8}$$

The maximum frequencies are typically lower than 100 kHz. Consequently, this necessitates a broadband amplification of the detector signal that is linear for all frequencies.

Typically, several interferograms are co-added for improving the signal-to-noise ratio (SNR). To do this, a reference point must be defined for the signals to be averaged and digitized. This can be done via a so-called white-light interferogram having a sharply defined maximum and which is detected, in addition to the laser-sine signal, with another photodiode. The collimated light originates from a small light bulb and is mostly fed into a parallel-operated Michelson interferometer. Alternatively, the connection to the individual interferograms can be achieved via laser interference signals, which are parallel registered by two photodiodes. Regarding this, the reference laser beam is split, e.g. by a $\lambda/4$-plate consisting of double-refracting material, after which both partial beams show a phase difference of $\lambda/4$. By various phase differences between both laser signals in the forward and backward path (so-called quadrate detection [10]), the absolute positions as well as the direction of the mirror movement can be determined. In this way, interferograms can also be recorded in both scan directions, whereby the measurement recording is further optimized with respect to time.

Because of experimental limits, each interferogram can only be recorded to a limited maximal retardation. The spectral resolution here is inversely proportional to the maximum retardation. This can be elucidated by means of Fig. 3-16b: the interferogram of a spectrum with two emission lines can be presented, as is shown, as a product of two cosine signals; for one, the periodicity is given by the mean of both wavenumbers $(\tilde{v}_1+\tilde{v}_2)/2$, for the other, by the difference $(\tilde{v}_1-\tilde{v}_2)/2$. In order to separate both spectral lines, which means the same as determining $\Delta\tilde{v}=\tilde{v}_1-\tilde{v}_2$, the interferogram – (presuming the same optical paths in both interferometer arms – must be measured at least until the first oscillation minimum. The smaller $\Delta\tilde{v}$ is, the greater must be the maximum retardation x_{max}. In general, the following applies for the spectral resolution:

$$\Delta\tilde{v} = 1/(2x_{max}) \qquad (3.9)$$

For a spectrum with little fine structure, as corresponds to an empty, evacuated or a dry-nitrogen purged spectrometer, only a small resolution is necessary, for example, in order to present spectral details. Its interferogram can thus be interrupted at a shorter path (Fig. 3-18). In contrast, the spectrometer filled with laboratory air gives an interferogram with clear structures also at greater retardations due to the H_2O and CO_2 absorptions. What is important is that, in contrast to dispersive spectrometers, the resolution here is constant over wide spectral ranges, because no grating and slit alteration is necessary.

With few exceptions, FT spectrometers are conceived as single beam instruments. Figure 3-19 illustrates a commercial routine instrument. Most of the spectrometer including the interferometer is encased. The position of the interferometer mirrors deviates from the 90% setting of the classical Michelson interferometer. This has been chosen for reasons of broadband, high

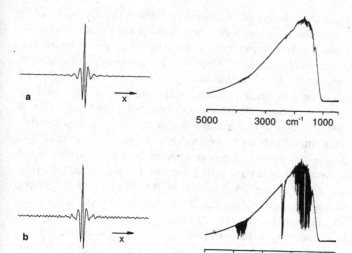

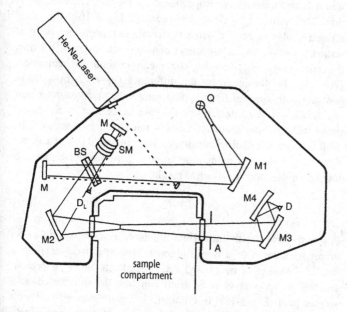

Fig. 3-18. Real interferograms (sections around the maximum) with corresponding single beam spectra (spectrometer with Globar, CaF_2/ Fe_2O_3 beamsplitter and DTGS detector; spectral resolution 0.5 cm^{-1}).
a) evacuated spectrometer
b) spectrometer aerated with laboratory air.

Fig. 3-19. Optical diagram of a commercial routine spectrometer (Perkin Elmer Paragon 1000).

Q	radiation source
BS	beamsplitter
SM	scan mechanism
A	aperture
D	detector
D_L	laser detector
M1–M4	focusing mirrors
M	plane mirror in the interferometer.

beamsplitter efficiency, whereby also polarization effects and a higher energy throughput are important with given beamsplitter size. In this case, the interferogram is not produced, as otherwise usual, by linear shifting of the movable mirror, e.g. on an extremely precise airshaft, but by rotation of a mirror pair (see Fig. 3-20). The essential advantage of the rotating system is the fact that the adjustment of the exiting beam remains constant during the rotation. Rotating systems are very successful, if one lists the various interferometer modules used in commercial FT spectrometers. For further details, please refer to literature [10].

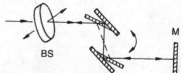

Fig. 3-20. Change of the optical pathlength with a rotating pair of mirrors (BS beamsplitter, M fixed plane mirror).

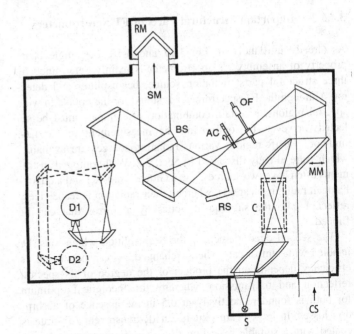

Fig. 3-21. High-resolution mobile FT-IR spectrometer, Bruker 120M.

S	radiation source
CS	collimated entry radiation
C	cell for calibration purposes
AC	aperture changer
OF	changer for optical filter
SM	linear scan mechanism
RM	retro reflector mirror
BS	beamsplitter
D1, D2	detectors
MM	movable plane mirror.

Especially for FT spectrometers with high spectral resolution, improved measurement quality can be obtained by dynamic alignment of one of the interferometer mirrors. For this, the wave front of the collimated, monochromatic beam is monitored in the interferometer at several points symmetrically to the center. If phase changes occur between the interference patterns on the respective detectors, an automatic correction on the fixed mirror can be done via the microprocessor by means of Piezo elements. For example, the spectrometer models of the Bomem company, which are based on this, are applicable for measurements with high spectral resolution up to the VIS- and UV-range.

An alternative to the dynamic adjustment is the use of retro reflectors instead of plane mirrors, as are found, for example, in high-resolution and mobile spectrometers of the Bruker Optics company. The instrument presented in Fig. 3-21 is used successfully for atmospheric spectroscopy. Excluding the position of the gas cell, there is no special sample compartment planned here. By means of a movable plane mirror, either atmospheric absorption and emission measurements or even calibration measurements are possible by using the spectrometer's own radiation source. For recording absorption spectra of large air layers, the sun or even the moon, for example, may be used as an external radiation source.

3.4.2.2   Important Structural Parts of FT Spectrometers

As already introduced in Fig. 3-17 and 3-18, the single beam property of an empty FT spectrometer is mainly determined by three structural parts: radiation source, beamsplitter and detector. Although the FT spectrometer is suited for the uptake in wide spectral regions, a certain combination of elements must be selected, however, for optimizing the measurement in the far-, mid- or near-IR region. Various radiation sources are available, which were already discussed in Sect. 3.2. Well-equipped instruments often have two sources, e.g. a Globar (mid-IR) and a quartz-halogen radiator (near-IR), whereby their radiation is respectively reflected onto the subsequent optics by a mirror that can be flipped.

The second cited element is the beamsplitter, which must be handled with utmost care when exchanged. As can be shown, its efficiency depends on the product of the degree of its spectral reflection and transmission, whereby the theoretical optimum for both is found respectively at 0.5 in the absence of absorption losses. In most beamsplitters, an IR-transparent substrate is coated with a suitable reflection film (e.g. germanium or potassium bromide). In order to keep the optical path in both interferometer arms symmetrical, a compensating plate made of the same substrate material is additionally incorporated. In contrast, freestanding Mylar polymer films (polyethylene terephthalate) of various thicknesses are used in far-IR as beamsplitters. The usable regions are listed in Tab. 3-3. For certain types of interferometers, an automatic beamsplitter change may also be made possible, which is favorable for the computer controlled spectrometer operation.

As the detectors were already discussed in Sect. 3.3, only the frequently used types should be reiterated here. The most widely used detectors are the pyroelectric DTGS detectors, because these can be used at room temperature without any problem. Their windows, e.g. made of KBr, limit measurements to the mid-IR spectral region, or if consisting of polyethylene, to

Table 3-3. Usable spectral regions of various beamsplitters (in cm^{-1})

Beamsplitter	$\tilde{\nu}_{max}$	$\tilde{\nu}_{min}$
Quartz	15600	2700
CaF_2/Fe_2O_3	12000	1200
KBr/Ge	6500	450
CsI/Ge	7000	220
Mylar (6 µm)	500	50
Mylar (12 µm)	250	25
Mylar (25 µm)	120	20
Mylar (50 µm)	55	15

the far-IR spectral region. Also used are semiconductor detectors having a wavenumber-dependent, though clearly greater spectral detectivity, and which have a high response speed thereby making faster scan rates possible with the interferometer. Characteristic for them is their long-wave cut-off frequency, below which the detector no longer responds. With some photodetectors, the use of liquid nitrogen is necessary, as is the case for semiconductor detectors made of HgCdTe (MCT), which are mostly used in mid-IR. They are more sensitive than DTGS detectors and can be variously broadband selected. Semiconductor detectors made of InSb are suited for the near-IR region.

Moreover, there are further optical elements such as e.g. apertures in the optical beam path of a FT spectrometer. Ideally, the retardation in the interferometer should be equal for all collimated partial beams. However, this cannot be fulfilled due to the finite size of the radiation source or of its image, so that a certain beam divergence in the interferometer is unavoidable. Hence, somewhat different optical paths, i.e. phase differences, result for e.g. the partial beams emitted from the middle and from the edge of the source. The larger the emitting area is, the greater are the resulting, disturbing phase differences, which finally diminish the attainable spectral resolution. The tolerable degree of divergence in the interferometer depends on the desired spectral resolution and on the maximum wavenumber to be measured in the spectrum. This can be established via the detector element size or by a variable aperture, namely the Jacquinot stop. Its effect is similar to the slit system in the monochromator: with higher resolution, the optical light conductance, which determines the radiation flow in the spectrometer, is reduced with a narrower aperture. Lastly, the incident energy on the detector is inversely proportional to the square of the aperture diameter.

Furthermore, optical filters are also used for limiting the spectral ranges. For instance, a black polyethylene film can filter off visible light of the mercury vapor lamp for measurements in the far-IR region. A glass or quartz window is capable of absorbing a large fraction of the long-wave radiation of the mid-IR region. For high-resolution spectroscopy, bandpass filters are used to reduce the sampling frequency in the interferogram, with which the data quantity may be considerably decreased also for the Fourier transform. Instead of optical filters, electronic low- and highpass filters can also be used in the audiofrequency range, which, together, act as a bandpass (also see Fig. 3-17). However, one is mostly interested in reducing the incident radiation power on the detector that cannot be done with electronic or digital filters.

3.4.2.3 Advantages of FT Spectrometers over Dispersive Instruments

There were many publications in the past in which the efficiency of the various instrument types have been compared [14, 15]. FT spectrometers have a number of advantages over dispersive spectrometers:

a) *Multiplex Advantage:* This advantage is also named after the spectroscopist *Fellgett*. All wavelengths are measured simultaneously in the interferometer, while these are measured in the monochromator successively. With dominating detector noise typically observed in the mid- and far-IR region, FT spectrometers are superior for two reasons: at the same spectral measurement time and corresponding to the spectral elements $N = (\tilde{\nu}_{max} - \tilde{\nu}_{min})/\Delta\tilde{\nu}$, a measurement time per element that is a factor N longer can be realized, whereby the noise is reduced by a factor $\sqrt{N}$. Moreover, compared to dispersive instruments, FT spectrometers need a measurement time that is a factor of $1/N$ shorter for obtaining the same signal-to-noise ratio.

　　Very sensitive semiconductor detectors are available in the near-IR and especially in the visible spectral region, so that the detector noise no longer dominates here but rather the photon noise that is proportional to the square root of the signal intensity. Thus, the multiplex advantage is not effective in these spectral regions.

b) *Throughput Advantage:* This term is also known as the *Jacquinot advantage*. At the same spectral resolution, the light conductance of a FT spectrometer, with its circular apertures, can generally be higher than that of a dispersive instrument, which is equipped with two slits within the monochromator. At the same resolution capacity, interferometers show an optical conductance that is two orders of magnitude higher than grating spectrometers [16]. This was specifically demonstrated in a comparison of two at the time commercially available instruments [15]: under comparable measurement conditions, a 100-times greater radiation power is detectable in mid-IR at about 2000 cm^{-1}. This advantage is not constant due to the variable, angle-dependent dispersion of the diffraction grating, but decreases by a factor of $\tilde{\nu}^2$. The signal-to-noise ratio obtainable with a FT spectrometer was estimated by Mattson [17] in an interesting and worthwhile study, wherein the various parameters are discussed in detail.

c) *Connes Advantage:* The wavenumber stability of spectra obtained via an interferometer is clearly higher (also measured over longer periods) than for dispersive spectrometers. The reason for this is that the frequency scale of the FT instrument is linked to the He-Ne laser, which provides an internal reference for every interferogram.

Another considerable aspect is scattering to be neglected, because the sample position typically lies behind the modulating interferometer. In addition to this, the necessary spectrometer computer has long been advantageous, first allowing spectral processing to be feasible.

3.4.3 Acquisition of FT-IR Spectra

3.4.3.1 Steps in Spectral Calculation

This section introduces the various steps in getting from an interferogram to a spectrum. As was already shown in Sect. 3.4.2.1, the interferogram is available only with a limited length, bound to a certain attainable spectral resolution. One can think of the finite interferogram as originating as a product of the infinite function with a rectangle, the so-called *boxcar truncation function*, which is unity within the experimentally available interferogram interval and zero outside of this interval. This function and its Fourier transform, the sinc function, are shown in Fig. 3-22. Likewise specified is its full line half-width. In dispersive spectrometers, the slit function (see Fig. 3-14) of the monochromator determines the line form, which is the *convolution* of the true spectrum with the slit function. Analogously, FT spectrometers likewise yield characteristic instrument functions, the so-called *instrument line shape* (ILS) functions, which depend on the weighting function of the interferogram, the so-called *apodization function*.

Especially when the true half-width of the absorption bands is smaller or comparable with the measured spectral resolution, measuring the interferogram only up to a certain point, i.e. a finite resolution, causes artificial side lobes (or pods). Although the highest spectral resolution is reached with the boxcar truncation, i.e. with no apodization, the side lobes are burdensome when closely neighboring lines should be quantitatively evaluated. Therefore, by multiplying the interferogram with an apodization function, these side lobes can be reduced, but at the cost of spectral resolution.

One of the most common apodization functions used in infrared FT spectrometry is a triangular apodization function and its Fourier transform, the sinc^2 ILS, which has exclusively positive side bands. Between no (boxcar) and strong (e.g. triangle) apodization, various trapezoidal weighting functions with variable side lengths are defined.

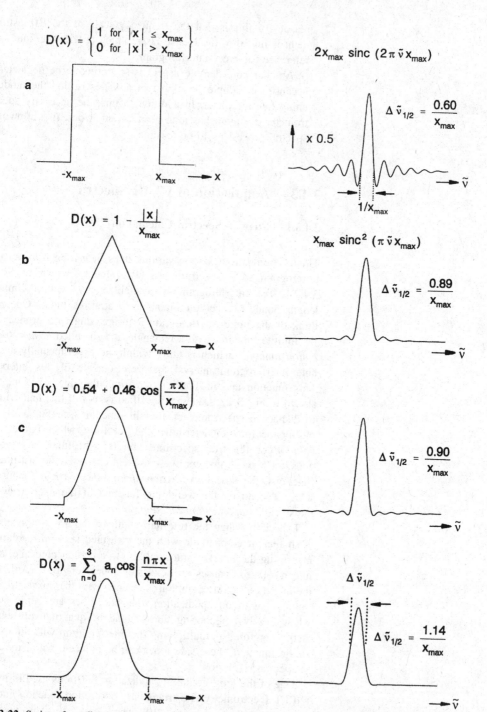

Fig. 3-22. Series of apodization functions and their corresponding instrument lineshape functions (Fourier transforms). a) Boxcar truncation; b) triangle, c) Happ-Genzel, d) 3-term Blackman-Harris apodization.

Most spectrometer operating programs also offer smoother alternative apodization functions, some of which are also shown in Fig. 3-22. Widely used is the Happ-Genzel apodization that can be seen as a special case of the Blackman-Harris functions. Aside from the 3-term apodization presented, a function having 4-terms is commonly used but which leads to a wider line shape (half-width $\Delta\tilde{v}_{1/2}=1.3$). Further possibilities of weighting an interferogram are given by the Norton-Beer functions, which likewise can be presented as a series development with various coefficients $D_{NB}(x) = \sum c_i (1-(x/x_{max})^2)i$. The half-widths of the corresponding spectrometer functions are $\Delta\tilde{v}_{1/2}=0.72$ for the weak, $\Delta\tilde{v}_{1/2}=0.84$ for the moderate and $\Delta\tilde{v}_{1/2}=0\%$ for the strong Norton-Beer apodization. The latter can generally be recommended, as well as the Blackman-Harris function with three terms, the ILS of which shows a somewhat larger half-width than the previous one. With these functions, the intensity of the side lobes decreases faster than the likewise favorable Happ-Genzel apodization.

Another aspect should be mentioned here: compared to using the non-weighting rectangle-function, an apodization reduces the average noise amplitude in the interferogram. Consequently, the triangle-function decreases, for example, the spectral noise by a factor of square root of 2.

As shown in 'Fig. 3-16, the interferogram to be measured should theoretically be symmetrical, so that one only needs to record it unilaterally. However, the digitization mostly does not occur exactly in the maximum and symmetrically to this, but is somewhat shifted, thereby resulting in a wavenumber-linear phase error for the various spectral elements. Moreover, further phase errors result through the optics (e.g. dispersion effects of the beamsplitter) and the electronics (including various response behaviors for the respective frequencies), which may make the interferograms even more asymmetrical than those shown in Fig. 3-18.

In many FT spectrometers, only unilateral interferograms are recorded due to a limited maximum retardation. In these cases, a part of the other side of the maximum is recorded in order to be able to calculate the spectral phase information from a short bilateral interferogram with correspondingly lower resolution. A prerequisite for this is that the spectral phase function can be well interpolated, which generally is the case.

The width of the bilateral interferogram can often be selected within certain limits by the user. For example, if a high photometric accuracy is intended in spectral regions with strong absorption, a wide interval should be chosen. The phase correction in the spectral domain is executed according to the discrete complex Fourier transform that is implemented in the spectrometer software without user interaction. Details and effects of the so-called Mertz phase correction and a comparison to another alternative algorithm can be found in the publication by Chase [18].

FFT (Fast Fourier transform) algorithms, which considerably decrease the number of otherwise necessary multiplications and trigonometric function calculations, are used exclusively for the discrete Fourier transform. Because computer time was extremely expensive in the earlier days of FT spectroscopy, these FFT algorithms constituted a breakthrough in FT technology. A disadvantage in e.g. the Cooley-Tukey procedure is that the number of data points to be transformed must be a power of 2 [19].

This condition can also be fulfilled by using a so-called zero filling. This is the adding of zeros to both ends of the interferogram before the Fourier transform. It results in an increased number of points in the spectrum corresponding to an interpolation. For example, if a doubling of the interferogram length is produced in this manner (zero fill factor, ZFF=2), an interpolation with respectively one intermediate point results in the spectrum domain. For maximum photometric accuracy, e.g. in gas analysis, a multi-zero fill (ZFF=8) is necessary. Figure 3-23 shows the effect of the interpolation on the line shape. Depending on how the points lie, an insufficient number of sampling points is partially linked to a considerable error in the determination of band maxima.

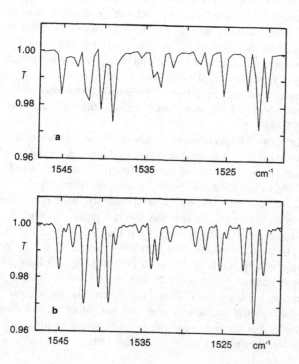

Fig. 3-23. Influence of the interpolation by zero fill on the transmittance spectrum of H_2O (spectral resolution 0.5 cm^{-1}, triangular apodization): a) no interpolation and b) 8-fold zerofill (produces seven additional data points).

3.4.3.2   Practical Tips for Measuring Spectra

After a suitable instrumental configuration has been selected (e.g. detector with a proper scan velocity), the alignment of the interferometer should be checked and optimized. Automatic routines are incorporated in some instruments, while especially for the older ones, the maximization of the interferogram signal has to be done manually, whereby typically the fixed mirror of the interferometer is adjusted.

Furthermore, suitable measurement parameters for recording the interferograms must be chosen. For standard spectra, these not only include the resolution, which also affects the size of the Jacquinot stop (most operation programs offer these suggestions), but also the number of interferograms that are to be accumulated. For strongly absorbing samples, the signal-to-noise ratio can be improved by co-adding several scans. As was already mentioned, the spectral noise can be reduced by a factor of square root of N, whereby N is the number of scans. For a fixed scan number and the same aperture, the noise amplitude is inversely proportional to the spectral resolution. In turn, the spectral resolution is inversely proportional to the interferogram length (proportional to the measurement time), so that a 4-fold longer measurement time at doubled spectral resolution is necessary in order to obtain the same signal-to-noise ratio. If the Jacquinot stop must be narrowed for the higher resolution, one even finds noise that is inversely proportional to the square root of the resolution.

For calculating the spectra from the interferograms, the choice of the apodization and the zero fill is necessary for interpolating.

Up to now, single beam spectra, which occur primarily in the spectral measurement, were predominately calculated, whether it is with sample or as a pure background spectrum. Because commercially obtained FT spectrometers are conceived as single beam instruments, it is necessary to measure the cited spectra consecutively, whereby the quotient of both gives the transmittance spectrum. The procedure is explained in Fig. 3-24. Indeed, there are also spectrometers having two or more equivalent sample compartments, which are accessible via deflection mirrors and are practical for different, fixed experiments to be set up. However, their optical performance is less satisfactory, so that a double beam option is not used for measuring the transmittance spectra.

Most often, the sample is put into the sample compartment after a background spectrum had already been recorded and stored. Before being opened up, the sample compartment in many instruments can be closed off from the rest of the spectrometer area so as not to affect the dry atmosphere in the inter-

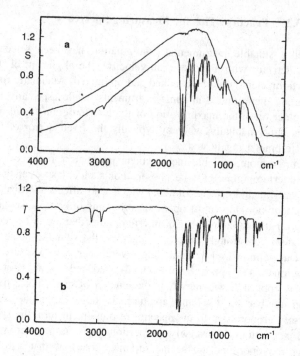

Fig. 3-24. Calculation of the transmittance from sample and background single beam spectra. a) spectrum of caffeine (measured as KBR pellet) and background spectrum with a reference pellet (shifted in the ordinate), b) quotient from both spectra.

nal part of the instrument, whenever capsuling with IR-transparent windows is not provided for. After the sample is placed and the sample compartment lid closed, one should allow enough time for renewed purging with nitrogen or clean dry air.

Instead, an automatic sample changer is advantageous, which is equipped as a sample sled with two identical cell holders and allows various positions to be reached. Thus, for example, the sample and reference cells can be subsequently analyzed without having to involve the spectrometer. An intermittent interferogram recording is also possible in order to enlarge the time proximity of sample and reference measurement. This is necessary when maximum signal-to-noise ratios have to be acquired also in those spectral regions, in which disturbances by vapor and carbon dioxide from the laboratory atmosphere are expected. Other sample changers having up to 64 different positions on a sample wheel are available for the software-operated routine measurement of pellets.

In order to test the efficiency of the spectrometer with respect to obtainable signal-to-noise ratios, so-called 100% lines are recorded that can be represented as the quotient of two successively taken single beam spectra of the empty sample compartment. On the one hand, this spectrum gives information about the stability of the instrument: after the start-up of the spectrometer, it lasts a little time until a perfect horizontal line with a slowly changing noise amplitude can be produced. Constant purging of the instrument can be recognized by the absence of obvious residue lines of H_2O and CO_2. On the other hand, a 100%-line documents obtainable spectral signal-to-noise ratios. Because of the single beam property, the signal-to-noise ratio is reduced (the spectral intensities are standardized to 100% for this presentation) based on the decreasing intensity towards the borders of the spectrum. This dependency is not found in dispersive spectrometers that are equipped with slit programs. For documenting the satisfactory functioning of a FT spectrometer, one should save a 100%-line, recorded under set standard conditions, for later comparisons.

Dull polystyrene films, which should be free of interference effects (humps in the spectrum), were recommended for checking the reproducibility of the wavenumber and absorbance scale. For such tests, certified reference materials exist from the National Physical Laboratory (NPL) in England as well as from the National Institute for Standards and Technology (NIST) in the USA. The aspect of instrument standardization via the measurement of the sample standard will be increasingly important in the future.

The wavenumber stability is linked with the Connes advantage of the FT-IR spectrometer and has already been cited in Sect. 3.4.2.3. The instrument manufacturers generally specify accuracies better than 0.01 cm^{-1}. Besides decreased resolution, a wavenumber-linear frequency shifting towards lower values also occurs due to the slight non-parallelism of the radiation beam in the interferometer. A minimal movement of the He-Ne laser beam compared to the IR beam in the optical beam path of the interferometer may cause a change. Because the relative difference to the exact wavenumber is about proportional to the relative error in the laser wavenumber, a calibration through comparison with a few absorption bands of standard gases [7] is not complicated to do. Hawkins et al. [20] give an example for the obtainable accuracy with a commercial FT spectrometer for gas spectra with a spectral resolution of 0.06 cm^{-1}.

The high wavenumber reproducibility of the interferometer is advantageous when evaluating extremely small signals, which sit on stronger absorption bands of e.g. solvents. Otherwise, during the spectral subtraction, so-called "derivative signals" are produced from the solvent spectrum, which may impair the evaluation [21].

3.4.3.3 Photometric Accuracy of FT Spectrometers

The high reproducibility of spectral recording with FT spectrometers has often been praised. The absolute photometric accuracy plays such an essential role, that a comparison of FT spectra with secondary standard spectra of various liquids is much discussed in literature [22]. In principle, the quantitative analysis is affected here, although one can fall back to the standpoint of sufficient reproducibility, if, for the calibration, one tolerates signals that are non-linearly dependent on the concentration. These aspects, in particular effects by apodization and insufficient resolution with respect to the band half-width, will be handled in detail in Chapter 7.

FT spectroscopy has inherent errors with different effects which are caused, for example, by the thermal radiation of the sample, whereby this enters the interferometer and, doubly modulated, contributes towards the detector signal [23], or by defocusing via the change of the optical path upon sample placement as compared to the empty spectrometer. Still other effects are discussed by Hirschfeld [24]. A survey of the error sources to be expected and their diagnosis is also given by Guelachvili [25].

An unfavorable case entails the situation when the applied detectors detect incident radiation intensity non-linearly. This is especially a problem, because the interferogram demands extremely high dynamics due to its maximum being in the zero position. The pyroelectric DTGS detector is known for its high linearity, so that no photometric inaccuracies occur with this, e.g. in the characterization of optical filters [26].

In contrast, the HgCdTe (MCT) detector, which is mostly operated as a photoconductor with constant current, behaves extremely non-linearly at higher radiation intensities. The direct effect is a compression of the interferogram around the maximum. Chase [27] elucidated the effects in the spectral domain, and an example of a spectrum was already presented in Fig. 3-17. The cutoff wavenumber of the applied MCT detector lies at ca. 530 cm^{-1}, but falsely, intensity below this limit is pretended. In addition, phase errors of absorption bands occur in the long-wave region. By using a smaller interferometer aperture, this effect can naturally be reduced but, unfortunately, the signal-to-noise ratio is relinquished through this. A drastic example of how the effect of non-linearity can act is shown in Fig. 3-25 in which the spectra of the SF_6-sample, recorded with DTGS and MCT, can be compared. In contrast, if the sample and background spectrum can be measured under comparable conditions, because e.g. diluted aqueous solutions are to be analyzed and a pure water sample is used as the reference, the photometric errors for the resulting absorbance spectrum can be eliminated in good approximation [28].

Aside from various hardware solutions for eliminating the non-linear detector behavior, an algorithm was developed and patented, with which the occurring non-linearity in the interferogram can be estimated via the respective single beam spectrum [29].

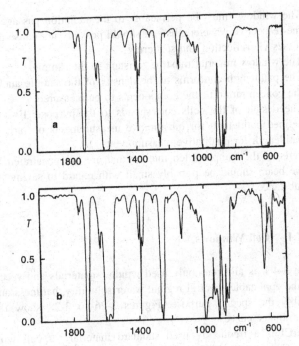

Fig. 3-25. Effects of detector non-linearity on the transmittance spectrum of an SF_6-sample (1000 hPa, 10 cm, 25 °C).
a) correct spectrum, measured with a DTGS-detector,
b) deformed spectrum, acquired with a MCT-detector and large aperture.

After correction of the interferogram, the usual further processing with Fourier transform etc. delivers comparable spectra having clearly less noise than those obtainable with a DTGS detector.

3.5   Standard Accessories

The following section covers possibilities for placing the sample into the IR beam in order to attain optimal transmission measurements. This primarily concerns all types of cells, holders for solid samples (films, pellets) and condensers for concentrating the beam onto a small sample volume (microtechnique). Optical accessories for measuring reflectance spectra are elucidated in Chapter 5.

3.5.1   Cells

Cells are containers suited for bringing liquid or gaseous samples with a defined pathlength into the beam. The following prerequisites must be met for these:

- The windows must be permeable to the radiation passage at the respective wavelengths and should possibly not cause any losses via reflection and scattering.
- The window material must be resistant to the sample.
- The pathlength conforms to the density of the sample and to the concentration of the components to be measured.
- The design of the cells corresponds to the purpose: Exactly defined pathlength for quantitative measurements or variable pathlength for qualitative measurements.
- Parts of the sample-filled interior that are not penetrated by the beam should be possibly small with regard to saving the substance.

3.5.1.1   Cell Windows

Table 3-4 lists all commonly used window materials and specifies the applicable spectral region, water solubility, hardness and possibly the special purpose. Figures 3-26 to 3-28 show the transmittance curves.

NaCl is a frequently used standard material for cell windows. One resorts to **KBr** and primarily to expensive **CsI**, when the spectral region to be examined requires this.

Infrasil cells with windows made of synthetic quartz are very robust compared to other materials but, unfortunately, are only permeable until 3.5 µm (2850 cm^{-1}). However, this material is advantageous for measurements in the near-IR region, for studies of fundamental vibrational bands in the region around 3000 cm^{-1} as well as for making pressure cells.

KRS-5, a mixed crystal of 42% thallium bromide and 58% thallium iodide, is a red soft crystal that is particularly suited for studies in the long-wave spectral region and for aqueous solutions because of its insolubility in water and its broad transmission range. Only alkaline samples and, particularly, phosphates attack the material and leave a yellow residue behind. Because **KRS-5 is toxic**, caution is necessary. Upon handling the material, predominantly upon grinding and polishing, rubber gloves must be worn and careful attention should be paid to discarding wastes. Because the material shows cold flow under pressure, the pathlength of fixed assembled cells must be frequently checked.

Irtran windows are glasslike, sintered crystals under pressure which show a relatively large hardness and are resistant to thermal and mechanical shocks. Besides the materials listed in Table 3-4, there are:

- **Irtran-1** (MgF$_2$, permeable until 7.5 µm or 1330 cm^{-1}),
- **Irtran-4** (ZnSe, permeable until 20 µm or 500 cm^{-1}) and

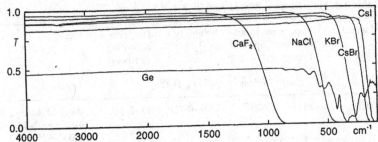

Fig. 3-26. Transmittance of cell windows, window materials: CsI, CsBr, KBr, NaCl, CaF$_2$ and Ge; pathlengths: Ge 3 mm, CsBr 4 mm, all others 5 mm.

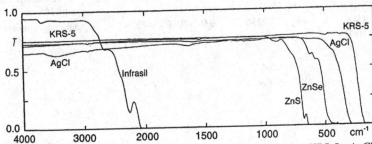

Fig. 3-27. Transmittance of cell windows, window materials: KRS-5, AgCl, ZnSe, ZnS and Infrasil; pathlengths: AgCl 1 mm, Infrasil and ZnS 2 mm, as well as ZnSe and KRS-5 5 mm.

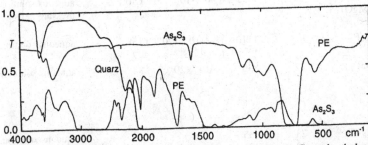

Fig. 3-28. Transmittance of cell windows, window materials: As$_2$S$_3$, polyethylene and quartz; pathlengths: As$_2$S$_3$, 2.5 mm, polyethylene and quartz 1 mm.

- **Irtran-6** (CdTe, permeable until 28 μm or 350 cm^{-1}), which is permeable in a broad range but is very expensive.
- **Irtran-3** is CaF$_2$ that is described as such in Tab. 3-4. This quite expensive material is suited for studies at high pressure as well as at high and low temperatures. Although sapphire is much more resistant to pressure, its applicability is limited, however, due to its very short-wave transmission limit.

If tools for producing KBr pellets are available, cell windows for a single use can be made by oneself. A pair of such pellets

Table 3-4. Material for cell windows [30, 31]

Material	Long-wave Limit of Usable Spectral Range		Solubility in H₂O	Hardness according to Knoop	Use	Remarks
	λ (μm)	$\tilde{\nu}$ (cm⁻¹)	g/100 g H₂O			
NaCl	16	625	35.0 (0 °C)	15–2–18.2	standard windows for the range 2–16 μm (5000–625 cm⁻¹)	inexpensive, poorly soluble in alcohol when water-free; halogen exchange
KBr	25	400	53.5 (0 °C) hygroscopic	6.5	for measurements until 400 cm⁻¹	soluble in alcohol; halogen exchange
CsI	60	165	44 (0 °C)	soft	for measurements until 165 cm⁻¹	soluble in alcohol; halogen exchange
KRS-5 (TlBrI)	40	250	0.02 (20 °C)	40	aqueous solutions	42% TlBr, 58% TlI; sensitive to various org. solvents; very soft; high refraction index (2.37 at 100 cm⁻¹); high losses by reflection (28.4% at 100 cm⁻¹ for 2 surfaces) **toxic!, gloves!**
Irtran-2 (ZnS)	14	710	0.69·10⁻⁴ (18 °C)	250	aqueous solution; acids, bases; higher pressure; high temp. (up to 800 °C)	very robust, insensitive to temperature changes; mechanically stable
Irtran-5 (MgO)	8.5	1175	6.2·10⁻⁴	640	aqueous solution; pressure cells	sensitive to acids and ammonium salt solutions
CaF₂	9	1110	2·10⁻³ (20 °C)	158	aqueous solution	soluble in ammonium salt solutions
BaF₂	12	830	0.17 (20 °C)	82	aqueous solution	soluble in acids and ammonium salt solutions; sulfate- and phosphate ions form residue
AgCl	23	435			aqueous solution, corrosive liquids incl. HF, deep-temperature cells	sensitive to light! do not polish! very soft, soluble in ammonia, Na₂S₂O₃ and hydrocarbons; inexpensive
AgBr	38	285	insoluble	9.5		
Al₂O₃ (sapphire)	5.5	1800	9.8·10⁻⁵	1370	high pressures	extremely hard; difficult to grind
SiO₂ (Infrasil)	4.0	2500	insoluble	ca. 470	near-IR; pressure cells	very robust; sensitive to alkali
Polyethylene	1000	10	insoluble		aqueous solution; org. sovents, acids etc. in long-wave and far-IR, robust; cheapest water-insoluble material	over 500 cm⁻¹ only usable as film to protect other window materials; only sufficiently permeable below 500 cm⁻¹

made of KBr, KI, CsI or RbI with an embedded liquid sample, placed in a pellet holder, is well suited for producing qualitative overview spectra (see Sect. 3.5.1.3 and 4.2.1.4).

Further information about optical materials for infrared spectroscopy is found in Miller and Stace [30].

3.5.1.2 Handling and Care of Cell Windows

As shown in Table 3-4, the cell windows consisting of single alkali halogenide crystals dissolve well in water and are hygroscopic. Hence, care must be taken upon storing and handling them.

The windows are best stored in an exsiccator or in another airtight closable container having a corresponding amount of blue gel that must be occasionally replaced.

Upon handling the windows, especially those of the very expensive CsI, the wearing of at least rubber fingers is advisable. However, gloves must always be worn when handling KRS-5 windows because of their toxicity!

The careful and responsible chemist should always make it a point to wear rubber gloves when manually coating the cells because of possible skin-damaging and other toxic properties of samples having physiological effects that are often unknown. Moreover, the responsible spectroscopist may make sure that co-workers under his/her responsibility also follow this safety precaution!

Before a cell is coated, a possible water content of the sample must be taken into account or, if needed, be determined. Water contents of up to 1% are no problem for alkali halogenide cells. At higher concentrations, especially for aqueous solutions, water-compatible materials such as KRS-5, Irtran, CaF_2, BaF_2, AgCl or Infrasil must be used, accounting for the wavelength range of interest. For some water-resistant materials, the refraction index is relatively high, thereby resulting in disturbing interference effects through multiple reflections within the cell (see Tab. 3-4).

A clouding of the alkali halogenide windows caused by air moisture or water-containing samples is no problem, because the crystal surfaces can be repolished in the following way:

A piece of soft window leather or a cotton cloth is stretched over a flat surface, e.g. a ca. 10–15 mm high metallic or glass cylinder with a smoothly polished surface and a diameter of ca. 10 cm and held in place by a correspondingly sized clasping ring or rubber ring. Very slight clouding can be removed onto the smooth dry leather by polishing the window using circular motions.

A better effect is attained when the leather is dampened with dry isopropanol or toluene (fluid film between leather and surface!). In this case, the solvent residue must be removed by buffing on a dry part of the leather. The polishing tool must always be clean and dust-free.

If a window shows stronger clouding or slight scratches, better polishing can be reached when the leather is dusted with e.g. polishing red (iron (III) oxide powder) before it is wetted. This case also requires subsequent cleaning in solvent and buffing on the dry leather. For strongly scratched window surfaces, some water may be added to the isopropanol in order to improve the effect. Nonetheless, surfaces treated in this way are no longer exactly flat and are still applicable only for qualitative measurements.

3.5.1.3   Dismountable Cells

The simplest type of cell (Fig. 3-29) consists of two, generally circular cell windows (diameter e.g. 25 mm) separated by a spacer of aluminum, lead or Teflon having a thickness of between 10 µm and 500 µm, depending on the concentration and the intensity of the spectrum expected. In extreme cases, the spacer may be a few millimeters thick, which is typical especially for near-IR spectroscopy. The spacer may also be waived for extremely small pathlengths.

The pathlength, dictated by the spacer, is not precisely defined but rather depends on the acting force, after the screwing together, and on the amount and viscosity of the excess sample quantity entering between the spacer and the window. Hence, dismountable cells are only suitable for qualitative measurements (see Sect. 3.5.10.1 for using empty pellets as cell windows).

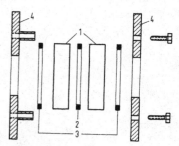

Fig. 3-29. Dismountable cell.
1) Cell windows
2) spacer ring
3) intermediate ring
4) holder.

3.5.1.4   Cells with Defined Pathlength

This term describes cells, which are assembled once and then are frequently reused. Like dismountable cells, fixed cells consist of two windows with a suitably thick spacer placed in between, but one of the windows is furnished with two boreholes for coating the cell. The boreholes continue into the holder and respectively end in a small socket that can be closed with a Teflon stopper (Fig. 3-30).

By using intact spacers and windows, fixed cells in a closed state are tight enough for solvents having boiling points above 50–60 °C. One must consider, however, that the sample temperature increases by radiation absorption that, in turn, causes a corresponding increase of the partial pressure of the cell contents. This may result in the sample or solution partially or complete-

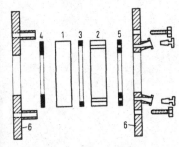

Fig. 3-30. Cell for quantitative measurements on liquids.
1) Cell window, not bored through
2) cell window, bored through
3) spacer ring
4) intermediate ring
5) intermediate ring, bored through;
6) holder.

ly evaporating through leaks between the windows and spacer. For such purposes, the instrument producers market "sealed" liquid cells produced with freshly amalgamated (by dipping into mercury) lead spacers. Such cells are also sufficiently tight for use of lower boiling solvents (e.g. *n*-pentane, chloroform, carbon disulfide). Even assembled cells with Teflon spacers are just as tight.

Fixed cells are mainly applied for quantitative measurements that depend on accurately knowing and maintaining the pathlength. See Sect. 7.10.2 for determining the pathlength. However, it mostly depends on the pathlength remaining constant during a measurement series, consisting of calibration- and measurement samples. This is generally guaranteed when the absence of moisture in the samples is assured. It is appropriate, however, to check this prerequisite by measuring an absorption band, e.g. of cyclohexane, before and after the entire measurement series (see Sect. 7.2.4.1).

3.5.1.5 Cells with Variable Pathlength

An elegant and indispensable technique for achieving smaller pathlengths or higher concentrations is the use of a cell that allows a change of the pathlength in the mounted and full state (Fig. 3-31).

By means of a fine winding, the distance between the window surfaces can be altered by a turning motion of the outer cell parts. The setting of defined pathlengths is simplified by an attached measure, mostly with a nonius. If very small pathlengths are needed, one first fills the sample at a higher pathlength and then reduces this to the extent required.

Fig. 3-31. Cell with variable pathlength.

3.5.1.6   Gas Cells

Corresponding to the lower density of gases, the pathlength needed here is considerably larger and generally amounts to 5 or 10 cm. A gas cell consists of a glass cylinder with two closable entries (vacuum-tight glass spigots) and flatly parallel polished ends. Because the beam path towards the source diverges, an inner diameter of about 45 mm is necessary and thus a window diameter of 50 mm.

The windows may be sealed to the glass body by using a suitable rubber-elastic material; for very aggressive gases, also with Teflon, if necessary. The cell is mounted by means of a corresponding holder and two rubber rings for protecting the windows. The coating takes place according to Sect. 4.4.1 by using a vacuum apparatus.

If trace substances are to be detected in a weakly absorbing or IR-inactive gas, an increase of the pathlength beyond 10 cm is necessary. Because the sample compartment generally accommodates cell lengths of maximally 10 cm, a multi-reflection cell according to *White* is used for reaching higher pathlengths.

Within the cell room, the beam is deflected by a plane mirror horizontally by 90° and, by using a mirror system, reflected back and forth many times until it finally normally leaves the ordered cell exit window after having traveled a defined pathlength (Fig. 3-32). In this manner, pathlength of up to 40 m and more can be reached. The high number of reflections on the mirror surfaces, however, results in considerable radiation losses within the cell. The most frequently used longpath cell with a 1 m pathlength is still easy to use with relatively low losses and only four reflections. Cells with variable pathlengths of up to 10 or 40 m are also applicable with current FT spectrometers and allow the measurement of traces down to, in the extreme case, 0.1 ppm and lower.

Fig. 3-32. Long-path gas cell (schematic).

3.5.1.7   Holder for Pellets and Films

KBr pellets with a diameter of 13 mm are made with the usual instruments for producing KBr pellets for the study of solid substances. These pellets are put into simple holders, which can be bought from the instrument manufacturer. The pellet is oriented in such a way that it is near the focus where the spreading of the radiation beam reaches its minimum.

There are special holders for freestanding films where the sample is fixed with a magnet ring. Depending on the size and solidity of the available material, the film can also be mounted between two cell windows or in a KBr pellet holder. The easiest way is placing the film onto a rectangular perforated piece of cardboard which fits into the slits intended for the standard cell holders. Polystyrene film that is generally included as a test sample is put on such a piece of cardboard.

3.5.1.8   Microcells and Beam Condensers

For very small amounts of sample, correspondingly small microcells are earmarked for gases and liquids as well as tools for producing micropellets. Whereas micro gas cells are, at best, precisely fitted only to the convergence of the radiation beam thereby preventing dead volume, some microcells for liquids and micropellets are so small that they would screen the radiation beam. By means of optical units housed in the sample compartment, the radiation beam can be focused on the site of the microcell. In this way, sample volumes of a few microliters or micropellets of one-half millimeter diameter are measurable upon careful adjustment of the sample in the beam condenser. This unit is also known under the name "micro-illuminator" and may consist of a mirror or lens system. The latter has the disadvantage of increased absorption nearby the transmission limit of the lens material (NaCl or KBr) and, hence, its use is uncommon.

Ultra-micro liquid cells, placeable into the pellet holder, allow the investigation of liquid volumes down to $0.3 \, \mu l$ without optical aids.

3.5.1.9   Cells for Measurements at Increased
and Reduced Temperature

In connection with studies about the effect of thermal influences on material structure by using the IR spectrum, information about a large number of heatable and coolable cells was published. Because such a review would overextend the framework of this book, only a few fundamental aspects are mentioned here. A survey of the most important heating and cooling cells with a detailed reference overview is found in Miller and Stace [32]. Moreover, accessory retailers offer up to $250 \, °C$-heatable cells for gases and liquids as well as cells that can be cooled down to the temperature of liquid helium.

In measurements of heated samples, one must always heed an important condition: the eigenradiation of the substance. Every material found in the beam path which has a higher temperature than that of its surroundings yields, in addition to the radiation of the light source, a radiation fraction corresponding to its temperature (see Sect. 5.4). This radiation fraction can already be noticeable at $50 \, °C$ and may lead to errors in intensity measurements. Because, according to the Kirchhoff law, the ra-

tio of the emission to absorption is independent of the substance
and is only a function of temperature and wavelength, the substance located in the beam path selectively emits in the regions
of its absorption band upon an increase in temperature. The
sample compartment in FT spectrometers typically lies behind
the radiation-modulating interferometer, so that the sample
emission falls non-modulated on the detector and therefore is
not detected. Nevertheless, errors by double modulation are possible (see Sect. 3.4.4.3).

For low demands, heating cells can also be produced by
using heating bands. For sealing up the sample compartment,
silicon rubber rings or Teflon rings have been shown to be better than sealings with hardened two-component glues. Stronger
temperature changes usually result in a break on the interface,
because the hardened glue mass does not give with the thermal
expansion of the window material.

Moreover, NaCl and KBr windows are quite insensitive to
temperature changes when the heating or cooling occurs slowly.
However, hot cell windows should never be exposed to the bare
metal of a cold tweezer (tweezer tip should be warmed up
somewhat or covered with a Teflon tube). In general, the upper
limiting temperature for heating cells is not only determined by
the window material but also by the sealing material applied
and lies in the range of ca. 250 °C for silicon rubber and Teflon.

Studies at lower temperatures must be conduced in especially
designed cryostats, the IR-permeable cell windows of which are
protected from tarnishing by placement in an evacuated external
cell [32]. A special technique is used in the matrix-isolation
spectroscopy (see Sect. 5.5.1).

3.5.1.10 Cells for Studies under Pressure

Cells which can withstand high and maximum pressures are
needed not only for spectroscopic-theoretical research targets
but, by all means, also for practical problems such as e.g. the
mode of action of a catalyst system of a high-pressure reaction
[33]. For several window materials, Table 3-4 (Sect. 3.5.1.1)
shows the hardness according to Knoop as a measure for the applicability of pressure cells. No doubt, sapphire is best suited
but, unfortunately, is permeable only to $1800\ cm^{-1}$ (5.5 μm). Irtran and calcium fluoride are also usable materials for pressures
that are not excessive.

Pressure cells, except those for relatively moderate pressures,
are hardly available on the market (see Sect. 5.5.3). Miller and
Stace [32] give an overview of the most important cell designs
with a large number of literature references. In addition, Tinker
and Morris [34] describe a heatable pressure cell.

3.6   Spectral Processing

The spectra produced by the spectrometer are typically stored digitally. Here lie the enormous possibilities of spectral processing which are partially offered by the spectrometer operation programs for routine work. It is sensible to somewhat consider what is behind the algorithms in order to understand their prerequisites and limits. The reader should refer to more detailed literature for a more intensive handling of this material (e.g. [12, 35, 36]).

A special problem is presented by the import and export of spectral data measured on spectrometers of various manufacturers. Because of the different data formats, in the past compatibility problems always reoccurred. A few years ago, a consensus could be reached on a format recommended by the Joint Committee on Atomic and Molecular Physical Data (JCAMP) [37], in which a spectrum inclusive for the recording of relevant instrument parameters as well as other information may be stored as a text file in ASCII-format. Corresponding programs for converting the files are made available by the spectrometer software and instrument producers. In the meantime, this format was further developed to likewise implement molecular structural information [38]. Other types of spectroscopy such as nuclear magnetic resonance (NMR) and mass spectrometry (MS) also use the JCAMP-DX format. Davies [39] gives a survey of the present state of activities.

Some of the options concern the presentation of the spectra as a discrete point function. The interpolation for FT spectra via zerofill in the Fourier domain was already handled (see Sect. 3.4.3.1). Similar possibilities exist naturally for already existing spectra by using polynomial functions to be adapted. With this, other scan wavenumbers can also be realized in order to obtain, for example, spectra compatible in the abscissa values. On this basis, most software packages offer a conversion of wavenumbers to equidistant wavelengths that are moreover used in NIR spectroscopy for presenting spectra.

Another aim concerns the conversion of transmittance values into absorbance values and vice versa that the user often meets. In addition, other transformations e.g. according to Kubelka-Munk and Kramers-Kronig will be handled during the introduction of reflection techniques (see Sect. 5.1).

Baseline correction with linear or parabolic functions where the number of baseline points can be set represents an important processing of spectra. Uncompensated reflection losses e.g. by cell window surfaces, often cause sloping baselines, or wavenumber-dependent scattering effects occur with opaque pellets (see Sect. 4.2.1.2, Fig. 4-1). A standardization of ordinate values can be undertaken with the selected baseline – in transmittance units, this is then represented by $T = 1$ (absorbance $A = 0$). Upon using this kind of spectral processing, however, one should be careful in selecting the number of data points. Further methods for baseline correction will still be cited when filtering via of the Fourier domain (interferogram) is discussed (see Sect. 3.6.3).

The spectral background may also include disturbances through bands of solvents or impurities, which can be eliminated by subtraction. The simplest method constitutes using a corresponding reference for compensation in the recording of the background spectrum. Often successful, however, is only a scaled subtraction of the disturbing substance bands in the absorbance spectrum. In many cases, the user has the possibility here of conducting the subtraction interactively, whereby one can vary the scaling factor and remove a linear ordinate offset. Spectral artifacts that are especially observable at high signal-to-noise ratios and corresponding ordinate spreading [21], however, may form by band shifting and other effects.

3.6.1   Data Reduction and Spectral Simulation

A data reduction can be undertaken with existing spectra in such a way that the spectrum can be relatively completely described with a few parameters [36]. With the exception of gas spectra that may show a complex rotational-vibrational fine structure, the spectra of substances in a condensed phase can be simulated by a small number of simply described bands. The individual bands are often described alone by the band position, namely the abscissa value of the relevant band maximum (for absorbance spectra) as well as by the half-intensity width. The latter denotes the width of a band at the site of half of the maximum absorbance.

Furthermore, the integral intensity plays a role, for example, in the quantitative analysis or when it concerns the calculation of transition probabilities of vibrational states, which are interpretable via molecular models described via structure and vibrational dynamics. Besides many definitions with various units [40], one uses the following practical equation for the band integral

$$A = \int a(\tilde{\nu})\mathrm{d}\tilde{\nu} = 1/(cb) \int \log\{I_0/I(\tilde{\nu})\}\mathrm{d}\tilde{\nu} \qquad (3.10)$$

$$[\mathrm{L\ mol^{-1}\ cm^{-2}}]$$

where b is the pathlength and c, the fractional amount of the substance. For determining the integral intensity, the band can be integrated numerically or be analytically calculated via a suitable function.

Suitable functions for band fitting can be based theoretically [41, 42]. One of these functions is the Lorentz function

$$F_L(\tilde{v}) = \frac{A_{max}[\Gamma/2]^2}{(\tilde{v} - \tilde{v}_0)^2 + (\Gamma/2)^2} \qquad (3.11)$$

which is applicable e.g. for the profiles of rotational-vibrational lines of gases, when the broadening is mainly caused by collision processes (so-called pressure broadening; A_{max} is the maximum band absorbance, Γ the full line half-width). The integral of a Lorentz curve can be specified by:

$$\int F_L(\tilde{v})d\tilde{v} = \pi A_{max}\Gamma/2 \; .$$

In contrast, in the lower pressure region of gases, the Doppler broadening predominates, which is caused by the Maxwell velocity distribution of the molecules and which leads to Gauß curves:

$$F_G(\tilde{v}) = A_{max}e^{-(\ln 2)(\tilde{v}-\tilde{v}_0)^2/(\Gamma/2)^2} \qquad (3.12)$$

In Gauß curves, the line intensity at the band flanks decreases much more rapidly than in Lorentz curves having the same half-widths (see Fig. 3-33). This is favorable for neighboring lines, which might otherwise hinder the determination of the maximum or the numerical integration of the band area. On the other hand, the integral of the Gauß curve can be represented by:

$$\int F_G(\tilde{v})d\tilde{v} = (\pi/\ln 2)^{1/2}A_{max}\Gamma/2 \; .$$

When collision effects gain importance with increasing pressure, Voigt curves are formed which can be described by a convolution of the Gauß- and Lorentz function.

These band shapes can also be used for condensed phases. Product or sum functions are often applied to be able to model the spectra better. An example is given by:

$$F_s(\tilde{v}) = LF_L(\tilde{v}) + (1 - L)F_G(\tilde{v}) \qquad (3.13)$$

where L is the Lorentz term with $0 \leq L \leq 1$, A_{max}, and F_L and F_G are respectively equal to Γ.

The band fitting becomes more complex when various bands overlap one another (see Fig. 3-34). In addition to the spectral parameters of the individual bands, the number of components making up the spectrum, of course, is important. On the one hand, one can estimate this by visual inspection and, on the other hand, one

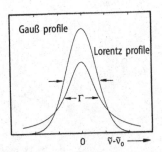

Fig. 3-33. Gauß- and Lorentz profile with the same half-width and band area.

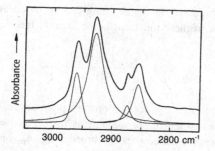

Fig. 3-34. Baseline-corrected spectrum of a solution of hexane, cyclohexane and dimethylformamide in CCl$_4$, as well as its separation into four single bands (ordinate-shifted).

can use resolution-improving methods which are practical for determining the position and which are discussed in more detail in Section 3.6.3. For calculating the band parameters, one first starts out with approximate values and optimizes iteratively until the sum of the squares of the deviation between the experimental and simulated spectrum reaches a minimum. In the correct model, the difference spectrum between both should only show noise, otherwise the model has deficiencies e.g. with respect to the number of bands, their symmetry or the baseline. Comprehensive papers on curve fitting and their limits were published by Gans [35] and Maddams [43].

3.6.2   Spectral Smoothing

Background noise overlaps all experimental spectra, so that they may be interpreted only to a limited extent. Previous smoothing of spectra via digital filters is helpful for processing information better. As Fourier transform is available, corresponding operations in the Fourier domain are applicable.

In order to increase the signal-to-noise ratio, several spectra can be determined, analogous to the procedure in recording interferograms by coaddition. As we nonetheless have seen, a smoothing of the spectra is possible by multiplying the interferogram with a weighting function (see Sect. 3.4.3.1). A shortening of the interferogram via a narrower boxcar function eliminates higher frequency fractions in the spectrum, thereby reducing noise; analogously, the resolution becomes worse.

As we have seen earlier, this multiplication in the Fourier domain corresponds to a convolution operation in the spectral region, which will be elucidated in Fig. 3-35. In the numerical example, the convolution function was not normalized for the sake of simplicity (the integral (area) or the sum of the convolution coefficients is otherwise equal to 1; this is necessary in order to keep the band area constant).

Such a folding or convolution is also performed in the best known smoothing method according to Savitzky and Golay

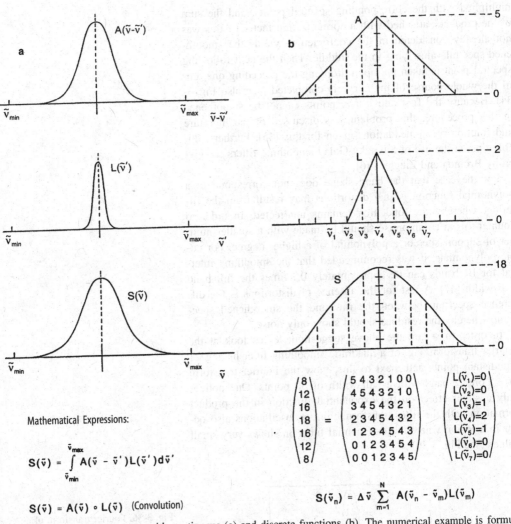

Mathematical Expressions:

$$S(\tilde{v}) = \int_{\tilde{v}_{min}}^{\tilde{v}_{max}} A(\tilde{v} - \tilde{v}')L(\tilde{v}')d\tilde{v}'$$

$$S(\tilde{v}) = A(\tilde{v}) \circ L(\tilde{v}) \quad \text{(Convolution)}$$

$$S(\tilde{v}_n) = \Delta\tilde{v}\sum_{m=1}^{N} A(\tilde{v}_n - \tilde{v}_m)L(\tilde{v}_m)$$

Fig. 3-35. Convolution process with continuous (a) and discrete functions (b). The numerical example is formulated by using the matrix form (A is hereby e.g. the instrumental line function, L the true line function and S the measured spectrum).

[44], based on the fitting of polynomials, among others, of the 3rd or 5th degree, according to the method of the smallest square of the deviation. The results with a polynomial of even and next-higher order, for example, squared and cubed, are identical. A various number of data points may be used; typically applied are values of between 5 and 25 (2 m+1) for which corresponding tables of convolution coefficients (discrete convolution function) already exist (e.g. [12]). One should note that the tables listed in the original publication [44] contain errors. However, recursion formulas also exist with which the coefficients can be easily calculated. The coefficients must be respectively

multiplied with the corresponding spectral points, and the sum of the products, attached with a normalization factor (if this was not already considered in the coefficients), yields the smoothened spectral value lying in the middle. Then the next following spectral point is taken up, upon dropping the preceding one, until the whole course of the curve is completed (see also Fig. 3-34). Because the first and last m points cannot be smoothened in this procedure, this problem was discussed in the literature and solution recommendations given for this [45]. Further valuable tips on the use of Savitzky-Golay smoothing filters are given by Bromba and Ziegler [46].

For the case that the band shape does not correspond to a polynomial function, signal distortions may result from the filtering, whereby the total area remains unaffected. In order to counter this, a compromise should be made with a smaller number of support sites or a polynomial of a higher degree. For cubic polynomials, it was recommended that the smoothing interval for IR bands can be approximately 0.7-times the full band half-width [47]. A test for the absence of distortions is the difference spectrum between the raw and the smoothened spectrum, whereby the residues should show only noise.

In order to secure the above suggestion, let us look at the Fourier transform, e.g. of a quadratic smoothing filter having 21 equidistant points and, next to this, view the Fourier transform of a Lorentz band with a half-width of 30 points. One notices only a slight effect on the exponential function in the product formation with the filter transformation. Its oscillations also occur in a region where the exponential function shows very small values (see Fig. 3-36).

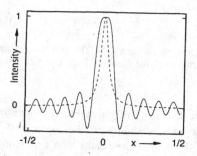

Fig. 3-36. Fourier transform of a Savitzky-Golay smoothing filter (quadratic polynomial, 21 smoothing points) and a Lorentz band with a half-width corresponding to 30 points (dashed-line function); the "half-life" of the exponential function is inversely proportional to the half-width of the spectral band.

3.6.3 Derivative Spectroscopy

Derivative spectra can be calculated using the Savitzky-Golay algorithm. This allows a simultaneous smoothing of the spectrum, which reduces the noise generated by derivation. Derivative spectra show a few advantages over the raw data. For example, linear baseline functions can be eliminated with the first derivative and quadratic ones with a second derivative. For higher polynomial orders of the baseline, further derivatives are accordingly needed to eliminate it. Because of the resulting baseline correction, band fittings based on spectra of the second derivative were recommended [48]. For calculating derivative spectra, a very good signal-to-noise ratio in the original spectrum is generally necessary, in particular when higher derivatives are intended [49].

Another area for the analysis of spectra in the first and higher derivatives lies in band separation, i.e. in the setting of band maxima, the points of inflection and the number of components. Inherent for the derivative spectroscopy is that the band shape becomes smaller the higher the selected derivative is. A disadvantage of this technique is that negative "band overtones" occur that hinder the spectral interpretation. The second derivative is often used producing satisfactory results (of course, depending on the number of convolution baseline points as discussed in the previous section). Fig. 3-37 shows an application of the second derivative on the known multi-component spectrum for which a band separation was displayed (Fig. 3-34).

Both discussed advantages, namely baseline correction and narrower band shapes are applied, for example, in gas analysis by means of diode laser spectroscopy. The second derivatives here are experimentally obtained via a fast modulation in the wavenumber axis. Talsky gives a literature review on this [49].

The derivative spectra can be calculated via the Fourier domain. As shown by the definition of the Fourier integral (e.g. [59]), corresponding multiplicative filter functions $D_n(x) = (-1)^{n/2}(2\pi x)^n$ can be found, where n denotes the even derivative order, respectively. The Fourier transform of a Savitzky-Golay filter for calculating the second derivative shows, starting from zero, a similar increase as the quadratic direct derivative operator for the interferogram, but the former displays the same side lobes around zero as was shown in Fig. 3-36. These "apodization functions" were compared with others, which incorporate comparable resolution improvements for the spectra but result in

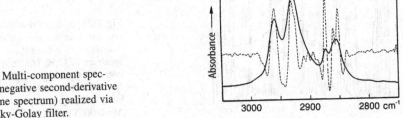

Fig. 3-37. Multi-component spectrum and negative second-derivative (dashed-line spectrum) realized via the Savitzky-Golay filter.

a greater similarity of the processed spectrum with the initial data. These methods are discussed in the next section.

3.6.4   Spectral Deconvolution (Resolution Enhancement)

With insufficient spectral resolution, the experimental spectra are decisively affected by the instrumental line functions – slit function for dispersive spectrometers or Fourier transform of the apodization function for FT spectrometers (see Fig. 3-35). In the past, various attempts have been undertaken to deconvolute such a shape in the spectral domain. As is illustrated in the figure above, for discrete functions, the convolution can be formulated as a linear system of equations; the deconvolution is to be resolved according to the true band function. Because the system of equations is poorly conditioned, instead of a direct matrix inversion, iterative techniques are applied having corresponding boundary conditions between the individual steps (e.g. noise reduction via smoothing, non-negativity of the spectral absorption). Figure 3-38 shows the resolution improvement achieved in this way for a C_2H_6-spectrum in the C–H stretching region [51].

A review of the various algorithms described in the literature for the spectral domain is found in the monograph by Jansson [52]. Likewise, alternative and calculatively intensive techniques (e.g. the concept for maximizing entropy often used in the NMR spectroscopy) are introduced, but which cannot be handled further.

In most software packages for processing spectral data, another method is implemented, namely the self-deconvolution via simple steps in the Fourier domain [53]. The theory on this is simple when the bands are represented as Lorentz curves and instrument effects can be neglected. As was already seen in Fig. 3-36, such a band shape in the Fourier domain can be depicted as a decreasing exponential function $e^{-|x|\Gamma/2}$. A multiplication of the Fourier transforms of the spectrum to be processed

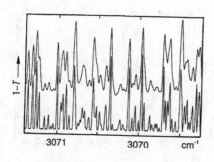

Fig. 3-38. Iterative resolution improvement in the spectral domain. Above: section from the absorbance spectrum of C_2H_6 (spectral resolution of the grating spectrometer $\Delta\tilde{\nu} = 0.025$ cm^{-1}). Below: deconvoluted spectrum with $\Delta\tilde{\nu} = 0.0015$ cm^{-1} [51].

with the inverse function $e^{|x|\Gamma/2}$ yields the deconvolution from the original band shape with the result that the effects of the terminated interferogram (boxcar apodization) appear after renewed Fourier transform. In order to avoid the side lobes in the band flanks, one multiplies with additional, suitable apodization functions (see Sect. 3.4.3.1). In addition to this, the noise in the experimental spectra allows the exponential amplification of the decreasing signal in the Fourier domain only until a certain degree and dependently on the abscissa value.

Estimating the band half-widths can prove to be difficult in the existing band multiplet because of overlapping. A further complexity arises in that the bandwidth can vary depending on the vibrational form. If the exponential amplification factor is now chosen to be smaller, thereby corresponding to a narrower estimated bandwidth, a self-apodization remains in the "interferogram". In contrast, an overestimation of the half-widths causes considerable lobes in the band flanks.

Figure 3-39 shows the result of a self-deconvolution that likewise demands spectra of high quality, i.e. with a low noise level. Interesting is the direct comparison with the derivative spectrum in Fig. 3-37 with which the differences of both methods for improving resolution are apparent. The band at 2775 cm^{-1} shows the lowest half-width compared to the remaining bands, thereby signifying a mild overdeconvolution. In general, one must not be misled in overdoing the resolution improvement. A critical estimation can also be reached here by inspection of the results.

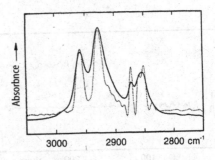

Fig. 3-39. Multi-component spectrum and the spectral result after self-deconvolution (dotted curve).

3.6.5 Further Practical Methods for Spectral Processing

Besides disturbing baselines, transmission measurements of thin, plane-parallel layers may show a periodic baseline modulation caused by multiple reflections. Samples can be, for example, films or liquids in transmission cells (experimental possibilities on preventing interference effects are described in Sect. 4.2.4.4).

The maxima of these interference patterns are found for the following wavenumbers

$$\tilde{v}_m = m/(2nd) \qquad (3.14)$$

where $m = 0, 1, 2, \ldots$ is the order of interference, n the refraction index of the layer and d its thickness. For a known thickness, wavenumber-dependent refraction indices of substances in the IR spectral region can also be measured hereby [54].

In order to compensate for this mostly undesired baseline modulation, the subtraction corresponding to the fitted sine functions was recommended in the literature. However, following their Fourier transform or already in the interferogram, such spectra provide, at best, localized signals that may be classified to the spectral sine function. This is apparent upon observing the first "Fourier transform pair" in Fig. 3-16. More advantageously, such signals in the Fourier domain can be used for precisely determining the pathlength of empty liquid cells [55] (also see Sect. 7.1.2) or can be eliminated with very little expense, e.g. set at zero or other suitable manipulations [56]. The transformation of such purified interferograms can yield usable spectra. However, the interference signals in the Fourier domain are often (otherwise also called "spikes") more widely distributed. Figure 3-40 depicts the interferogram and absorbance spectrum of acetonitrile. As the reader can figure alone by using the above equation, a mean refraction index of about $n = 1.38$ can be estimated in the mid-IR region from the distance between the principle and first side maximum.

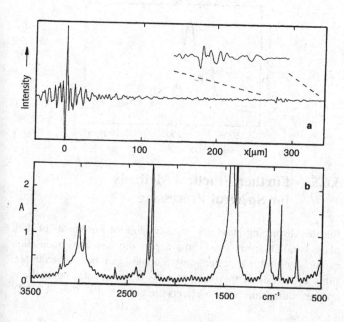

Fig. 3-40. Interferogram of the single beam spectrum (a), as well as the corresponding absorbance spectrum (b) of acetonitrile measured in a 100 µm cell (ZnSe-window) [21].

Earlier, when electronics were not so dependable, one had to expect temporary disturbances (the so-called 5 o'clock effect played a role here, namely when most laboratory instruments were turned off); undesired voltage peaks caused "spikes" in the interferogram, resulting in enormous modulations of the baseline. If the interferogram was still available, the measurement could be saved by simple interpolation between the apparently unaffected neighboring data points. The same effects are discernible when, for example, so-called "gain-ranging" is used in the FT spectroscopy, which allots a lower amplification for the central region of the interferogram than for the remaining part in order to improve the signal-to-noise ratio as a whole. If the electronics are not well tuned, steps may occur in the turnover point for the higher amplification that have effects similar to "spikes".

On the other hand, spikes are also found in the spectrum when, for example, overtones of net buzzing were not filtered off satisfactorily. These are shown equidistantly in the spectrum. An experimental possibility of eliminating them is (if this is still permissible) to increase the mirror velocity in order to widen the distance between the frequency spectrum to be measured and the disturbing frequencies (also see Eq. 3.8, Sect. 3.4.2.1). However, if spikes appear randomly in the spectrum, as occurs e.g. during the measurement with extremely sensitive charge-coupled device (CCD) detectors in near-IR, one can hereby use statistical estimation techniques for eliminating the extreme values [57].

References

[1]   Blank, R.E., Wakefield II, T., *Anal. Chem.* **1979**, 51, 50.
[2]   Brügel, W., *Einführung in die Ultrarotspektroskopy*, 4th Edition, Darmstadt: Dr. Dietrich Steinkopff, **1969**, p. 92.
[3]   Franzis, R.J., Zwafink, J., Widmann, W., *GIT* **1979**, 23(6), 500.
[4]   Bergmann, L., Schaefer, Cl., *Lehrbuch der Experimentalphysik*, Vol. 3, 7th Ed., *Optik*, Berlin: Walter de Gruyter, **1978**, p. 343.
[5]   Thelen, A., *Appl. Optics* **1965**, 4, 977, 983.
[6]   Williams, V.Z., *Rev. Sci. Instr.* **1948**, 19, 135.
[7]   Cole, A.R.H., *Table of Wavenumbers for the Calibration of Infrared Spectrometers*, 2nd Ed., Int. Unionof Pure and Applied Chemistry (IUPAC), Oxford: Pergamon Press, **1977**.
[8]   Kauppinen, J., Kärkkainen, T., Kyrö, E., *J. Mol. Spectrosc.* **1978**, 71, 15.
[9]   Molt, K., *Nachr. Chem. Tech. Lab.* **1994**, 42, 914.
[10] Johnston, S., *Fourier Transform Infrared – A constantly Evolving Technology*, New York: Ellis Horwood, **1991**.
[11] Bergland, G.D., *IEEE Spectrum* **1969**, 6, 41.
[12] Ziessow, D., *On-line Rechner in der Chemie*, Berlin: Walter de Gruyter, **1973**.

[13] Lam, R.B., Wieboldt, R.C, Isenhour, T.L., *Anal. Chem.* **1981**, 53, 889A.

[14] Sheppard, N., Greenler, R.G., Griffiths, P.R., *Appl. Spectrosc.* **1977**, 31, 448.

[15] Griffiths, P.R., Soane, H.J., Hannah, R.W., *Appl. Spectrosc.* **1977**, 31, 485.

[16] Schrader, B., Niggemann, W., Belz, H.H., Schallert, B. *Z. Chem.* **1981**, 21, 249.

[17] Mattson, D.R., *Appl. Spectrosc.* **1978**, 32, 335.

[18] Chase, D.B., *Appl. Spectrosc.* **1982**, 36, 240.

[19] Cooley, J.W., Tukey, J.W., *Math Comp.* **1965**, 19, 297.

[20] Hawkins, R.L., Hoke, M.L., Shaw, J.H., *Appl. Spectrosc.* **1983**, 37,134.

[21] Heise, H.M., *Fresenius' J. Anal. Chem.* **1994**, 350, 505.

[22] Bertie, J.E., Keefe, C.D., Jones, R.N., Nantsch, H.H., Moffatt, D.J., *Appl. Spectrosc.* **1991**, 45, 1233.

[23] Tanner, D.B., McCall, R.P., *Appl. Optics* **1984**, 23, 2363.

[24] Hirschfeld, T., *Quantitative FT-IR: A Detailed Look at the Problems Involved*, in: J.R. Ferraro, L.J. Basile (eds.), Fourier Transform Infrared Spectroscopy, Vol. II., New York: Academic Press, **1979**, p. 193.

[25] Guelachvili, G., Distortions in Fourier Spectra and Diagnosis, in: Vanasse, G.A. (ed.), Spectrometric Techniques, Vol. II, New York: Academic Press, **1981**, p. 1.

[26] Compton, D.A.C, Drab, J., Barr, H.S., *Appl. Optics*, **1990**, 29, 2908.

[27] Chase, D.B., *Appl. Spectrosc.* **1984**, 38, 491.

[28] Heise, H.M., Marbach, R., Koschinsky, Th., Gries, F.A., *Appl. Spectrosc.* **1994**, 48, 85.

[29] Keens, A., Simon, A., *Proc. Soc. Photo.-Opt. Instrum. Eng.* **1994**, 2089, 222.

[30] Miller, R.G.J., Stace, B.C., *Laboratory Methods in Infrared Spectroscopy*, 2nd Ed. London: Heyden & Sons, **1972**, p. 25 ff, p. 152.

[31] Nonnenmacher, G., Perkin-Elmer-Tips 31 UR, Bodenseewerk Perkin-Elmer & Co. GmbH, Überlingen, **1966**.

[32] loc. cit. [30], p. 129-148.

[33] Ferraro, J.R., Basile, L.J., *Appl. Spectrosc.* **1974**, 28, 505.

[34] Tinker, H.W., Morris, D.E., *Developments in Applied Spectroscopy*, Vol. 10, Plenum Press, **1972**, p. 123.

[35] Gans, P., *Data Fitting in the Chemical Sciences*, Chichester: J. Wiley & Sons, **1992**.

[36] Horak, M., Vitek, A., *Interpretation and Processing of Vibrational Spectra*, Chichester: J. Wiley & Sons, **1978**.

[37] McDonald, R.S., Wilks, P.A., *J. Appl. Spectrosc.* **1988**, 42, 151.

[38] Gasteiger, J., Hendriks, B.M.P., Hoever, P., Jochum, C., Somberg, H., *Appl. Spectrosc.* **1991**, 45, 4.

[39] Davies, A.N., *Analyst* **1994**, 119, 539.

[40] Pugh, L.A., Rao, K.N., *Intensitites from Infrared Spectra*, in: Rao, K.N. (ed.), Molecular Spectroscopy: Modern Research, New York: Academic Press, **1976**,165.

[41] Seshadri, K.S., Jones, R.N., *Spectrochim. Acta* **1963**, 19, 1013.

[42] Steele, D., Yarwood, J., *Spectroscopy and Relaxation of Molecular Liquids*, Amsterdam: Elsevier Science Publ., **1991**.

[43] Maddams, W.F., *Appl. Spectrosc.* **1980**, 34, 245.

[44] Savitzky, J.A., Golay, M.J.E., *Anal. Chem.* **1964**, 36, 1627.

[45] Gorry, P.A., *Anal. Chem.* **1990**,62, 570.

[46] Bromba, M.U.A., Ziegler, H., *Anal. Chem.* **1981**, 53, 1583.

[47] Edwards, T.H., Willson, P.D., *Appl. Spectrosc.* **1974**, 28, 541.

[48] Holler, F., Burns, D.H., Callis, J.B., *Appl. Spectrosc.* **1989**, 43, 877.
[49] Talsky, G., *Derivative Spectrophotometry*, Weinheim: VCH, **1994**.
[50] Cameron, D.G., Moffatt, D.J., *Appl. Spectrosc.* **1987**, 41, 539.
[51] Cole, A.R.H., Cross, K.J., Cugley, J.A., Heise, H.M., *J. Mol. Spectrosc.* **1980**, 83, 233.
[52] Jansson, P.A., *Deconvolution with Applications in Spectroscopy*, Orlando: Academic Press, **1984**.
[53] Kauppinen, J.K., *Fourier Self-Deconvolution in Spectroscopy*, in: Vanasse, G.A. (ed.), Spectrometric Techniques, New York: Academic Press, **1983**, p. 199.
[54] Heise, H.M., *Appl. Spectrosc.* **1987,** 41, 88.
[55] Heise, H.M., *Fresenius' J. Anal. Chem.* **1993,** 346, 604.
[56] Heise, H.M., *Proc. Soc. Photo.-Opt. Instrum. Eng.* **1985,** 553, 247.
[57] Hill, W., *Appl. Spectrosc.* **1994**, 47, 2171.

Detailed Monographs

Stewart, J.E., *Infrared Spectroscopy*, New York: Marcel Dekker, **1970**.
Bell, R.J., *Introductory Fourier Transform Spectroscopy*, New York: Academic Press, **1972**.
Griffiths, P.R., de Haseth, J.A., *Fourier Transform Infrared Spectrometry*, New York: John Wiley & Sons, **1986**.
Steel, W.H., *Interferometry*, 2nd Ed., Cambridge: Cambridge University Press, **1983**.
Laurin, T.C. (ed.), *The Photonics Design & Applications Handbook*, 38th Ed. Pittsfield: Lauring Publ., **1992**.
Dereniak, E.L., Crowe, D.G., *Optical Radiation Detectors*, New York: John Wiley & Sons, **1984**.

4   Sample Preparation

Only very rarely is the sample placed into the beam without being pretreated in some manner. The form, aggregate state or concentration of the sample must usually be altered so that it can be measured in one of the cells described in Sect. 3.5. This chapter covers the most important methods for sample preparation.

4.1   General Remarks

4.1.1   Sample Quantity

According to the Beer's law $I = I_0\,e^{-abc}$, the transmittance $T = I/I_0$ is a function of the absorptivity a and the pathlength b of the sample and, for solutions, also a function of the concentration c of the absorbing sample (see Sect. 7.1). The intensity of a spectrum, i.e. the transmittance in the regions of the absorption maxima can be accordingly influenced by the corresponding pathlength selected. The relative change of the absorbance $\Delta A/A$ reaches a minimum at $T = 0.367$ or $A = 0.435$; therefore, quantitative measurements are most precise when the band used for the determination reaches a maximum at about this value (see Sect. 7.3.3).

In order to obtain a spectrum of moderate intensity, pathlengths ranging from 0.01 to 0.05 mm (10–50 µm) suffice for solid and liquid, undiluted samples, and correspondingly higher pathlengths are needed for solutions depending on the concentration. The minimum sample volume is yielded from the pathlength and the area irradiated by the sample beam and, thus, the required sample quantity. For an area of 20×4 mm and a pathlength of 0.025 mm resulting in a volume of 2 µl, the necessary sample amount is about 2 mg. Considerably larger amounts are often necessary because of the dead volumes mostly present. If one uses the smaller dimension of the converged beam near the focus for KBr pellets, the required amount of sample is then reduced to 1 mg; upon using a beam condenser (see Sect. 3.5.1.8), about 10 ng of sample are still sufficient in the extreme case.

Because ultimately the number of molecules hit by the beam as it passes through the sample is relevant for the transmission, the

molar volume of the sample is decisive. Consequently, the measurement of gases requires layers that are a 1000-times larger.

4.1.2 Purity Requirements

4.1.2.1 Spectra of Pure Compounds, Reference Spectra

In general, the sample to be measured does not have to be extremely pure. A purity of 99% is often sufficient, because only
very strong bands of impurities having a concentration of <1%
are still observable. Traces of <0.1% are only determinable under favorable experimental conditions and do not hinder the
spectrum of the main components.

Similar thinking applies for the production of calibration
mixtures for quantitative analyses by considering the fact that
the relative standard deviation of the measurement method, at
best, lies at ±0.4%

These prerequisites, however, are not applicable when trace
determinations ought to be made for which the highest requirements on e.g. solvent purity are demanded. This applies in particular to high-boiling and non-volatile impurities in solvents
used for separating and enriching small amounts of sample. The
currently obtainable spectral signal-to-noise ratios, indeed, allow
the measurement of trace concentrations and very small sample
quantities.

4.1.3 Moisture

Section 3.5.1.1 indicated the sensitivity of commonly used cell
windows to water. Hence, samples and solvents must be carefully dried. If the sample cannot be dried because of danger of
its composition being altered or if adsorption to a drying agent
is impossible, a corresponding cell material (see Tab. 3-7) must
be chosen. Water contents of <1% do not pose a problem for
NaCl and KBr windows.

4.1.4 Precautionary Measures

The most important precautionary measures (see e.g. [1]) in
handling samples and solvents are:
- Wear safety glasses to protect yourself against any spraying.
- Work under a fume hood upon handling unknown samples
  and many solvents (chlorinated hydrocarbons, carbon sulfide,
  benzene, toluene, etc. (also see Tab. 4-1).

- Wear rubber gloves upon handling caustic or other samples with an unknown physiological effect.

Do not let the smallness of the sample quantities prevent you from heeding these precautionary measures. Experience has shown that even the smallest sample quantities can harm the eyes (also see Sect. 4.3.2.4, Tab. 4-1).

4.2 Solids

The following sample preparation methods are available for solids:
- Preparation of a solution (see Sect. 4.3.2),
- KBr pellet technique (see Sect. 4.2.1),
- Preparation of a suspension in oil (see Sect. 4.2.2),
- Application of a thin crystalline layer on the cell window from the melt (see Sect. 4.2.3),
- Production of a film by pressing in a heated state, cast from solution or cutting (polymers; see Sect. 4.2.4),
- Use of the ATR method for unmeltable and insoluble polymers with a smooth surface (see Sect. 5.1.2).

The selection of a suitable method depends on the state of sample, on its physical properties (melting point, solubility) and on the measurement purpose (qualitative overview spectrum or quantitative determination).

4.2.1 KBr Pellet Technique

4.2.1.1 Materials for Producing Pellets

Alkali halogenides have the property of exhibiting *cold flow* at pressures of about 0.7 to 1.0 GPa (10^5 Pa corresponds to 1 bar). At these pressures, the material sinters and can be formed into a transparent tablet resembling a single crystal.

The most commonly used of them is potassium bromide that, like the KBr windows consisting of real single crystals, is permeable down to a wavenumber of $400 \, cm^{-1}$. Sodium chloride also shows cold flow but is hardly used because of its limited transmission range down to $650 \, cm^{-1}$. Potassium iodide and cesium iodide are suited for measurements in long-wave IR regions down to $200 \, cm^{-1}$. Furthermore, thallium bromide, silver chloride and polyethylene can be used for special purposes.

The KBr pellet technique is excellently suited for qualitative measurements on sufficiently impact-resistant, powdered solids. The method, however, is poorly suited for quantitative measurements. One must consider interactions between the sample and the embedding medium that may lead to spectral changes (see Sect. 4.2.1.7).

For remedying disturbances by the Christiansen effect (see Sect. 4.2.1.6), KCl or KI is occasionally advantageous. The permeability of the various materials has already been shown in Fig. 3-21 to 3-23.

The following describes the cleaning of the material and the production of the pellets by using KBr. The procedure with other materials is analogous.

4.2.1.2 Handling of Potassium Bromide

KBr is known to be hygroscopic, especially when it has been finely pulverized for producing pellets, thereby resulting in a large surface area. Adsorbing moisture leads to disturbing bands at ca. 3450 cm^{-1} and 1640 cm^{-1} as well as to a decline of the background above 1000 cm^{-1} (see Fig. 4-1). The KBr can be treated for a few hours at about $200 \,^{\circ}\text{C}$ to prevent this disturbance. Better is drying in a vacuum-drying oven at $150 \,^{\circ}\text{C}$ after previous pulverization. KBr treated in this manner can be stored in the exsiccator over silica gel (blue gel). More frequent removal requires it to be occasionally dried later on. It is therefore advantageous to divide the KBr into smaller portions.

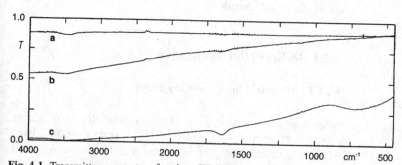

Fig. 4-1. Transmittance spectra of various KBr pellets:
a) Normal pellet,
b) Opaque pellet with low moisture (the weak finely structured bands are attributed to insufficient compensation of the absorption by CO_2 and H_2O in the spectrometer during the sample and background measurement),
c) Pellet with extreme moisture.

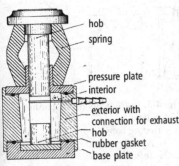

- hob
- spring
- pressure plate
- interior
- exterior with connection for exhaust
- hob
- rubber gasket
- base plate

Fig. 4-2. KBr-pressing tool (from an advertisement brochure of Paul Weber™).

4.2.1.3 Pressing Tools

Hydraulic presses and pressing moulds for producing pellets with a 13-mm diameter can be obtained from most instrument manufacturers and accessory suppliers.

Figure 4-2 shows the sectional diagram of a pressing tool. The inner section of the press mould, with a cylinder where the tablet is pressed between the bottom and upper stamp, consists of two cone halves held in place by the outer press mould. The pressed tablet can be removed by dismantling the inner section.

The lateral supports serve for evacuating before and during the pressing procedure. This is important, because otherwise, entering air would lead to premature recrystallization of the tablet.

4.2.1.4 Pellet Production

0.5 to 1.5 mg sample is finely ground in a small flat agate mortar and, after addition of ca. 300 mg KBr powder, intimately, but carefully mixed with this. Adsorbed water is minimized by grinding the KBr as little as possible. The sample is ground separately and then mixed with KBr, after which it is ground as little as possible to achieve good mixing.

The mixture produced in this way is introduced into the assembled, but still open press mould. By a careful one-time placement and pressing of the upper stamp on the still loose and irregularly layered powder, the powder becomes equally distributed on the bottom stamp and is precompressed.

With an incorporated rubber hollow spring, the upper stamp is now finally placed on, the pressing tool put into the hydraulic press and the vacuum pump attached. With the running pump, the KBr sample mixture is exposed to a force of ca. 10^5 N, corresponding to a pressure of ca. 0.75 GPa. The required pressure, based on the stamp area of the hydraulic press, is generally labeled on the manometer of the special presses for the KBr press technique.

The sinter process needs some time; therefore, the end pressure is held for ca. 2 minutes. Maintaining the right pressure is important. Too little pressure leads to an incomplete sintered and, hence, opaque tablet which has a tendency towards re-crystallization. Excessive pressure damages the pressing tool for which there are recommended load limits. After relieving the press and releasing the vacuum connection, the pressing tool is dismantled, the pellet carefully removed with tweezers and placed into the pellet holder (see Sect. 3.5.1.7).

After pressing, the press mould must be carefully cleaned, since the residues of the hygroscopic KBr powder soon becomes moist and is very corrosive in this state.

The following procedure is suited for the routine preparation of a larger sample series: the sample is added to the KBr powder and placed, together with two agate balls, into a thick-walled agate capsule (Fig. 4-3 a). The volume of the capsule is so measured that the layering constitutes a maximum of 2/3 of the inner chamber. The closed tube is then fitted into an electromagnetic vibrator and vibrated for 3 to 5 minutes. Under the influence of the agate balls, the sample and KBr powder are mixed and finely pulverized. Upon transfer of the mixture into the press mould, one should think of removing the agate balls. Especially useful for this is e.g. a glass funnel having a net of glass rods melted in (Fig. 4-3 b). Metal sieves are unsuitable because of the strong corrosivity of moist KBr.

Sometimes very hard, shock-resistant or overly soft compounds cannot be finely pulverized in the manner cited above. In this case, very rough kernels of sample can be seen in the pellet by bare eye. Irregularly weak absorption bands are the result. In such cases, it is helpful to first finely grind the sample before it is mixed with KBr. This can be done by using the above vibrating mill under cooling with liquid nitrogen with a somewhat larger amount of sample (ca. 50–100 mg) or by using a small agate mortar.

A blank pellet may be used for compensating the background and the water bands. With double-beam grating spectrometers, the blank pellet is placed directly into the reference beam and, with FT spectrometers, the background measurement typically takes place with it. In order to possibly obtain good compensation of the water bands, the blank pellet should have been prepared shortly before or after production of the sample pellet.

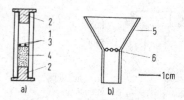

Fig. 4-3. Instruments for preparing KBr pellets.
a) Grinding capsule
   1) Agate cylinder
   2) Teflon stopper
   3) Agate balls
   4) Potassium bromide powder and sample
b) Funnel with glass grating for separating the agate balls
   5) Glass funnel
   6) Glass grating

4.2.1.5 Micropellets

Even when sample quantities of less than 0.8 mg are only available, a good spectrum can be obtained by producing a pellet with a correspondingly smaller radius. Special pressing tools for making micropellets are commercially available. Their design corresponds to the standard pressing tool, but the pellet is pressed into a central bore of a 1 mm thick steel disc having a diameter of 13 mm. The bore and upper stamp with the corresponding stamp lead must have the respective diameter. Commonly used are diameters of 1.5 and 0.5 mm.

In the production of a micropellet, a smaller force is, of course, sufficient for obtaining the necessary pressure of 0.75 GPa ($75 \cdot 10^3$ N/cm^2). Corresponding to the smaller surface area, this is ca. 1300 N for a diameter of 1.5 mm, as the following calculation example shows:

$$K = p \cdot F = p \cdot \pi r^2 = 75 \cdot 10^3 \cdot 3.14$$
$$\times (2.5 \cdot 10^{-2})^2 = 147.3 \, \text{N}$$

$$(4.1)$$

where p = pressure in N/cm^2
  F = surface area in cm^2
  K = force in N
  r = radius in cm.

This condition must be absolutely heeded, because otherwise the sensitive upper stamp would be destroyed. The correct pressures are best obtained with the spindle press, which fits the micro-pressing tool. In the production of a pellet, having a diameter of 0.5 mm, the compression of the upper stamp by hand is sufficient.

The micropellet remains in the bore of the steel plate and is inserted with it into the holder of the micro-illuminator (see Sect. 3.5.1.9).

The micropellets with a diameter of 1.5 mm in their steel templates may also be placed into the standard pellet holder at the beam site intended for macropellets. Although the steel templates cover ca. 95% of the beam, the reserves of higher performance instruments, however, suffice for obtaining a usable spectrum with the remaining 5% of the energy. To compensate for the loss in beam energy, a larger number of interferograms may be accumulated.

Upon using micropellets, the required amount of sample can be reduced from 5 to 10 µg (4 mg KBr) and even down below 1 µg (0.4 mg KBr) with careful handling (see e.g. [5]). The detection limits, partially down to 10 ng, depend also, of course, on the intensity of the absorption bands. The demand for cleanliness during the sample preparation and for purity of the compounds obviously increases with the reduced detection limits.

Also see Sect. 4.3.2.6 regarding the transfer of solutions to the micropellets. A typical method uses a porous wedge of pressed KBr (wick-stick, Fig. 4-4) that is put into the solution. Upon slow evaporation of the solvent, the dissolved sample becomes accumulated at the tip, which is then broken off and may be subsequently processed to a micropellet [6].

If no micropressing tool is available, one can place, according to Patt and Kuhn [7], a funnel-shaped lead ring into the normal pressing tool and, by using 50 µg sample and 15–20 mg KBr, produce a pellet having a diameter of ca. 3 mm in the lead mask. A rectangular piece of cardboard (1.9×1.9 mm) may also serve as a template for a semi-micropellet [8].

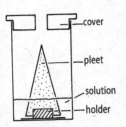

cover
pleet
solution
holder

Fig. 4-4. "Wick-Stick" apparatus.

4.2.1.6 Disturbance by the Christiansen Effect

According to the laws of wave optics, all samples show irregular dispersion in the region of strong absorption bands, i.e. the refraction index, which normally constantly increases towards higher wavenumbers, is subjected to strong changes in these spectral regions. The course of the absorbance of a sample embedded in KBr or in another matrix, however, is, in its turn, affected by the difference of the refraction index between the sample and the embedding matrix as well as by the grain size of the sample. If the refraction index difference between the sample and the matrix is especially large to begin with, the absorption bands show a typical asymmetry resulting from the abnormal dispersion (Fig. 4-5 a).

The higher frequency (short-wave) flank is very steep, and the higher frequency foot of the band often shows a higher transmittance than that corresponding to the spectrum baseline. Moreover, the lower frequency (long-wave) band flank displays tailing. This

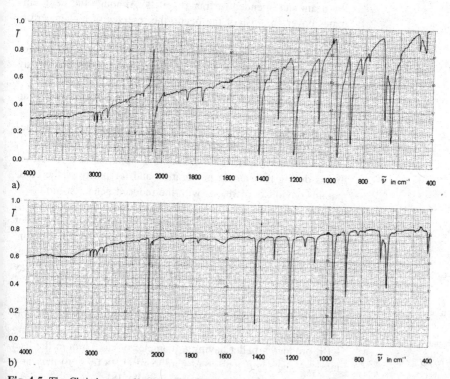

Fig. 4-5. The Christiansen effect.
a) IR spectrum of trans-butene-2-dithioisocyanate in KBr, pronounced Christiansen effect.
b) In contrast to a), the Christiansen effect was avoided by a longer grinding of the sample.

phenomenon is called the *Christiansen effect*. In addition, a more or less sharp decline of the baseline towards higher wavenumbers is observed as well as a shifting of the band maxima.

The Christiansen effect is lower the smaller the refraction index difference between the sample and the embedding matrix is and the smaller the grain size of the sample is. This disturbance can be avoided accordingly by finer grinding (Fig. 4-5b) (longer grinding period, pre-grinding of the sample without KBr or repeated pulverization of the pellet) or by selecting another embedding matrix (KCl or KI instead of KBr).

4.2.1.7 Disturbances by Reaction with the Embedding Medium

As the KBr pellet technique is a practical and uncomplicated method, it is preferred over other methods for preparing solids. However, one should not use this thoughtlessly, because KBr and all other alkaline halogenides are highly polar compounds, which may interact physically or chemically with the embedded sample. As the result of such interactions, specific differences of the spectra of the same sample may be observed, depending on whether the sample is embedded in KBr or measured by another means (solution, oil suspension, film). The changes can be so large that the compound cannot be identified by comparison with the spectrum of the same compound but prepared in another way. Some of the reasons for the falsification of the spectra in KBr-prepared samples are given below:

- Change of the crystal modification (Fig. 4-6),
- Hydration by the moisture attaching to the KBr,
- Saponification of esters and acid chlorides,
- Halogen substitution of functional groups,
- Decomposition of unstable compounds,
- Ion exchange [10],
- CO_2-inclusion [11].

In case difficulties arise during the identification of a sample prepared in KBr, one should try to apply another preparation technique (solution, oil suspension, melt). One will often find that the spectrum of the KBr pellet shows specific properties that may lead to failure in the identity testing through spectral comparison or even to misinterpretations. The spectra of a copper phthalocyanine complex, illustrated in Fig. 4-6, clearly deviate from each other at the marked wavenumbers. The β-modification is retained in the paraffin oil preparation (see Sect. 4.2.2), whereas it is converted into the α-modification upon grinding in KBr. Figure 4-7 exemplifies the spectral change in a sample embedded in KBr due to traces of moisture.

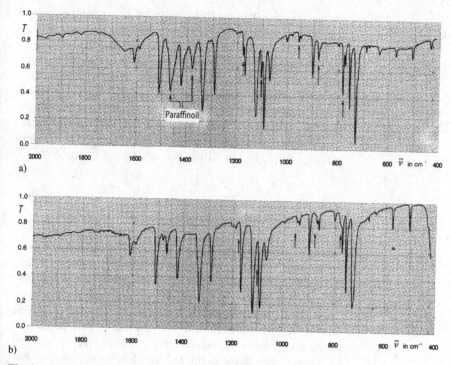

a)

b)

Fig. 4-6. IR spectrum of copper phthalocyanine (β-modification). Dispersion in paraffin oil, b) KBr pellet.

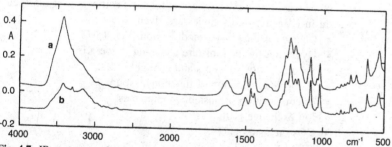

Fig. 4.7. IR spectrum of calcium-2,5-dihydroxybenzosulfonate. a) Measurement of the KBr pellet directly after the preparation. b) Measurement of the same pellet one day later (traces of moisture caused an enormous scattering; therefore, the spectrum was baseline-corrected and shifted in the ordinate axis for comparison.

4.2.1.8 Quantitative Analyses

The KBr pellet technique is only poorly suited for quantitative analyses for two reasons:

1. The pathlength is poorly reproducible even with careful KBr-weighing because of uncontrollable sample losses in the pressing tool.

2. The absorbance in the band maxima depends on the degree of dispersion of the sample [10]. By using a vibrating mill, the intensity of the absorption bands can be increased by a longer grinding period. Due to lower scattering effects, the spectral baseline, e.g. in absorbance units, lies lower than when the pulverization and homogenization of the sample with KBr occurred in the agate mortar. Hence, the grinding process must possibly be reproducible. Moreover, the addition of an internal standard, such as $K_3Fe(CN)_6$ is occasionally recommended.

Also, see Section 7.2.4.2 for performing semi quantitative measurements in KBr.

4.2.2 Suspensions in Oil

The following described preparation technique for solid samples is older than the KBr-pressing technique. It was an extremely effective technique earlier and it has been unjustly forgotten. This method is based on the production of a suspension of finely pulverized sample in paraffin oil (Nujol®) or hexachlorobutadiene. Its only and decisive disadvantage is the occurrence of absorption bands, attributed to the liquid phase, which overlap the spectrum of the suspended sample (Fig. 4-8). The interesting spectral region can be kept free of disturbing bands by suitably selecting the liquid phase.

However, the method is advantageous in that no expensive aids are needed. Moreover, upon mainly using the non-polar paraffin oil, no disturbances have to be expected as would be the case in using the polar KBr (see Sect. 4.2.1.7). Most of all, samples which are sensitive to air and moisture can be hereby prepared without being damaged.

At best, one uses two rough (matted) glass plates for preparing the oil suspension. These are made in the following way:

Two glass plates (diameter 6–7 cm, thickness 3–4 mm) are moistened, covered with very fine sea sand (600 mesh) and polished matt by rubbing each plate against the other. The last polish takes place for 4 to 5 minutes with still finer sand with a grain size of about 2.5 µm (222 mesh).

For producing the oil suspension, one drop of oil is placed on the matted plates, 5–7 mg of the sample to be examined is added to this and rubbed 2 minutes between the plates. The clear suspension is carefully pushed together with a razor blade and is then transferred with the razor blade or a spatula onto the window of a dismountable cell and covered with a second window with a slight turning motion so that a bubble-free film is formed between the windows.

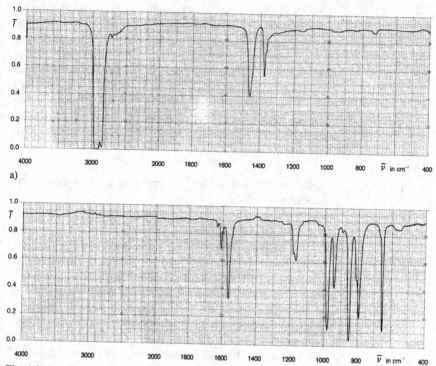

Fig. 4-8. Dispersing agents for preparing solid samples. a) Paraffin oil,
b) 1,3-Hexachlorobutadiene.

One can also grind the suspension in an agate mortar. With
using the glass plates, it is also possible to process waxy material when a little alkali halogenide powder is added to the suspension [12].

4.2.3 Films from the Melt

Crystalline or amorphous solids, which melt in the air without
being decomposed, can be applied onto the cell window from
the melt as a solid thin film. This method is suited for cases in
which every effect of a solvent or a dispersing agent should be
excluded. This preparation technique is also suited for viscous
and waxy samples.

Two windows of a dismountable cell are placed on the cold
end of a Kofler-Heizbank® (heater with a temperature gradient
for determining the melting point). Possibly after placing a suitable spacer ring, the window nearer to the cold end is given just
enough of the sample. By using tweezers, the ends of which are
covered with a Teflon tube, the windows are slowly pushed
towards the higher temperatures so that the windows, which

poorly conduct heat, are gradually heated. When the surface of the colder window has reached the melting temperature of the sample, the sample liquefies. At this point, the hotter window is grabbed with the tweezers and placed on top of the colder one. It requires a little aptitude to produce a bubble-free layer. When this is obtained, the pair of windows is slowly pushed to the cold end, whereby the melt solidifies to a film.

Especially with slow and shock-free cooling, well-crystallized compounds solidify occasionally with a more or less preferred orientation of the crystal surfaces. This may cause disturbances when the optics of the spectrometer partially polarizes the radiation. Depending on the position of the partially oriented crystallate relative to the polarization orientation of the radiation, the bands of parallel or perpendicularly polarized vibrations display intensity differences (see Sect. 5.6).

Low-melting and poorly crystallizing compounds often stay longer in the liquid state as an under-cooled melt between the cooled windows. Because the spectra of the same compound in different aggregate states differ from one another, sometimes even very strongly, this may lead to errors in the sample identification through comparison of spectra (Fig. 4-9). If crystalliza-

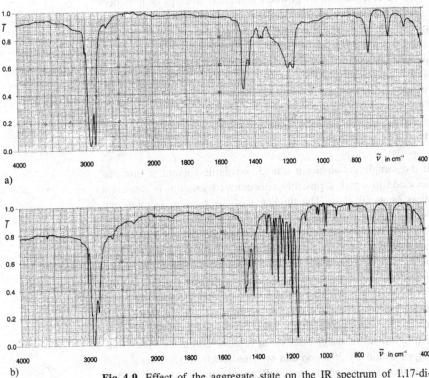

Fig. 4-9. Effect of the aggregate state on the IR spectrum of 1,17-diiodoheptadecane: a) liquid, b) solid.

tion occurs during the measurement, the background shifts and non-reproducible absorption bands may suddenly appear. It is therefore advisable to make sure that the sample is crystallized before the measurement begins.

4.2.4 Films

Polymer films can be produced according to three different methods:
- for thermoplastic polymers, by pressing between two heated plane-parallel metal plates,
- for soluble polymers, by casting from solution,
- for insoluble and unmeltable polymers, by producing micro-tomic sections.

4.2.4.1 Pressing of Films

Warm, pliable (thermoplastic) polymers can be easily pressed to films between two heatable metal plates under moderate pressure. Heatable cases for the KBr-presses are hereby useful.

A suitable amount of powder, granulate or chips of polymers is placed between two metal plates (e.g. aluminum) and is then heated to melting temperature between the heatable pressing units and pressed to a film under slight pressure. There are commercial accessories for this which allow temperatures of up to 400 °C and with which films of fixed pathlengths ranging from 15 µm to 500 µm can be manufactured.

4.2.4.2 Cast Films

If the sample is soluble in a readily volatile solvent, a film can be made in which a possibly concentrated solution is cast on a glass or metal plate. The thickness of the running solution can be influenced by tilting the plate at a certain inflection. After the film is somewhat dried, the solvent is possibly evaporated completely at moderate temperature (drying oven, Kofler-Heiz-bank®), or if necessary, in a vacuum. Then, the film can be peeled off the plate.

According to Wehling [13], a thin cast film can be removed intact from a glass plate or from another base by outlining it to the desired size with a needle and, at one of the tears, hosing it with a stream of water directed at a sharp 60° angle. The films bulges within 20–30 s and can be lifted off from the plate.

If the casting is done on a reflecting metallic foundation, the film does not need to be peeled off. In this case, the measurement is performed by using a reflection accessory (see Sect. 5.1.1).

One may also directly pour the solution onto the cell window, which is advantageous in producing very thin films, which are difficult to be peeled off from the base in an intact state. Of course, this variation is not limited to polymers, i.e. non-decomposed meltable monomers can be measured in this way as a film in the solid state.

With this preparation technique, one must account for the possibility of solvent bands still occurring in the spectrum. Irrespective of this disadvantage, however, one can make very thin films in this way, e.g. for studying especially intensive absorption bands.

4.2.4.3 Microtomic Sections

If a polymeric material cannot be processed to a film with a suitable pathlength by using any of these aforementioned methods, one can use a microtome. Preparation areas of several square centimeters with pathlengths of between 20 and 50 μm are nevertheless unusual for the technology of producing microtomic sections and are seldom obtained, especially considering the necessary area. One possibly has to be satisfied with smaller film pieces and fasten these e.g. on a KBr pellet holder or on a carrier plate of a micropellet.

If the surface area of the sample allows it, using the ATR technique is better (see Sect. 5.1.2).

4.2.4.4 General Tips for Producing Films

For producing films by pressing or casting, the surface may become very smooth, that can possibly result in interferences during the measurement. Wavelength-dependent intensity maxima and minima, created by multiple reflection of the beam on the inner surfaces of the films and the succession of which depends on the thickness of the films, overlap the spectrum. Although one can use them for exactly determining the thickness (see Sect. 7.1.2), they considerably disturb the qualitative interpretation of the spectrum.

The interferences can be avoided by scrunching the film before measurement and again smoothing it or by roughing its surface on a suitable instrument (e.g. clean file or polishing canvas).

Another method using a polarizer is applicable when perpendicular incident radiation is waived and the measurement takes place in the Brewster angle [14]. For eliminating spectral interference patterns, a simple reflection accessory was recommended with which the transmitted and reflected radiation is recombined, so that the resulting transmission-reflection spectra can be recorded without baseline disturbances [15]. Calculating techniques, especially in applying FT-IR spectroscopy, are likewise possible. In the Fourier domain (e.g. interferogram), the spectral interference patterns yield locally limited disturbances, which can be eliminated. A survey of this is found in [16].

The pathlength is measured easiest with a caliper, whereby one must take care for a constant pressing of the measurement sensor. Well suited is a caliper with a pressing attachment that automatically provides for a reproducible pressure application.

If quantitative analyses that are more precise should be performed on plastics (e.g. determination of the content of softeners, stabilizers, components of copolymers etc), the use of high-precision instruments is recommended for determining the pathlength. Such electronic precision measurement instruments have an inductive measurement element and sensitive measurement sensor with a scale or a digital indicator and several measurement regions. The occurring measurement forces as well as the ratio of the increase of measurement force to the change of the measurement path are minimal.

4.2.5 Films as Sample Substrate

It was mentioned in Sect. 4.2.1.4 that KBr pellets are also applicable as a substrate for various samples. Another possibility is the use of microporous thin films of polyethylene (PE) or polytetrafluoroethylene (PTFE) that are commercially available. The films are mounted in cardboard that are compatible with the standard sample holders.

Various fluids, pastes or lubricants can be considered as samples. About 15 to 100 µl sample are needed. Volatile solvents already quickly evaporate on the porous surface at room temperature. The PE films are heat-resistant for a short period at 100 °C, which can be important in the evaporation of less volatile solvents. Such films also allow highly viscous samples to be warmed up for better application. After pretreating the film with a polar solvent like acetone, liquids having a high surface tension can moisten the substrate better.

With the exception of the aliphatic CH-stretching region, the spectral range of 4000 to 400 cm^{-1} can be well used with the PE film (see Fig. 4-10). The prerequisite is that a reference spectrum is recorded by the unused film for compensating the existing absorption bands. In particular, this film material can be applied excellently for qualitative investigations in which the fingerprint region is important. In the case that the region

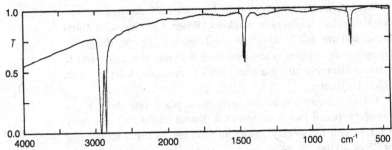

Fig. 4-10. Transmittance spectrum of a microporous polyethylene film.

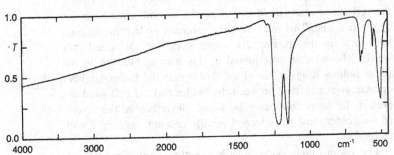

Fig. 4-11. Transmittance spectrum of a microporous polytetrafluoroethylene film.

around 3000 cm^{-1} ought to be evaluated, PTFE films can be used instead, since their free spectral region extends from 4000 to about 1300 cm^{-1} (see Fig. 4-11). This material is further advantageous in that it is temperature-resistant until 200 °C.

4.3   Liquids and Solutions

4.3.1   Liquids

If the sample to be investigated is liquid and clearly transparent, generally no further preparation is necessary, especially for qualitative studies. Possible colorings of the sample are based on a light absorption in the visible region and do not disturb the IR region. However, clouding caused by solid particles (dispersion) or by a finely distributed liquid phase (emulsion) is disturbing in that they scatter the incident radiation by diffraction and reflection thereby causing strong background absorption. In these cases, a homogeneous phase must be provided for by filtration or centrifugation. The residue can be examined after washing and drying.

Water contents of greater than 1% up to 2% damage the surfaces of alkali halogenide windows. When a drying with anhydrous sodium sulfate is unadvisable because of the danger of changing the sample by absorption or reaction, one can resort to a water-insoluble cell material (KRS-5, calcium fluoride, silver chloride, Irtran).

For *qualitative studies*, it suffices to place one drop of the sample between two windows of a dismountable cell (see Sect. 3.5.1.3), and for weakly absorbing samples, to use a spacer ring of 25 or 50 µm.

After laying the spacer ring onto the first window and applying enough sample corresponding to the respective pathlength, the cell chamber is closed by laying the second window on top of the first one. This must be done carefully so that no air bubbles form in the interior. Excessive sample is displaced outwards. The window pair joined in this way is fastened in the metal holder. Rings made of an elastic material (silicon rubber or analogous material) between the holder and the cell windows protect the latter from easy breaking. Nevertheless, the screws of the holder must be tightened equally and not too tight ("with feeling").

For viscous and poorly volatile samples, very thin films are made by repeated sandwiching the windows together and then pulling them apart after having first applied just enough of a drop of sample without a spacer ring. If the sample is insensitive to air and is sufficiently viscous, the measurement can subsequently take place in which only this window is placed in the holder. This technique again leads to a reduction of the pathlength which is occasionally advantageous for strongly absorbing viscous samples.

For *quantitative measurements*, cells with a defined pathlength ("fixed cell", see Sect. 3.5.1.4) of ca. 25 µm are used, while those of 50 or 100 µm thickness are used only for very non-polar samples (hydrocarbons). Determinations of trace amounts of sample may require even greater pathlengths.

For filling a fixed cell, one lays it slightly at an angle on the work surface and inserts the sample by means of a pipette into the deeper lying bore. The capillary forces acting in this thin layer are normally sufficient in order to pull the liquid from below up into the upward inclining cell chamber. For larger pathlengths, the hydrostatic pressure of the fluid located in the filling ports causes it to spread into the cell chamber. The cell is closed in the same tilted position starting with the bottom opening and, then, the upper opening. If the cell were closed in the opposite way, the air found in the upper filling port would be easily pushed back into the sample compartment.

Using a fixed cell is impossible for viscous samples. In this case, a solution must be made. For determining main components in liquids, it is likewise advisable to produce solutions and to use correspondingly higher pathlengths, because the relative error by uncontrolled alterations of the pathlength (e.g. by heat expansion) might be considerable for the low pathlengths required for pure liquids.

4.3.2 Solutions

The production of a solution is an important preparation technique in IR spectroscopy. It is used for transferring solid samples into the liquid aggregate state, for reducing the viscosity of samples but, most of all (also for liquid samples), for decreasing the concentration to enable the use of larger and thus better reproducible pathlengths for quantitative measurements. Furthermore, studies using solvents of various polarities and with different concentrations are important for identifying the constitution of compounds capable of forming hydrogen-bridge bonds (see Sect. 4.3.2.5).

4.3.2.1 Solvents

Due to the fundamental laws of IR radiation absorption, every molecule shows an IR absorption spectrum, with the exception of homonuclear elements in the gaseous state. Therefore, there is no solvent that is free of absorption bands in the entire IR spectral region[1]. The number of possible solvents for many compounds, which is nevertheless strongly limited by the solvent properties, becomes even more narrowed:

Figure 4-12 lists the most significant solvents and illustrates the regions of sufficient permeability. The corresponding spectra for these are found in the Appendix (see Sect. 10.2). The regions labeled as permeable are not fully free of bands, as one can see from the individual spectra. Scaled subtraction with solvent spectra, which were measured with the same fixed cell used for the sample, allows removing the disturbing solvent bands. Here, the absorbance of the solvent spectrum is mathematically trimmed by using the suitable scaling factor so that no positive or negative "overlaps" remain in the difference spectrum. In the region of strong absorption bands, a dependable compensation is no longer possible anymore and considerable artifacts may result [18].

---

[1] Exception: liquefied gases as solvents [17]. The low solubility of organic compounds in non-polar media at around 100 K (sample fraction 10^4–10^{-6}%) can be compensated by high pathlengths. The cell designs are experimentally complicated.

For quantitative measurements, the solvent should still show at least a transmittance of 0.05 in the region of the bands to be evaluated. Lower transmittances of the solvent are certainly permissible for qualitative measurements.

4.3.2.2 Purity of the Solvents

The most common solvents used in IR spectroscopy are listed in Figure 4-12. Chloroform generally contains 0.8% ethanol as a stabilizer but which can be easily removed chromatographically by means of silica gel or aluminum oxide columns. Unstabilized chloroform is only limitedly stable and tends to split off chlorine and hydrogen chloride.

4.3.2.3 Concentration Ranges

Because of the limited sensitivity of the measurement method and the eigenabsorption of the solvents, relatively high concentrations are generally selected. In order to obtain optimum transmittances (Sect. 4.1.1) with good reproducible pathlengths of at least 0.1 mm or, better, 0.2 mm, the 5–10-fold dilutions of the sample, i.e. concentrations of between 20% and 10%, are necessary. Such high concentrations are mainly then necessary, when one has to fall back on solvents due to their favorable solutizing properties but those show a strong eigenabsorption. If the sample is soluble in less polar solvents with fewer bands, higher

Fig. 4-12. Transmittance of the most common solvents for IR spectroscopy. In the black-labeled regions, the transmittance is at least 0.2 at a pathlength of b=150 μm.

pathlengths (0.5–2 mm) and, accordingly, lower concentrations (5–1%) can be applied. This concentration range is better, because here disturbances through interactions between the sample molecules are considerably suppressed and are negligible. This range is generally in the linear region of Beer's law (see Sect. 7.1). At higher concentrations, however, bands may appear which are specific for hydrogen-bridge bonds. Moreover, at a higher concentration, the relationship between the concentration and absorbance often is no longer linear due to interactions between the sample and solvent.

4.3.2.4 Production of Solutions

Because always only small amounts of solutions are needed, i.e. mostly 5–10 ml and possibly only 1–2 ml or less, the gravimetric production of the solution is preferred over the volumetric method. In general, the sample is first weighed and then the solvent. If the sample is readily volatile, however, errors may occur through evaporation losses. These can be extensively avoided if the solvent is prepared and the sample is weighed afterwards. For higher concentrations and higher volatile solvents, the danger of evaporation losses of the solvent must be considered, especially for quantitative measurements. For the same reason, the headspace over the solution must be kept possibly small; the containers for producing the solutions therefore should not be selected too large and must be well closable.

Be careful when working with hazardous solvents (e.g. carbon disulfide, benzene, toluene, carbon tetrachloride, dioxane)! Despite the relatively small amounts, damaging concentrations readily occur in the workplace atmosphere (see Tab. 4-1 on this). Therefore, upon making solutions and filling cells, one always should work under the hood. Because carbon tetrachloride is supposedly carcinogenic, it should be replaced by 1,1,2-tetrachlorotrifluoroethane.

4.3.2.5 Solvent Effects

Not only during the production of pellets but also with solutions can one expect changes of the spectrum opposed to that of the pure compound. The changes may be based either on interactions between the sample and solvent or on interactions between the molecules of the sample themselves. Although these effects can disturb the identification by comparing spectra, they can provide, however, useful additional information for structural investigations.

Table 4.1. Maximum workplace concentrations (MAK values) of some common solvents for IR spectroscopy

	MAK Value in Air	
	ml/m^3 (ppm)	mg/m^3
Acetonitrile	40	70
Benzene*	TGC**	
Chloroform***	10	50
Cyclohexane	300	1050
Dichloromethane***	100	360
Diethyl ether	400	1200
Dimethylformamide	10	30
Dioxane***	50	180
Ethanol	1000	1900
n-Hexane	50	180
Methanol	200	260
iso-Propanol	400	980
Carbon disulfide	10	30
Tetrachloroethene***	50	345
Carbon tetrachloride***	10	65
Tetrahydorfurane	200	590
Toluene	50	190

* Carcinogen, i.e. compound that causes malignant tumors in humans.
** Technical guidance concentration: 1 ml/m^3, corresponding to 3.2 mg/m^3.
*** Suspected of having carcinogenic potential.

Interaction between the Sample and Solvent

Changes of the spectrum resulting from interactions between the molecules of the sample and those of the solvent are to be greater feared, the more polar the solvent and the sample are. Such phenomena may be attributed to dipole interactions, chelate formation (hydrogen–bridge bonds), π-complexes or other types of complexes.

Weak interactions lead only to concentration-dependent alteration of the absorptivity, so that the linearity of the relationship between concentration and absorbance no longer exists. This phenomenon is so frequent that quantitative measurements, for which only moderate precision (standard deviation $< \pm 10\%$) would suffice, never should be performed without previous calibration, which means testing the linearity of $A = f(c)$, i.e. testing the applicability of Beer's law.

Stronger interactions between the sample and solvent may also cause shifting of the band maxima of the dissolved sample by a few wavenumbers.

If a complex is formed, completely new bands that are typical for this complex may occur. The best known and the easiest ones to recognize and to interpret in the spectrum are the hydrogen-bridge bonds. Hydroxy compounds, dissolved in low concentration (<1%) in unpolar solvents, show a narrow OH-stretching vibrational band at ca. 3640 cm^{-1}. If a solvent capable of forming hydrogen-bridge bonds is selected, e.g. acetone, this band widens and shifts, depending on the strength of the association, towards lower wavenumbers, e.g. for acetone, by 140 cm^{-1} to 3500 cm^{-1}.

Interactions between the Molecules of the Dissolved Sample
The most frequently observed hydrogen-bridge bonds belong to this type of interactions. The OH-stretching vibrational band of hydroxy compounds present in high dilution (<1%, see above) disappears upon increasing concentration and gradually favors a considerably broader band located at a lower wavenumber which distinguishes the complex formed by hydrogen-bridge bonds between two or more oxy-groups of different molecules (see Sect. 6.4.9). An extreme case is the dimeric complexes of carboxylic acids (see Sect. 6.4.11.3).

Intramolecular Interactions
The previous sections discussed the interactions between different molecules, i.e. intermolecular interactions. However, hydrogen-bridge bonds may occur within the same molecule (e.g. in *o*-nitrophenol). These show a more or less strongly widened OH-band shifted towards lower wavenumbers (to 400 cm^{-1} and greater), which can be affected neither by a dilution nor by a change of the polarity of the solvent. Such observations are very important for structural studies (see Sect. 6.5.6).

Hence, in case of doubt (especially upon using stronger polar solvents), it is advisable to examine whether the solvent may effect the spectrum of the studied sample by systematically altering the concentration and possibly changing the solvent.

4.3.2.6 Microtechniques for Solutions

By using microcells (see Sect. 3.5.1.8) that can be placed in the sample beam instead of the pellet holder, liquid quantities of a few microliters are measurable. For reducing the losses by dead volumes and by adhesion to the container walls, in such cases the measurement of solutions is also preferred when low-viscous liquids are concerned. For example, one can concentrate extracts of chromatographically separated fractions (see Sect. 5.7.2) to the extent that the solution amount is just enough for filling a microcell of the corresponding pathlength. In this way, the necessary amount of sample to be effective can be reduced to a few tenths of a micro-Liter [19]. Even smaller substance quantities can be measured by using micropellets (see Sect. 4.2.1.5), e.g. by evaporating the solvent from a solution to the corresponding amount of KBr powder, a method which is applicable for solid and high-boiling liquid substances.

4.4   Gases

The preparation of gaseous samples for IR spectroscopy demands another working method than those discussed in the previous sections. Quantitative measurements can be done with gases to the same extent as for liquids or solids. With suitably high resolution, the spectra allow additional conclusions to be made on the molecular design through interpretation of the rotational fine structure.

4.4.1   Filling of Gas Cells

Gas samples are delivered and stored in either metallic pressure cylinders or in cylindrical gas collection containers made of glass (Fig. 4-13). Alternatively, there are aluminum or Teflon bags with a septum as the sampling location. A gas sample is transferred over to the cell mainly from a gas-collection container by using a vacuum apparatus, which allows an exact pressure measurement, possibly not directly from the pressure cylinder.

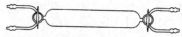

Fig. 4-13. Gas collection container.

Figure 4-14 shows the principle of the apparatus for filling gas cells. It consists of a conduit system (glass tube, 6 mm diameter) with two attachment possibilities for the cell and the collection container, as well as a manometer M1 for measuring the exact pressure in the cell, a cold trap and two manometers M2 and M3 for measuring the pressure in the cold trap or in the gas-collection container.

After the conduit system is evacuated, the sample is transferred to the cell by opening the corresponding valves until the desired partial pressure is attained in the cell. After attaching a nitrogen container (gas-collection container), the sample can be filled with nitrogen in the same way until the desired partial pressure (e.g. 500 hPa) is reached. Solvent vapors or condensable gases can be added for calibration mixtures without any problem when the sample, after cooling with liquid nitrogen, is again slowly warmed. One must note that these substances may be adsorbed by the cell surface, so that e.g. dynamic calibrations methods are preferably selected over the static method presented here. Examples for measuring gases and solvent vapors as well as literature on the dynamic techniques can be found in [20].

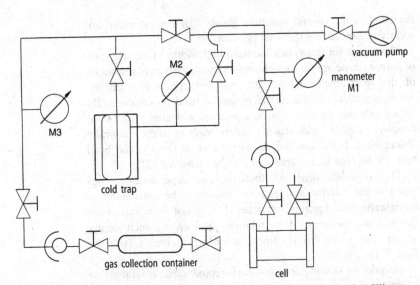

Fig. 4-14. Schematic diagram of a vacuum apparatus for filling gas cells.

The cold trap can also be used for freeing the samples from inert gases (e.g. air) by deep cooling, for pumping off as well as for recovering the sample.

4.4.2 Total Pressure Dependency of the Absorptivity in Gas Spectra

Figure 4-15 shows the CH-stretching vibration of methane, measured under two different experimental conditions. In spectrum (a), only methane was found in the cell with a partial pressure of 21.5 hPa. Prior to measurement (b), nitrogen was added up to a total pressure of 530 hPa. The considerably higher intensity of

Fig. 4-15. C-H stretching vibration of methane at 3020 cm^{-1} (methane-partial pressure 21.5 hPa):
a) Without addition of nitrogen,
b) Filled with nitrogen to 530 hPa.

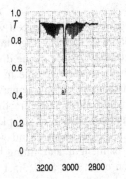
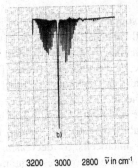

the band in the second spectrum shows that the measured absorptivity depends on the total pressure.

The reason for this is that the band half-widths in the spectrum of pure methane are very much smaller than the spectral resolution of the spectrometer. These are so-called convolution effects, which considerably reduce the measured band absorbances. By filling with nitrogen, we obtain a pressure widening of the rotational-vibrational bands that ultimately leads to larger maximum absorbances. If the spectral resolution can be fitted to the band half-widths, true line shapes can also be measured [21].

The frequently observed, total pressure dependency, recognized in the cited example, must absolutely be considered in quantitative measurements of gases. It may not be mistaken for the nonlinear behavior of the relationship $A = f(c)$, which ideally would have to be strictly linear according to Beer's law (see Sect. 7.1).

In order to obtain clear and well-reproducible conditions in quantitative measurements and regarding qualitative determinations, to avoid disturbances in the intensity relationships of different bands (due to variously strong total pressure-dependencies), one should fundamentally analyze gases in spectroscopy by using the same total pressure. It does not matter what total pressure is selected. It just has to be the same only for the measurement and the corresponding calibration.

Consequently, for the measurement of a calibration series for determining the concentration dependency of the absorbance, at first, the respectively desired partial pressures of the gases to be calibrated are to be inserted into the cell and subsequently the chosen total pressure is to be set by filling with nitrogen. For the calculation, the partial pressure of the calibration gas or of the sample is important, since the measurement imprecision of this affects the reproducibility of the result. In contrast, the precision with which the total pressure is set by filling with nitrogen is not very important.

4.4.3 Micromethods for Gases

The volume of 150–180 ml of a normal cylindrical gas cell with a pathlength of 10 cm is relatively large, because a large internal diameter (ca. 45 mm) must be selected because of the divergence of the radiation beam. The resulting dead volume in the cell and the remaining residual amount in the sample container for the method of pressure compensation allow large fractions of the sample to be unused. If only a small portion of sample is available, the cell volume can be reduced to about 25 ml with using microcells by decreasing the pathlength and fitting the cell to the convergence of the radiation beam (see Sect. 3.5.1.8).

Moreover, losses through the residual volume remaining in the sample container can be avoided by selecting the sample container to be possibly small. Condensing the sample into a small cold trap melted onto the cell is optimal, whereby the volume of the cold trap is negligibly small compared to the cell volume.

Whether the micro gas cell or a possibly available long-path cell is chosen for small sample portions, depends on the ratio of the cell volume/pathlength (see Sect. 4.4.4).

4.4.4 Trace Gas Studies

According to Beer's law, the absorbance increases, and thus the detectability, proportionally to the pathlength. This results in an improved detectability that, however, is not linear to the pathlength due to e.g. the increasing radiation losses for a larger number of reflections within a long-path cell (also see Sect. 7.4.2). The use of larger pathlengths, however, is generally prohibited in the IR spectroscopy, since the matrix, i.e. the remaining main and secondary components present, including the eventually present solvent, likewise show higher absorbances with increasing pathlength and thus overlap the spectrum of the trace component. Only a small selection of solvents depicts larger band-free regions, which allow measurements in pathlength of a few millimeters in the liquid phase (see Sect. 4.3.2.1 and 4.3.2.3).

However, if a gas sample consists mostly of IR-inactive components (nitrogen, oxygen, hydrogen and other dipole-free gases), of course the pathlength and, consequently, the detectability, can be considerably increased for the IR-active secondary components and trace substances in the sample. This also applies for wide ranges of di- and triatomic molecules (HCl, NO, CO etc.) so that trace analyses can be successfully performed here with corresponding long-path cells (see Sect. 3.5.1.6).

The coating of long-path cells takes place as described in Sect. 4.4.1. If air samples are to be investigated, only the very band-rich spectrum of steam disturbs in the region between 1900 and 1200 cm^{-1} as well as between 3900 and 3600 cm^{-1}. The absorption bands of CO_2 also disturb to a lesser extent (see Sect. 2.2.2.2). Because drying methods are particularly unsuited for trace studies because of the danger of trace components adsorbing to the drying agent, one can resort to using scaled subtraction by means of suitable steam spectra.

The determination limits obtainable by using long-path cells for pathlengths of 20 m lie in the lower ppb-range for components having strong absorption bands (e.g. CCl_4). These limits are naturally correspondingly higher (around 100 ppb) for other components such as NH_3 or SO_2 [20].

Here, we would like to point out the special problems that result by the adsorption to the vessel walls. The more polar the substance is, the greater the danger is that the trace substance is already adsorbed to the sample vessel or, at the latest, is adsorbed to the cell (metal wall!) and is partially or totally missing for the measurement. Short residence times in the sample vessel and cell before the measurement, frequent filling for the purpose of saturating the container walls and repeating the measurement as well as calibration measurements with test mixtures in the concentration range of interest are all measures with which the validity of a negative finding or the accuracy of a found value can be checked. Complete certainty in trace studies, however, is only provided by dynamic methods in which a stream of the gas to be analyzed is directed through the cell for a longer period. In this way, the vessel walls are saturated and a constant and correct measurement value is found after a particular time has elapsed. Dynamic calibration measurements on trace components or air (e.g. in environmental analysis) can be conducted by using, for instance, suitable dosing pumps.

Long-path cells are not only suited for trace gas determinations. Despite their large volume, they may also be used at times advantageously for measurements on small sample quantities. The task in the measurement of small substance portions is obtaining a possibly high absorbance for a small prescribed sample mass.

Inserting $c = m/V$ into Beer's law $A = a \cdot b \cdot c$, where m is the mass of the sample portion, and V denotes the volume of the cell, one obtains

$$A = a m \frac{b}{V} \tag{4.2}$$

Because the absorptivity a (base 10) is a constant and m (in our example) is given, A depends only on the quotient of the pathlength and the volume (b/V).

Table 4-2 gives these quotients for several commercially available cell types of various volumes and pathlengths. One recognizes here that the microcell is very effective because of its small volume but that a sample portion of the same mass in the larger volume of the mini-long-path cell, at pathlengths of greater than 2.4 m, can be assayed with still higher sensitivity. The normal cell actually offers the worse use of the given substance amount. However, not taken into account here are the effect of raising the absorbance at increased pressure in the microcell with unfitted spectral resolution, the counter effect of the narrower band half-widths (hence, higher maximum absorbance, important for measurements with IR diode lasers (see Sect.

Table 4.2. Quotient of the pathlength and volume for several cell types of various volumes and pathlengths

Cell Type	Cell Volume V in cm^3	Pathlength b in cm	Quotient b/V in cm^{-2}
Microcell	33	10	0.33
Normal cell	160	10	0.06
Ultramini-long-path cell	120	240	2.00
Mini-long-path cell	530	120	0.23
(variable in steps of 1.2 m)		240	0.45
		600	1.13
Long-path cell	8500	200	0.02
(variable in steps of 2 m)		–	–
		3200	0.38
Super long-path cell	25000	11000	0.44

8.2)) at lower pressure in the long-path cell, as well as possible adsorption phenomena on the large surface of the long-path cell.

For IR spectroscopy, Infrared Analysis, Inc. (Anaheim, U.S.A.), among others, produces long-path cells that exploit the multiple reflections within the cell (also see Sect. 3.5.1.6).

For atmospheric measurements (an overview on this is found in [22, 23], one can even do without cells by selecting the sample pathlength e.g. as the distance between a radiation source positioned far away and a spectrometer with telescope optics. Arrangements with retro reflectors are also possible, whereby the radiation source and the spectrometer then may stand next to each other. Still longer pathlengths have been realized in which the sun was used as the radiation source.

References

[1] *Sicheres Arbeiten in chemischen Laboratorien*, ed. Bundesverband der Unfallversicherungsträger der öffentlichen Hand e.V. (BA-GUV) in collaboration with the German Chemical Society and the Employment Accident Insurance Fund of the Chemical Industry (BG Chemie), **1988**

[2] *Unfallverhütungsvorschriften der Berufsgenossenschaft der chemischen Industrie.* Heidelberg: Jedermann-Verlag Dr. Otto Pfeffer

[3] Günzler, H., Borsdorf, R., Danzer, K. et al. (ed.), *Analytiker Taschenbuch*, Vol. 11. Berlin: Springer-Verlag, p. 216 ff

[4] *Maximale Arbeitsplatzkonzentrationen und biologische Arbeitsstofftoleranzwerte der Senatskommission zur Prüfung gesundheitsschädlicher Arbeitsstoffe der Deutschen Forschungsgemeinschaft.* Weinheim: VCH-Verlagsgesellschaft, **1994**

[5] Otto, A., Bode, U., Heise, H.M., *Fresenius Z. Anal. Chem.* **1988**, 331, 376

[6] Garner, H.R., Packer, H., *Appl. Spectrosc.* **1968**, 22, 122

[7] Patt, P., Kuhn, P., *Tips 9UR*. Bodenseewerk Perkin-Elmer and Co. GmbH, Überlingen/Bodensee, **1962**

[8] Mebane, A.D., *Anal Chem*. **1956**, 27, 37A

[9] Milne, J.W., *Spectrochim. Acta* **1976**, 32A, 1347

[10] Ataman, O.Y., Mark, H.B. Jr., *Appl. Spectrosc. Rev.* **1977**, 13, 1

[11] Keresztury, G., Insze, M., Seti, F., Imre, L., *Spectrochim. Acta* **1980**, 36A, 1007

[12] Miller, R.G., Stace, B.C., *Laboratory Methods in Infrared Spectroscopy*. London: Heyden & Son, 2nd ed., **1972**, 115.

[13] Wehling, W., *Perkin-Elmer Tips 37 UR*. **1968**

[14] Harrick, N.J., *Appl. Spectrosc.* **1977**, 31, 548

[15] Farrington, P.J., Hill, D.J.T., O'Donnell, J.H., Pomery, P.J., *Appl. Spectrosc.* **1990**, 44, 901

[16] Heise, H.M., *Proc. Soc. Photo.-Opt. Instrum. Eng.* **1985**, 553, 247.

[17] Bulanin, M.O., *J. Mol. Struct.* **1973**, 19, 59

[18] Heise, H.M., *Fresenius Z. Anal. Chem.* **1994**, 350, 505

[19] Bode, U., Heise, H.M., *Mikrochim. Acta I*, **1988**, 143

[20] Heise, H.M., Kirchner, H.-H., Richter, W., *Fresenius Z. Anal. Chem.* **1985**, 322, 397

[21] Heise, H.M., *Infrarotspektrometrische Gasanalytik – Verfahren und Anwendungen*, in: Günzler H et al. (ed.), Analytiker Taschenbuch, Vol. 9. Berlin: Springer-Verlag, **1990**, 331

[22] Hanst, P.L., *Fresenius Z. Anal. Chem.* **1986**, 324, 579

[23] Hanst, P.L., Hanst, S.T., *Gas Measurements in the Fundamental Infrared Region*, in: Sigrist M.W. (ed.), Air Monitoring by Spectroscopic Techniques. New York: John Wiley & Sons, **1994**, 335

5   Special Sample Techniques

Almost all samples can be brought into a suitable form for IR spectroscopic studies with the preparation methods yet described. The following section points out other important possibilities that can be applied for special cases.

5.1   Reflection Methods

Aside from the standard procedure of measuring samples in transmission, numerous other techniques using various reflection accessories were developed in recent years to also measure difficult samples in the laboratory, thereby considerably reducing the effort involved in the sample preparation. A particular aspect is that reflection spectroscopic methods often allow a non-invasive study of samples.

In all, two categories can be differentiated: for external reflectance, irradiation occurs e.g. from the medium air to an optically denser sample, while for internal reflectance, interactions of the electromagnetic radiation on the interface between the sample and a medium with a higher refraction index are observed. Techniques applying external reflection ought to be first introduced here.

5.1.1   Measurement of External Reflectance

5.1.1.1   Fresnel Reflection

Regarding reflection on real surfaces, one must note that this is often classified between the limits of mirror reflection, also called 'Fresnel reflection', which obeys the laws of geometric optics and is assigned to diffuse reflectance. Scattering processes are responsible for the latter in which refraction, diffraction and mirror reflection on small particles may be involved. In this section, however, we are interested in processes on plane surfaces.

For the measurement of oriented reflection, various commercial accessories are available which are placed in the sample

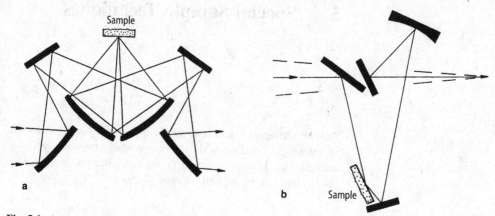

Fig. 5-1. Accessories for reflection measurements under a fixed (a) and variable (b) angle of incidence.

compartment of the spectrometer. Figure 5-1 shows two different arrangements that are mainly used. In the second attachment, the reflection angle can be widely varied so that spectroscopy can take place even when the incident radiation grazes the sample which can be helpful for certain uses that will be discussed later.

Spectra of oriented sample reflection differ considerably from those recorded in transmission. This is shown in Fig. 5-2. The Fresnel reflection on an interface defined by the sample and the air above it is determined by the refraction indices of

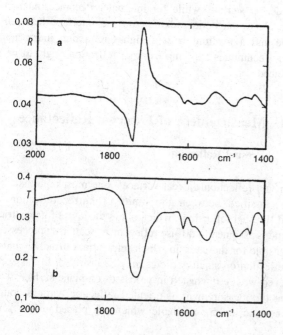

Fig. 5-2. Reflection spectrum (a) and transmission spectrum (b) of a polyester varnish (for the latter, a film was analyzed by IR spectroscopy on a KBr window).

both layers. Some fundamentals ought to be cited here about this. The complex refraction index is described by

$$\hat{n}(\tilde{v}) = n(\tilde{v}) + ik(\tilde{v}) \tag{5.1}$$

where $n(\tilde{v})$ is the refraction absorption index and $k(\tilde{v})$ the absorption index, whereby both are correlated by the so-called Kramers-Kronig transformation (see e.g. [1], $i = \sqrt{-1}$ for the imaginary part). Figure 5-3 illustrates the shape of the function of both optical constants. The absorption index $k(\tilde{v})$ of an optical medium is linked to its absorption coefficient $a(\tilde{v})$ via the equation

$$k(\tilde{v}) = a(\tilde{v})/(4\pi\tilde{v}) \tag{5.2}$$

where a is defined through the transmittance of a sample with the thickness b:

$$T(\tilde{v}) = e^{-a(\tilde{v})b} = 10^{-a(\tilde{v})cb} \tag{5.3}$$

By using the molar absorptivity $a(\tilde{v})$ and the molar concentration, c, the corresponding absorption coefficients can be determined as $a(\tilde{v}) = \ln(10)a(\tilde{v})c$.

Both optical constants are important for calculating the degree of reflection. By applying the so-called Fresnel equations, which describe the transmission and reflection on interfaces, one obtains the following expression by considering a non-ab-

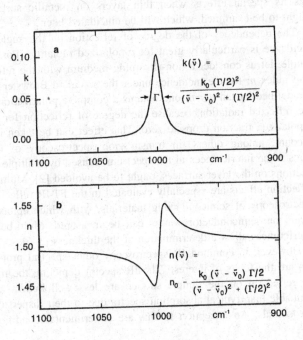

Fig. 5-3. Shape of the absorption index (a) and refraction index (b) for a Lorentz absorption band.

sorbing, optically thinner medium with a refraction index n_1 for perpendicular irradiation:

$$R(\tilde{v}) = \frac{\{n_2(\tilde{v}) - n_1(\tilde{v})\}^2 + k_2^2(\tilde{v})}{\{n_2(\tilde{v}) + n_1(\tilde{v})\}^2 + k_2^2(\tilde{v})} \tag{5.4}$$

For metals, the degree of reflection attains values close to $R = 1$ because of the very high absorption indices present. The absorption properties are apparent when the metals exist as a fine powder, which then may be counted among the "most black" substances. A similar case exists for minerals having strong absorption bands, in the region where likewise a high degree of reflection is found. This phenomenon is also called the reststrahlen effect and is used for reflection filters in order to obtain simple optical band passes for certain wavenumber regions.

The polarization of the electromagnetic radiation is very important, especially for the oriented reflection. The electrical vector of the wave hereby vibrates in one plane in the direction of propagation, for which the case of linear polarization is described. For non-perpendicular incidence for reflection, the degrees of reflection differ depending on whether parallel or perpendicularly polarized radiation was applied. The letters p and s have been agreed upon for characterizing the fractions. Hence, for p-polarized radiation, the electrical vector is found in the same plane defined by the incident and reflected radiation beam, and in the other case, perpendicular to this plane. This has its special effects when thin layers on metallic surfaces ought to be examined, which will be elucidated later.

The dependency of the degree of reflection on the angle of incidence is particularly great for p-polarized radiation. For example, let us consider a non-absorbing medium with the refraction index n: for the incidence under the so-called Brewster angle defined by $\tan a_p = n$, one obtains a complete polarization of the reflected radiation, because the degree of reflection for the p-polarized fraction drops to zero. This effect can be taken into account, among others, in transmission spectroscopy of thin films when interferences in the spectrum, caused by multiple reflections on the layer surfaces, ought to be avoided [2]. Multiple reflection effects are especially evaluated in the FT-IR reflection spectroscopy of semiconducting materials, with which epitaxial layers on semiconductor wafers can be measured. Multi-beam interferences allow a determination of the thickness [3–6].

However, in common IR spectroscopy other spectral properties are the focus of interest. The IR spectra depicting the mirror reflection on the sample surface are less well known and evaluable than typical absorption spectra due to their dispersion-like signals. An exception to this are experimental conditions

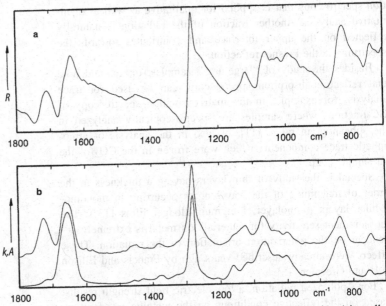

Fig. 5-4. Reflectance spectrum of a modern violin coating (nitrocellulose, oriented reflection with 6° angle of incidence measured *in situ*) (a) and its Kramer-Kronig transform (upper spectrum) compared to the absorbance spectrum of a commercial nitrocellulose coating prepared as a dry film on a KBr pellet (lower spectrum, (b)).

e.g. with parallel-polarized radiation near the Brewster angle, for which spectra result that are highly similar to absorption spectra, because the quadratic absorption index dominates hereby (see also Eq. 5.4) [7].

A more efficient procedure applies the aforementioned Kramers-Kronig transformation for converting reflectance spectra into absorbance spectra [8, 9]. Conversion programs are available in most spectrometer software packages. Such an analysis has been proven successful in investigating the coatings of historical works of art as well as of old violins [10]. An example regarding this is shown in Figure 5-4. The spectra obtained in this way can be used for a library search in order to identify materials.

5.1.1.2 Reflection-Absorption Spectroscopy

This measurement technique allows the study of thin layers located on metallic surfaces. In these measurements, the same type of accessories can be used as were described in the previous section. One part of the radiation penetrates the layer and is reflected by the metallic base layer whereby the sample to be examined is penetrated twice. This is called reflection-absorption spectroscopy that is applied for qualitative as well as quantitative analyses. Another fraction of the radiation is naturally reflected on the upper interface and contributes towards the spectrum via the Fresnel reflection.

Besides the study of "large-area" samples, one should note that reflection-absorption spectroscopy can be used for trace analysis: for example, in the matrix-isolation spectroscopy of GC-fractions where samples are spectroscopically analyzed in the subnanogram range [11], or also in the analysis of atmospheric trace components which were frozen in the CO_2 of the air sample [12].

Special is the study of thin layers having a thickness in the order of magnitude of the wavelengths occurring in monomolecular layers (monolayer, Langmuir-Blodget films [13]). The measurement sensitivity for materials on metal is extremely dependent on the polarization orientation of the radiation. These effects were already described years ago by Francis and Ellison [14] and Greenler [15].

For s-polarized radiation, a low electric field strength results, based on the reflective conditions on the metallic interface, so that only a weak interaction with the thin layer material results which is almost independent of the angle of incidence. The obtained signals, moreover, are not proportional to the pathlength. The situation is dramatically changed with polarized radiation which is parallel to the plane of incidence: especially for almost grazing incidence, one hereby obtains the highest electric field strengths on the metallic surface, so that amplification factors of up to 25, as opposed to normal transmission measurements, were precalculated [15]. In the practice, angles of incidences of between 75° and 89° are applied. One must take into account the fact that also the dielectric properties of the metal substrate affect the spectral quality [16]. It must also be noted that for such a measurement, especially vibrations with alterations of the dipole moment perpendicular to the metal surface are excited.

The investigation of thin polymer films and coatings on oxidized metallic surfaces is often a topic of interest as well as the study of surface treatments with e.g. lubricants and other organic materials [17, 18] and adsorbate coatings [19, 20]. Other objects of measurement are interfaces between electrodes and electrolytes (e.g. [21]) or interactions of adsorbed CO on metallic surfaces [22]. Other applications, also with polarization modulation as an improved measurement technique, are found in a review by Urban and Koenig [23].

5.1.2   Attenuated Total Reflectance

Attenuated total reflectance (ATR), also known as internal reflection spectroscopy (IRS) or multiple internal reflectance
(MIR), is a versatile, nondestructive technique for obtaining the
infrared spectrum of the surface of a material or the spectrum of
materials either too thick or too strongly absorbing to be analyzed by standard transmission spectroscopy. The technique
goes back to Newton [24] who, in studies of the total reflection
of light at the interface between two media of different refractive indices, discovered that an evanescent wave in the less
dense medium extends beyond the reflecting interface. Internal
reflection spectroscopy has been developed since 1959, when it
was reported that optical absorption spectra could conveniently
be obtained by measuring the interaction of the evanescent
wave with the external less dense medium [25, 26]. In this technique, the sample is placed in contact with the internal reflection element (IRE), the light is totally reflected, generally several times, and the sample interacts with the evanescent wave
(Fig. 5-5) resulting in the absorption of radiation by the sample
at each point of reflection.

    The internal reflection element is made from a material with
a high refractive index; zinc selenide (ZnSe), silicon (Si), germanium (Ge) and diamond are the most commonly used. To obtain total internal reflection the angle of the incident radiation
must exceed the critical angle θ_c [25]. The critical angle is defined as:

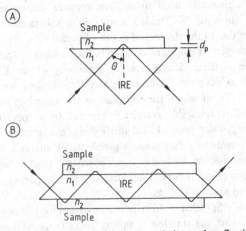

Fig. 5-5. Schematic representation of total internal reflection with:
A) Single reflection; B) multiple reflection IRE (internal reflection element). n_1 = refractive index of the internal reflection element;
n_2 = refractive index of the sample with $n_2 < n_1$; θ = angle of incidence;
d_p = depth of penetration.

$$\theta_c = \sin^{-1}\frac{n_2}{n_1} \qquad (5.5)$$

where n_1 is the refractive index of the internal reflection element and n_2 is the refractive index of the sample.

What makes ATR a powerful technique is the fact that the intensity of the evanescent wave decays exponentially with the distance from the surface of the internal reflection element. As the effective penetration depth is usually a fraction of a wavelength, total internal reflectance is generally insensitive to sample thickness and so permits thick or strongly absorbing samples to be analyzed. The depth of penetration d_p, defined as the distance required for the electrical field amplitude to fall to e^{-1} of its value at the interface, is given by:

$$d_p = \frac{\lambda_1}{2\pi(\sin^2\theta - n_{21}^2)^{1/2}} \qquad (5.6)$$

where $\lambda_1 = \lambda/n_1$ is the wavelength in the denser medium, and $n_{21} = n_2/n_1$ is the ratio of the refractive index of the less dense medium divided by that of the denser [24].

Although ATR and transmission spectra of the same sample closely resemble each other, differences are observed because of the dependency of the penetration depth on wavelength: longer wavelength radiation penetrates further into the sample, so that in an ATR spectrum bands at longer wavelengths are more intense than those at shorter ones (see Fig. 5-6).

The depth of penetration also depends on the angle of incidence; hence, an angle of 45°, which allows a large penetration depth, is generally used to analyze organic substances, rather than an angle of 60°, which results in a substantially weaker spectrum due to the decreased depth of penetration.

The degree of physical contact between sample and internal reflection element determines the sensitivity of an ATR spectrum. To achieve this, a horizontal ATR accessory such as the MVP[TM] [27], in which the top plate is the sampling surface, is used [28]; reproducible contact is ensured by a special sample clamp or powder press. Good quality spectra are thus obtained for many materials that present problems of analysis with routine transmission methods, e.g., powders, pastes, adhesives, coatings, rubbers, fibers, thick films, textiles, papers, greases, foams, and viscous liquids.

Harrick's SplitPea[TM] [27] is a nanosampling horizontal ATR accessory with the smallest sampling area of any ATR accessory – less than 250 μm in diameter for its Si ATR crystal (see Fig. 5-7). The SplitPea[TM] is configured to apply localized, measured pressure to produce superior contact between the sample and the ATR crystal. Thus making ATR nanosampling simple

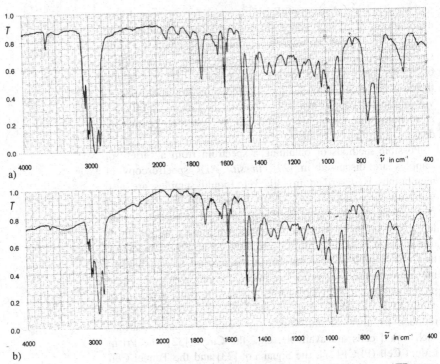

Fig. 5-6. Comparison of a transmission spectrum (a) with an ATR spectrum (b). Sample: styrene-butadiene copolymer.

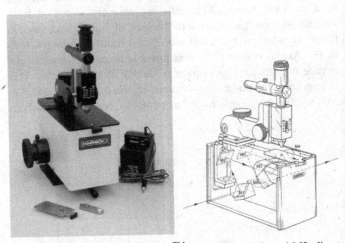

Fig. 5-7. Illustration of the SplitPea^{TM}. Two mirrors, M1 and M2, direct the beam to an ellipsoidal mirror, M3, which focuses the light onto the sample. The radiation reflected from the sample is collected by a second ellipsoid, M4. Mirrors M5 and M6 direct radiation reflected from M4 to the detector of the spectrometer. This configuration provides a six times linear reduction of the source image on the sampling surface (Reproduced by permission of Harrick Scientific Corporation, Ossining, NY 10562).

and straightforward, the SplitPea™ is ideal for quick and easy examination of a wide range of samples. These include: hard samples, like paint chips and combinatorial chemistry substrates; small samples, such as individual fibers and nanoliters of liquids; large samples, such as transparency film and defects thereon. For positioning samples on the sampling area, a 50X-viewing microscope is available. Overall, the famous SplitPea™ is an innovative alternative to infrared microscopes, beam condensers, and diamond cells.

Possible methods of obtaining a spectrum from a variety of samples are discussed in [29]; *in situ* ATR spectroscopy of membranes is described in [30, 31]. Liquid samples are also well suited to ATR analysis. Most liquids require a very short path length; aqueous samples, for instance, are measured at path lengths of no more than ca. 15 μm, which makes the design of transmission cells difficult because flow of liquids is hindered; they also exhibit interference fringes because of the small spacing between the high refractive index infrared windows. These problems are eliminated by using liquid ATR cells, a variant of solid ATR, in which the internal reflection element is surrounded by a vessel into which the liquid is poured. Various such liquid cells are available, e.g., the Circle [32], the Prism Liquid Cell (PLC) [27], the Squarecol [33] and the Tunnel Cell [34]. With optimized optical design and fixed internal reflection elements, these cells provide a highly reproducible path length, which permits the quantitative analysis of liquids and aqueous solutions [35]. Liquid ATR cells are uniquely suited to fully automatable, on-line process monitoring of liquids and viscous fluids. In addition, with suitable optical transfer modules such as the Axiom system [36], it is possible to operate liquid cells outside the sample compartment of the FT-IR spectrometer. This is important for on-line applications in which it is not practicable to pipe the sample to the spectrometer because the analysis

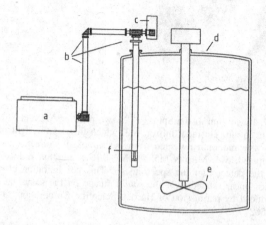

Fig. 5-8. Schematic drawing of an FT-IR measurement system utilizing the Deep Immersion Probe Model DPR-124 mounted in a batch reaction vessel [37]. a) FT-IR spectrometer; b) optical transfer elements; c) detector assembly; d) reaction vessel; e) mixing blade; f) ATR sensing head (reproduced by permission of Axiom Analytical, Inc., Irvine, CA 92614).

can be carried out at the most desirable location, even in a fume hood or on a process line. With an ATR sensing head, immersion probes [37] offer the possibility of *in situ* FT-IR spectroscopy, for monitoring batch process reactions (Fig. 5-8), laboratory process development, the identification of hazardous waste, the verification of the contents of storage drums, and the inspection of incoming raw materials.

5.1.3 Diffuse Reflectance

Diffuse reflectance is often used for powders and solids having a rough surface. The diffuse scattered radiation from the sample is possibly collected in a wide spatial angle. One can differentiate between two fractions of this radiation: one fraction is mirror reflected from uneven areas of the surface. The other fraction penetrates the sample, is partially absorbed, and via scattering processes in the interior, returns to the surface. Mostly, both reflection effects cannot be separated experimentally. This kind of diffuse reflectance (DR) is often named "remission"; the measurement technique for this has been already used for a long time in the ultraviolet and visible region [41].

In literature, diffuse reflectance is also known under the name "Kubelka-Munk reflection", because both scientists developed a theory on the radiation transport in scattering media [42]. The description of diffuse reflectance is based on a one-dimensional model, wherein the optical properties of the sample are given by two relevant constants, i.e. the absorption and the scattering coefficients. After integration of the fundamental differential equation, one obtains with R_∞, the degree of reflection on the sample surface at infinite thickness, the Kubelka-Munk (KM) equation that plays an important role for the quantitative evaluation of DR spectra:

$$f(R_\infty) = (1 - R_\infty)^2 / 2R_\infty = K/S \qquad (5.7)$$

with K denoting the absorption modulus and S, the scattering modulus (both in units of an inverse length), whereby the optical parameter is shown to be $K=2a$ (a absorption coefficient). Through this transformation of the degree of diffuse reflection and with a constant scattering modulus, one obtains a spectral function that is proportional to the normal absorbance spectra. The absorptive properties of the sample affect the penetration depth of the radiation, so it can happen that in regions of weak absorption, the layer base shines through and the prerequisite for the KM theory must be altered [43].

Especially for the measurement of diffuse reflection in the near-IR region, the function log $(1/R)$ is evaluated often analogously to the absorbance for quantitative evaluations, which, as Olinger and Griffiths [44] could show for an absorbing matrix, guarantees a better proportionality to the concentration here than the Kubelka-Munk function.

The integrating sphere, described by Ulbricht for measurements in the UV/VIS spectral region, belongs to the oldest arrangements [45]. The inner, diffuse scattering surface of the sphere is treated with gold vapor for measurements in mid-IR in order to guarantee a high degree of reflection. In contrast, for measurements in the near-IR region, Ulbricht spheres are used consisting of *spectralon*, a thermoplastic resin which is applied as a white standard because of its high degree of refection. With the Ulbricht sphere, the entire radiation reflected by the sample is integrated, thereby making absolute measurements of the degree of reflection possible [46].

The integrating sphere (see Fig. 5-9) is often applied for quantitative analysis in near-IR, whereby different foodstuffs and animals feeds can be directly studied for their protein, fat and water content, for example, without a complicated sample preparation. NIR spectroscopy hereby replaces the conventional and complicated wet chemistry in routine analysis and process control [47, 48]. Other arrangements, which allow a partial discrimination of the mirror reflection, apply elliptical collection mirrors (see Fig. 5-9).

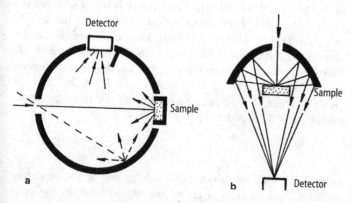

Fig. 5-9. Arrangements for reflectance measurements.
a) Ulbricht-sphere
b) segment of a rotational ellipsoid.

Several commercial accessories have long since been offered, whereby the schematic design of these mirror arrangements, intended for the use in the sample compartment of the spectrometer, is illustrated in Figure 5-10a. There are various designs, which collect the radiation in a respectively limited spatial angle. For one type of accessories, both mirrors are located above the sample in one plane. For reducing the mirror reflection, a so-called blocker placed on the sample has been recommended here [49].

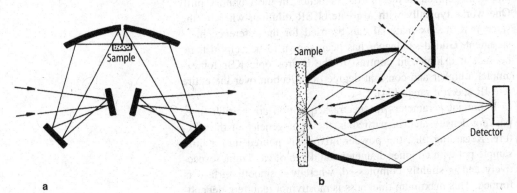

Fig. 5-10. Accessories for measuring diffuse reflection:
a) Sample compartment accessories,
b) reflection optics for measuring small regions on samples having a
large surface.

In a similar attachment ("praying mantis") [27], both mirrors
facing the sample are arranged offside of the optical axis in the
sample compartment, with which the Fresnel reflection can be
better discriminated. A commercial version allows studies, e.g.
of adsorptive processes of CO on rough metallic surfaces at high
temperatures (up to 750 °C) and pressures of up to 10 MPa by
means of DR spectroscopy [19]. On the other hand, the operation
is also possible in a vacuum at the temperature of liquid nitrogen.

One disadvantage of the commercial arrangements is that, often, only small samples can be examined. In order to relieve
this problem, arrangements with rotational ellipsoids were developed in which the illumination and collection optics are
found above the sample surface (see also Fig. 5-10b) [50, 51].
However, these accessories need their own detector, in contrast
to the reflection arrangements applicable in the sample compartment of the spectrometer.

Y-formed bundles of quartz fiber have proven themselves to
be advantageous especially for NIR spectrometry, e.g. for the input control of production raw materials. Hereby, a bundle of
fibers are used for illumination; the diffuse reflected radiation
from the sample is collected via a second bundle, whereby illu-
mination- and collection fibers are arranged mixed and equally
distributed in the measurement probe. The radiation from the
collection fiber bundle is focused on its own detector.

As a whole, the DR spectra are dependent on the various parameters of the sample preparation as well as on the geometric-
optical effects of the spectral measurement. The latter means
that the spectra vary depending on the accessories used [52, 53].

For measuring powder sample in the mid-IR region, these
samples are mostly mixed with KBr powder and are pulverized

and homogenized in the mortar, or better, in the vibratory mill. One works typically with a greater KBR dilution, whereby the same pure matrix material can be used for the reference measurement. Gilded sandpaper has been shown to be helpful here, because it exhibits, in contrast to the hygroscopic KBr for example, a higher and constant degree of reflection over the entire mid-IR spectral range [54].

Important parameters in spectra acquisition are particle size, packing density and, in particular, the homogeneity of the mixture. Regarding this, the powder mixture is poured into a little sample pot with an inner diameter and depth of ca. 5 mm, respectively and is slightly compressed, whereby a smooth surface is formed. This maximum thickness is mostly not required; depending on the absorption and scattering, already 200 μm or also 1.5–3 mm may suffice [55–57]. A simple accessory for doing reproducible sample preparation has been variously described [58, 59]. The reproducibility attainable with careful manual or automatic sample preparation lies at a relative standard deviation of 3% and better [60, 61]. It is important to note that the sample absorption, at small particle sizes, increases proportionally to the longer pulverization period, while the Fresnel reflection effects can be simultaneously reduced [62]. Further details with comprehensive literature data are found in Korte [63].

Substances dissolved in solvents can likewise be studied with diffuse reflectance after dropping of the solution onto KBr powder and then evaporating the solvent. In this way HPLC fractions, for example, can be examined by IR spectroscopy, whereby the preparation effort is very low. As an example, Figure 5-11 illustrates reflectance spectra obtained by this means. As a result, it is found that the DR measurement technique can be excellently used for microtrace analysis. Thus, by all means, detection limits in the nanogram range can be obtained [64].

For compatibility reasons, various diamond powders were recommended for preparing aqueous solutions. Several other materials were examined by Brackett and colleagues [65].

Some further uses of DR spectroscopy are named as follows. A small amount of substance can be rubbed off a solid compact sample with sandpaper in order to qualitatively analyze it by means of diffuse reflectance [66]. However, also undiluted samples can be directly measured in the mid-IR region. An example is the *in situ* spectroscopy of fractions on thin-layer plates that will be elucidated in Sect. 5.7.2. Other application examples include the measurement of textiles, varnish coatings, foams or paper, whereby the non-invasive *in situ* analysis is also applicable for rough surfaces [54].

Figure 5-12 depicts the spectrum of a polyester coating measured by diffuse reflectance compared with other measurements. It shows the specific differences expected for the respective

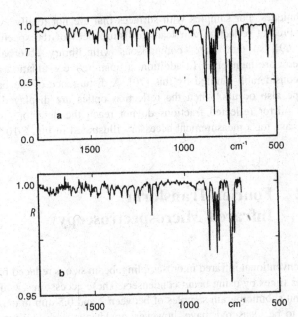

Fig. 5-11. Spectra of benzo[a]pyrene:
a) reference transmittance spectrum with 13 mm pellet,
b) diffuse reflectance spectrum of 2 µg in KBr powder (applied as a solution).

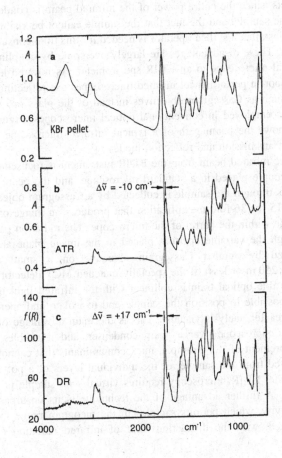

Fig. 5-12. Spectra of a polyester varnish, acquired with various spectroscopic measurement techniques:
a) Transmission as KBr pellet,
b) attenuated total reflection,
c) diffuse reflection.

techniques. For samples with surfaces that are not totally even, the Fresnel reflection can lead to disturbances of the spectrum [67–69], so that direct comparisons with library absorbance spectra are hampered. In addition, a quantitative evaluation can be considerably limited by this [70]. A disturbance of the band shape also occurs, when the reflection optics are displayed so that mirror reflected fractions do not reach the detector, as is the case for a measurement accessory illustrated in Fig. 5-10b.

5.2 Fourier Transform Infrared Microspectroscopy

In conventional infrared microsampling, beam size is reduced four to six times by using beam condensers. These accessories, which permit submicrogram samples of between 2 and 0.5 mm in diameter to be measured, have, however, drawbacks such as heating effects, since the entire power of the infrared beam is condensed on the sample, and the fact that the sample cannot be visibly located, so that a skilled operator is needed for this time-consuming work. These disadvantages are largely overcome by coupling an optical microscope to an FT-IR spectrometer. As infrared lenses are poor in performance, infrared microscopes use reflecting optics such as Cassegrain objectives instead of the glass or quartz lenses employed in conventional optical microscopes. Figure 5-13 shows the beam path of a typical infrared microscope with both transmission and reflection modes.

The infrared beam from the FT-IR instrument is focused onto the sample placed in a standard microscope, and the beam that passes through the sample is collected by a Cassegrain objective with 15- or 36-fold magnification that produces an image of the sample within the barrel of the microscope. The radiation passes through the variable aperture placed in the image plane and is focused by another Cassegrain objective on a small area (250×250 m or less) of the specially matched MCT detector. As the visible optical train is collinear with the infrared light path, it is possible to position the sample and to isolate and aperture the area for analysis visually. This is the main advantage of an FT-IR microscope over a beam condenser, and it enables the measurement of, for example, microcontaminants that cannot be removed from the sample, or the individual layers of a polymer laminate. FT-IR microscopy requires virtually no sample preparation. A further advantage of the technique is its unsurpassed sensitivity, which permits samples in the picogram range to be analyzed, with the diffraction limit of infrared radiation (10–

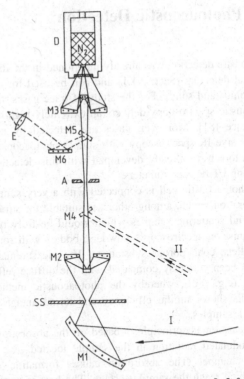

Fig. 5-13. Beam path of a typical infrared microscope. I = Infrared transmittance beam; II = infrared reflectance beam; M1 = condensing mirror; M2 and M3 = Cassegrain objectives; M4 = movable, semi-transparent mirror; M5 = movable mirror; M6 = mirror; SS = sample stage; A = aperture; E = eye-piece; D = MCT-detector, liquid N_2 cooled (Reproduced by permission of Bruker Optik GmbH, D-76275 Ettlingen, Germany).

20 m) as the limiting condition for the size of measurement spot. To minimize stray light due to diffraction being collected by the microscope, particularly when the size of the area investigated is approximately equal to the wavelength employed, a second aperture is introduced into the optics that focus the radiation onto the sample [71].

Uses of FT-IR microspectroscopy include general characterization of particulate matter, dichroic measurements with polarized light, polymer characterization, semiconductor measurements, the identification of contaminants, as well as forensic, biological, and pharmaceutical applications [72–74].

By using diamond cells with the beam condenser or microscope, the thickness of a sample (e.g., paint chips) can be adjusted by squeezing. An alternative to this is Harrick's Split-Pea™ [27], an ultrasmall sample analyzer [75], which allows nondestructive internal reflection studies of microgram and nanogram samples to be performed (see Sect. 5.1.2).

5.3   Photoacoustic Detection

Photoacoustic detection was already mentioned in the discussion
of thermal detectors (Sect. 3.1.1) and can be used for detecting
gases, liquids and solids. For the study of trace gases, the laser-
photoacoustic spectroscopy delivers an extremely high detection
performance [81]. Moreover, gases are detected by means of
non-dispersive IR spectroscopy with simple photoacoustic cells.
Gas monitors were already developed with this detection technique using FT-IR spectrometers.

The photoacoustic cell is connected with a very simple sample preparation which, among others, is suitable for strongly absorbing and scattering samples which would be fully ruled out
for transmission spectroscopy: powders, bodies with rough surfaces, pellets, coal, porous substances. The measurement technique has been variously compared with the diffuse reflectance
and ATR (e.g. [82]), whereby the photoacoustic method with
newer cells shows similar efficiency and its advantages for almost black samples.

The measurement principle is based on the absorption of in-
tensity-modulated radiation in the sample located in a sealed,
gas-tight chamber. The absorption causes formation of heat
waves, which reach the sample surface. The heat is partially released to the gases at the interface above the sample, whereby
the resulting pressure modulations can be detected with a sensitive microphone (see Fig. 5-14). The thermal expansion in the
sample leads to the development of an acoustic pressure wave
that can be measured in liquids and solids, for example, with
piezoelectric detectors.

As the theory is quite extensive, please refer to further literature for detailed information on this [83, 84]. Besides the optical depth of penetration, which is defined as the inverse of the
spectral absorption coefficient ($\mu_a = \alpha^{-1}$), the thermal properties
of the sample are important. The thermal diffusion distance
length μ_1 plays an important role hereby:

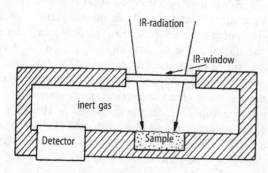

Fig. 5-14. Schematic diagram of a
photoacoustic sample cell.

$$\mu_t = \sqrt{\frac{2\lambda}{\varrho c v_s}} \tag{5.8}$$

where λ = thermal conductivity in $(\text{J cm}^{-1} \text{ s}^{-1} \text{ deg}^{-1})$
ϱ = density in (g cm^{-3})
c = specific heat in $(\text{J g}^{-1} \text{ deg}^{-1})$
v_s = modulation frequency in (s^{-1}).

As was shown in Sect. 3.4.3.1, the modulation frequency for fast-scanning FT spectrometers is proportional to the product of the wavenumber and the mirror velocity with which the probed sample depth can be varied thereby making deep-profile analyses also possible. An example for the study of laminar polymer layers as well as practical tips is given by Oelichmann [82].

The photoacoustic signal shows a relatively complex dependency on various factors that affect the signal intensity besides the sample absorption. Various case studies can be presented with respectively various optically and thermally dense samples, whereby slight saturation effects are also noted for strongly absorbing materials. A discussion of these effects and other applications are found in [85, 86]. Another parameter, for example in powders, is the particle size, which also determines the signal intensity [87].

An interesting development is the combination of phase modulation and step-scan FT spectrometers in order to obtain wavelength-independent modulation frequencies and, thus, the same spectral penetration depths [88].

5.4 Infrared Emission Spectroscopy

Each body with a temperature of $T > 0$ K emits photons to its surroundings. The radiation properties which are hereby important are described by the laws of Kirchhoff (absorption = emission), Stefan-Boltzmann (total radiation is proportional to T^4), Planck (spectral emission distribution) and Wien (radiation maximum is proportional to T^{-1}) with the supplementation that the radiation balance is formed by the sum of the emittance ε, reflectance R, and transmittance T, namely $\varepsilon + R + T = 1$. This last relationship implies that high-reflecting substances do no show any emission.

For performing emission-spectroscopic studies in the infrared range, the cited laws give the foundation for practical work. Therefore, according to Kirchhoff, the detector must be colder than the sample, because otherwise, the total system would be found in the radiation equilibrium state and no measurement signal would result. On the other hand, the temperatures of the sample (here the radiation source) and detector must be strictly kept at a constant temperature because of the T^4-law for reproducible measurements. Fortunately, the radiation maximum for $T = 300$ K lies at $\tilde{v} \approx 1000$ cm^{-1}, so that good emission spectra can be obtained in the range of $\tilde{v} \leq 1800$ cm^{-1} at typical laboratory temperatures. The steep decrease of the emissive power at short wavelengths, as described by Planck's law, is observable already in the C-H stretching vibration range to the extent that one can hardly see the CH-bands around 3000 cm^{-1}. This effect is exhibited by the emission spectrum of an organic substance in Fig. 5-15 in direct comparison with its transmittance spectrum: the signal positions all agree with one another, but the intensity relationships are totally different.

Compared to the principly same information obtained from transmission spectra, the technique of emission spectroscopy shows a few additional advantages but which require greater experimental effort:

- Contact with optical material is no longer necessary (pellet, ATR crystal).
- Studies are possible on strongly absorbing materials having a rough surface (corrosion processes, heterogeneous catalysts, films).
- Studies are possible on hot surfaces, of melts and in and on combustion gases.
- Phase transitions and solid-state reactions can be monitored.

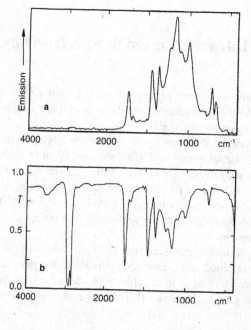

Fig. 5-15. IR emission (a) and transmittance (b) spectrum of an organic substance (wool fat) at 373 K (100 °C) for comparison. The radiation maxima lie at the same location as the transmittance minima. The C-H stretching vibration is practically non-existent in the emission spectrum. (The emission at $\tilde{v} = 672$ cm^{-1} is of unknown origin).

Very thin surface layers, composite materials or adsorbed substances come into question as samples. Various reports were made about the possibilities for surface analysis by means of emission spectroscopy [89, 90]. Other questions concern the contact-free measurement of gases, e.g. of flue gases from smokestacks (remote sensing) [91]. Moreover, emission spectra of the atmosphere, especially in the far-IR region, can be evaluated for quantitatively determining trace components.

A few particularities must be accounted for in measuring the sample emission: the emission spectrum also depends on the thickness. Normally, the sample is present as a thin layer on a base. For thicker samples, re-absorption may occur thereby causing the emission spectrum to be definitely more complicated, have less structure and to resemble the emission curve of a black radiator [92, 93]. The sample can also form gases, which re-absorb the radiation, or it can simply be thermally degraded.

In practice, an emission spectrum is acquired by using the heated sample, located in the emission port of the FT spectrometer, as the radiation source. A black radiator, having the same temperature as the sample, acts as the reference. If the spectrometer does not have any emission port, a sample arrangement is possible in the sample compartment, whereby the retro-reflection is used in the Michelson interferometer (efficiency of 50%) [94].

Spectra, which are represented as the quotient of the emission of the sample and of the black radiator, however, contain emission contributions from the sample environment and from fractions reflected by the sample. Information on methods and correction are found in a review by DeBlase and Compton [95]. A more interesting review is found in [96].

5.5 Measurements under Extreme Conditions

5.5.1 Cryogenics

Cells for low-temperature measurements were already mentioned in Sect. 3.5.1.9. IR studies at low temperature may yield spectral differences to measurements at ambient temperature, especially when phase changes occur. Transitions from the liquid to the solid amorphous state are characterized by reduction of the half-widths and band splitting. In addition, there are new aspects in the solid state: the forces change between the molecules at small intermolecular distances, the crystalline structure brings fully new viewpoints, hydrogen-bridge systems behave differently, and many molecules only exist in the solid state [97]. Moreover, the use of polarized radiation is linked to the solid. The spectra of e.g. carbohydrates [98] are surprisingly finely structured at low temperature. A summary review about low temperature research is found in Hermann et al. [99].

A variation of the low-temperature technique is used in the matrix-isolation (MI) spectroscopy. Since this technique was introduced by Whittle, Dows and Pimentel [100], various applications were described [101–104]. At first, short-lived decomposition products of labile compounds were analyzed by spectroscopy. In the experiments, for example, photochemically unstable substances are mixed at high dilution with an IR-transparent matrix gas and brought onto a deep-cooled target onto which the mixture solidifies. The investigated substance can then be cleaved e.g. photolytically. The fixed fragments incorporated in the matrix can be analyzed by spectroscopy as desired, i.e. in transmission, when the target (e.g. CsI) is IR-transparent, or with the reflection-absorption technique, when a metallic mirror was selected as the carrier. In this way, e.g. the quadratic structure of free cylcobutadiene was probably produced [105].

Ar or N_2 are often used as matrix gases. Then, He-cryostats are used for cooling with which the sample carrier can be cooled to temperatures down to 10 K. A prerequisite of this technique is the existence of a good high vacuum for thermally isolating the cooling finger, on the one hand, and also to exclude atmospheric contaminants which otherwise would be concentrated on the cryofinger (also see Fig. 5-16). A great excess of inert gas (1:500–1000 and greater) guarantees that the sample molecules are enclosed in the matrix, isolated from one another. The spectra are then free of interactive effects, with the exception of the matrix; hence, the narrow half-widths of the bands. Highly resolved, band fine structure may result from various "site trapping", i.e. the vibrational frequencies depend sensitively on the immediate cage surroundings. By careful thawing, one can determine diffusion and aggregation processes (metal carbonyl formation [106]).

One modification, the so-called pseudomatrix isolation, is a method mainly developed by Rochkind [107] for identifying components in gas mixtures. Here, the premixed sample is pulsed onto the target. The matrix-isolation technique simplifies the multi-component analysis considerably, because the rotational fine structure of the vibrational bands is not applicable, and also the so-called "hot" bands, transitions from excited vibrational states, no longer occur in the spectrum (see Figure 5-17) because of their low occupational density at low temperatures. It is interesting is that the vibrational band of N_2 appears at 2327 cm^{-1}, which otherwise is IR-inactive in the gaseous

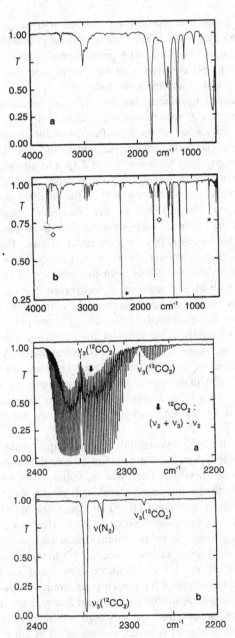

Fig. 5-16. IR spectra of liquid (a) and matrix-isolated acetone (b); the MI spectrum also contains absorption bands of H_2O (monomer, dimer and multimer) ($\diamond$) and CO_2 (*) as contaminants.

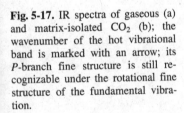

Fig. 5-17. IR spectra of gaseous (a) and matrix-isolated CO_2 (b); the wavenumber of the hot vibrational band is marked with an arrow; its *P*-branch fine structure is still recognizable under the rotational fine structure of the fundamental vibration.

phase. Its frequency concurs with the Raman vibrational band. According to Carr et al. [108], vibrations of matrix molecules can be activated by lattice defects, whereby ultimately a reduction in symmetry exists.

An important application area of the matrix-isolation technique is the classification of vibrational bands, whereby the narrow half-widths of the absorption bands is advantageous with overlaps because of the missing interaction between the molecules (see Fig. 5-16). There are excellent possibilities in the comparison of MI-spectra with gaseous phase data for structural clarification and description of new molecules (ions, radicals, complexes and of intermediate products).

Besides the quantitative analysis of components in complex gas mixtures, e.g. for atmospheric analysis [109], the MI spectroscopy plays a role in the identification of the smallest sample quantities as occurring after separation in the capillary column gas chromatography (also see Sect. 5.7.1). The advantage of this preparation method is that, on the one hand, the fractions with the matrix gas can be concentrated on the smallest sample areas (diameter ca. 150 µm), and on the other hand, that the half-widths of the bands are clearly smaller than those obtained with conventional spectroscopy of condensed phases. Both factors lead to higher maximum absorbances, so that lower detection limits can be obtained. Thus, with the coupling of gas chromatography with MI spectroscopy, it was shown that various dioxin isomers could still be differentiated in the lower nanogram range and be quantified [110]. IR spectroscopy shows its strengths especially in the differentiation of isomers in comparison with the frequently applied mass spectrometry. A commercially obtainable MI-IR system has been known under the name "Cryolect". An improvement of the chromatographic resolution over that with a flame ionization detector can be obtained with this system by selecting various scan velocities for the sample drum [111].

Despite the high detectability, this technique could not quite establish itself, because the handling is complex and, in practice, tedious. Therefore, there was an impetus towards simpler systems. With the instrument developed by Bio-Rad, which is marketed under the name GC-FT-IR, the fractions are evaporated onto a ZnSe-window cooled to 80 K. Here, too, as in the Cryoelect system, the entire chromatogram can be stored as a trace. IR microscope optics is necessary for the spectroscopic analysis of the ca. 100 µm-diameter crystallized sample spot. An online measurement of the chromatogram is possible, but may be measured again for a subsequent study of the individual fractions under optimized spectroscopic conditions, because the fraction chromatogram is saved, as long as no thawing and evaporation was programmed. The thereby obtainable detection limits partially lie at ca. 50 pg [112, 113]. Another advantage of this technique is that the spectra obtained from the condensed fractions may be compared with data from conventional spectral libraries, because the half-widths for the tracer technique do not considerably differ from those of normal spectra for solids, e.g. KBr pellet spectra, than is associated with the MI technique [114].

5.5.2 High-Temperature Spectroscopy

IR studies at high temperature are performed mainly with regard to phase transitions, catalyst behavior and kinetic questions. Another interesting area is studies on the intra- and intermolecular behavior of substances; a review is found in [115]. The studies always require especially designed cells. Gas cells are commercially available and are applicable at temperatures of up to 250 °C, although the boundary temperature for solid cells is 500–800 °C.

In general, the appearance of an IR spectrum definitely depends on the temperature. The characteristics for this are changes of the band shape, describable by the half-width, the maximum absorbance and the integral absorbance. At increased temperature, band widening with simultaneously increased intensity of the hot bands are ascertained (Fig. 5-18), from which also the shifting of band maxima may result. Also, refer to Fig. 5-17 regarding CO_2-spectral interpretation.

The possibilities of a contact-free temperature measurement using IR spectroscopy are interesting [115]. Because the occupational density of the molecular vibration and rotation level depends on the temperature (Boltzmann theorem), information about this parameter, temperature, can be obtained. In particular, for small molecules, the shifting of the maxima in the *P*- and *R*-branches of the rotational-vibrational bands can be used. Another possibility involves evaluating the temperature dependency of the integral band intensity of combination bands and over-

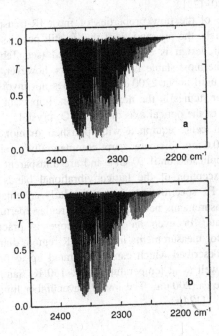

Fig. 5-18. Temperature dependency of the ν_3-rotational-vibrational band of CO_2; a) 200 °C, b) 25 °C. Other conditions: pathlength 10 cm, CO_2-pressure 8.1 hPa, spectral resolution 0.05 cm⁻¹.

tones. In contrast, the temperature dependency of fundamental vibrations is clearly lower. The emission measurement offers another possibility. For instance, the temperature dependency of the v_3-vibration of CO_2 was determined in the range of between 120 °C and 650 °C, in order to study the properties of a flame-emission detector for gas chromatography [116]. Moreover, the wavelength-dependent emission of a thermal radiator can be applied for determining temperature [117, 118].

A further application of higher temperatures is the coupling of thermogravimetric analysis (TGA) with IR spectroscopy. In the thermogravimetry, the change in mass of a substance is measured, which is subjected to a controlled temperature program. However, with this, no conclusions can be drawn about the gaseous reaction products. These gaseous products, however, are measurable by means of the IR spectrometer (EGA = evolved gas analysis) [119].

5.5.3 High-Pressure Measurements

Special cells, that allow IR-spectroscopic measurements under high pressures, were already cited in Sect. 3.5.1.10. Such investigations are used when the composition of samples under certain reaction conditions is the topic of interest or the behavior of molecules ought to be studied under extreme conditions. Intermolecular interactions, phase changes and various reactions, including thermal decomposition and polymerization, are investigated [120–123].

Because of the flow properties of most IR-transparent window materials, the pressure limit that is still permissible for the special cell design is quickly approached (see Tab. 3-7, Sect. 3.5.1.1). The most stable is sapphire. This, however, can be applied only until about 2200 cm^{-1} that does not involve any disturbance for studies in the near-IR region. Important here is the orientation of the optical axis of the Al_2O_3 crystal.

Diamond cells, equipped with polished diamond windows e.g. with 0.6 mm diameter, are commercially available. The cells are applicable until 10 GPa and are transparent in mid-IR, with the exception of the lattice vibrational bands at around 2200 cm^{-1}. FT spectrometers are particularly advantageous during the measurements because of their circular apertures. A micro-illuminator or even an IR-microscope is practical here. Especially for measurements in the far-IR region, diamond cells have been described which can be operated up to pressures of 16 GPa as well as at temperatures of 2–150 K. Sample diameters here are ca. 200 μm. The lower wavenumber limit was given at 20 cm^{-1} [124].

5.6 Measurements with Polarized Radiation

For natural radiation obtained from thermal radiators, for example, there is no preferred vibrational plane for the electric field vector. In Sect. 5.1.1.1, linearly polarized radiation was already introduced in which the electric vector is set in a particular plane of the direction of propagation. Elliptical polarized radiation is explained as a superimposition of two perpendicular, coherently vibrating vectors of the electric field that are phase-shifted to one another, thereby resulting in the sum vector moving like a helix in the direction of propagation. Circular-polarized radiation is a particular case of this, whereby both vectors occur with the same amplitude and a fixed phase difference of $\pi/2$, corresponding to $\lambda/4$.

Optically anisotropic crystals are known such as calcite, which show double refraction, whereby the refraction indices depend on whether the electric field vector is oriented parallel or perpendicular to the main axis of the crystal (optical axis). Regarding this, a beam can be separated into two partial bundles having different polarization planes that are perpendicular to each other (ordinary and extraordinary beam). However, polarized IR radiation is mostly produced e.g. with a polarizer consisting of a fine grating with parallel metal wires. The various absorption of radiation of parallel and perpendicular polarization is known as linear dichroism. The technique for measuring both of these optical properties, double refraction and dichroism by means of FT-IR spectrometers has been described by Jordanov et al. [125].

In oriented crystals, certain dipole moment changes lie in a fixed direction during the molecular vibration. Thus, such vibrations can be only excited when the IR radiation is also polarized in this direction (Fig. 5-19). The measured absorbance depends on the mutual orientation of the molecules and the polarization direction of the exciting radiation. As one can see, the transitional moment (proportional to the dipole moment change with regard to one molecular axis) of the C=O vibration at 1720 cm^{-1} of the mylar material is perpendicular to the remaining intensive molecular vibrations in the spectrum shown. Intensity studies can also be conducted depending on the crystal orientation to the polarization plane. However, one may not forget here that the spectrometer itself is partially a polarizer by its optical system.

Investigations on IR dichroism have been widely applied for oriented crystals and drawn polymers [126]. In polymers, the oriented crystallite likewise shows dichroic behavior after one

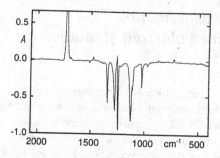

Fig. 5-19. Linear IR dichroism of polyethyleneterephthalate (Mylar, thickness 3.5 μm), obtained via difference measurement.

stretching. Hence, conclusions can be drawn about the position of the polymer chains and their order. Recently, new dynamic measurement techniques have been applied for such rheo-optical studies, namely the two-dimensional IR-correlation spectroscopy with which particular interactions between functional groups can be detected. Another advantage results from the improved spectral resolution that arises from applying a second dimension. A review of this is given by Noda et al. [127].

The interaction of polarized radiation with chiral substances is used for characterizing them. It is known that the solution of enantiomers, as is represented by glucose e.g., can rotate the plane of linearly polarized radiation (polarimetry). The latter can be imagined as being composed of two circularly polarized components, which move, however, in the opposite phase. Optical activity is manifested in that the refractions of the substance for left and right circularly polarized radiation differ.

The measurement of the so-called optical rotational dispersion (ORD) in the IR spectral range is possible for various substance groups. These include optically active substances as pure liquids or in solution. Cholesteric liquid crystals and polymers, as well as induced cholesteric solutions show stronger effects, and a third category is represented by optically active crystals. An example for measuring such spectra is given in [128]. An alternative measurement technique is possible and is based on determining the difference of the absorption coefficients for left or right circularly polarized radiation (circular dichroism, CD). By means of Kramers-Kronig transformation, it can be shown that ORD and CD spectra complement each other. Also, refer to [125] regarding the measurement of both effects via FT spectrometers. A review on spectroscopy with circularly polarized radiation is given by Nafie [129]. There is a variety of analytical applications including, for example, the determination of the conformation of biological molecules [130].

5.7 Combination of IR Spectroscopy with Chromatographic Methods

The qualitative interpretation of a spectrum, as well as the identification of pure substances through spectral comparison, is only successful when the sample is uniform, when namely all absorption bands of the spectrum definitely belong to the same substance. Although bands of solvents or of a few known accompanying components can be eliminated by compensation, the isolation of main and secondary components in a mixture is necessary for their elucidation. Possible separation methods include:

- Distillation and rectification,
- Gas chromatography,
- Thin-layer chromatography,
- Liquid chromatography.

Distillation and rectification typically need relatively large amounts of substances, with the exception of micro distillation apparatuses such as the rotating-strip column. Because of their low substance requirement and the high separation performance, chromatographic methods are better suited for isolating substances for IR spectroscopy. Liquid chromatography is mostly represented by HPLC (high pressure liquid chromatography). Supercritical fluid chromatography is another suitable technique, standing between gas chromatography and liquid chromatography. A detailed review on the coupling of chromatography and IR spectroscopy is given by White [131].

For combining the following described separation methods with IR spectroscopy, it partially depends on one being in the position to handle minute sample quantities. Beyermann [132] gives several ideas on this that are still useful nowadays.

5.7.1 Gas Chromatography/IR Spectroscopy

There are two possible techniques for GC/IR coupling: firstly, the separation of the GC-eluted sample by collecting the eluate and secondly, the direct transfer of the gaseous eluate into the IR cell with the possibility of an online measurement.

The simplest method for collecting a GC-separated fraction is cooling the gas stream exiting the gas chromatograph in a cold trap or directly collecting it in a microcell, which was described in different varieties e.g. in [133]. This working procedure, however, is not always successful. A rather universally applicable method is described by Witte and Dissinger [134], whereby the GC eluate is caught in a deep-frozen solvent.

The sample quantities of the fractions are generally in the microgram range by using packed columns. Today's capillary columns use much smaller sample portions, so that other techniques had to be developed for handling and analyzing them by spectroscopy. Hereby noted is the sample transfer via deactivated quartz capillaries as well as the use of cryogenic techniques in order to fix samples onto minute surfaces and to analyze them by spectroscopy (see Sect. 5.5.1). IR spectrometric detection limits of 50 pg have already been cited. With this, online studies of the chromatographic fractions are possible as well as a subsequent spectroscopy of the cryocondensed substances with an improved signal-to-noise ratio.

The problem of time-fitting in the spectrometric online measurement of the chromatograms, as in the case of GC-MS coupling, could be solved by the introduction of the fast FT-IR spectrometers, whereby another prerequisite hereby is the use of sensitive MCT detectors.

Here is an example for elucidating this procedure: at first, the separation of the substance mixture to be examined is prepared on the gas chromatograph (Fig. 5-20, uppermost chromatogram). Upon coupling with the IR spectrometer, the eluate flows first through an inner gilded glass tube, the so-called light pipe, through which the radiation of the interferometer is directed prior to detection of the fractions via e.g. a flame ionization

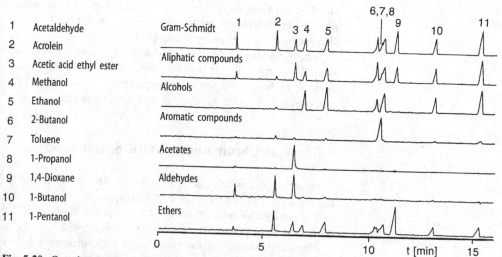

1	Acetaldehyde
2	Acrolein
3	Acetic acid ethyl ester
4	Methanol
5	Ethanol
6	2-Butanol
7	Toluene
8	1-Propanol
9	1,4-Dioxane
10	1-Butanol
11	1-Pentanol

Fig. 5-20. Gas chromatogram of a multi-component mixture; capillary column Carbowax 20M with a length of 50 m, 1.2 μm film thickness, 0.32 mm inner diameter, 2 ml/min He as the carrier gas; task 0.1 μl with split 1:50, all components with the same mass fraction, light-pipe length 12 cm, inner diameter 1 mm, operated at 200°C. Uppermost trace: Gram-Schmidt chromatogram, lower traces: chemigrams of the GC-separation (data on the selected IR windows found in text).

detector or an attached mass spectrometer. The interferograms are recorded in intervals of up to 0.1 s, converted and saved. Already a first rough evaluation can be made during this procedure. Regarding this, one can subdivide the IR spectral region into arbitrary regions.

Using Figure 5-20 as an example, specific spectral sections were chosen for aliphatics (2980–2920 cm^{-1}, 1478–1455 cm^{-1}, 1390–1380 cm^{-1}), for alcohols (3680–3660 cm^{-1}, 1390–1370 cm^{-1}, 1065–1050 cm^{-1}) for aromatics (3080–3020 cm^{-1}, 780–700 cm^{-1}) for aldehydes (2810–2805 cm^{-1}, 2715–2705 cm^{-1}, 1750–1745 cm^{-1}), for acetates (1770–1765 cm^{-1}, 1240–1237 cm^{-1}) and for ethers (1140–1085 cm^{-1}, 890–700 cm^{-1}). In the evaluation, it is determined how high the integrated absorbance in the individual channels is. This value is recorded continuously as the intensity signal. The so-called "chemigram" obtained, gives (read line-wise) the information about the substance groups in the chromatogram.

A specialty is the Gram-Schmidt trace [131, 135] that is calculated from the interferogram data and yields the total absorbance changes in the spectrum compared to a background spectrum that is defined before the chromatogram is taken. The Gram-Schmidt trace is by all means comparable mostly to a FID signal (for further details, see [136]).

In the subsequent post-processing of the stored data, formats are possible e.g. in the time-resolved presentation as shown in Fig. 5-21a. For the individually separated components, averaging can occur via the spectra within the half-widths of the corresponding chromatogram peaks in order to obtain better signal-to-noise ratios. Actual background spectra before and after the chromatogram peaks to be evaluated may be used for improved transmission spectra which are less disturbed by H_2O- and CO_2-absorption bands.

In the case at hand, the incompletely separated components 6 thru 8 can be mathematically subtracted from each other, or one integrates over the overlap-free fractions and thereby obtains the spectra of the individual fractions (Fig. 5-21b).

Generally, GC-separations with capillary columns can be performed without any problem. The detection limit by using a light pipe is at about 10 ng.

In the spectral interpretation, one must consider that the interpretation generally is best in absorbance form (see Fig. 5-21) and that the spectra are gas spectra, which partially deviate considerably from those of the liquid phase (Fig. 5-22).

For this reason and particularly for GC/IR coupling, there exist collections of similarly recorded gaseous phase spectra, which allow the identification of the eluate fractions. One of these collections is the EPA spectral library (Environmental Protection Agency, U.S.A.) with its about 3300 spectra which can

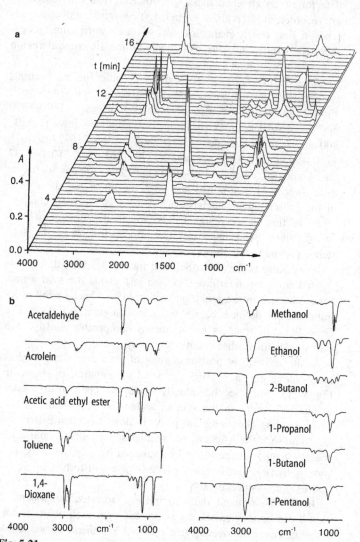

a
t [min]
16
12
8
4
A
0.4
0.2
0.0
4000   3000   2000   1500   1000   cm⁻¹

b
Acetaldehyde
Acrolein
Acetic acid ethyl ester
Toluene
1,4-Dioxane

Methanol
Ethanol
2-Butanol
1-Propanol
1-Butanol
1-Pentanol

4000   3000   cm⁻¹   1000   4000   3000   cm⁻¹   1000

Fig. 5-21.
a) Illustration of the IR spectra recorded during GC-separation as a
   function of time;
b) component spectra obtained via post-processing (spectral resolution
   8 cm⁻¹).

also be provided from the respective spectrometer distributors.
In addition, the library of Sadtler has 9200 entries (Vapor Phase
IR Standards) which, depending on the demand (evaporation
and chemical stability hereby play a role), had been recorded at
temperatures of between 25 °C and 300 °C. A published work on
the interpretation of gaseous phase spectra also including 500 library spectra is likewise available [138] (also see Sect. 9.1).

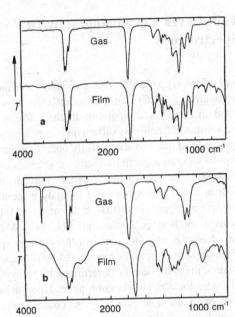

Fig. 5-22. Comparison of the transmittance spectra of a) non-polar (ester) and b) strongly polar substances (carboxylic acid) in the gaseous phase and as a liquid. With limitations, the spectra of the low-polar substances can be compared among each other in both aggregate states. Hydrogen bond-forming substances often have very different spectra (according to Welti [137]).

There is a variety of applications of GC/IR coupling. The study of solvent mixtures was already mentioned. Other application areas concern, among others, petrochemistry, the perfume industry and biomedicine; for example, see [136]. An extensive area is the study of natural substances and environmental samples [139]. Another use is the examination of pyrolytic products of polymers and humid materials. Kuckuk et al. [140] give an example for the structural identification by means of pyrolysis-GC-FT-IR.

In order to verify information for the component identification, parallel or series connections of the light pipe and mass spectrometers are operated. The Hewlett-Packard company markets an optimized FT-IR spectrometer with a heated gas cell, the so-called IR-detector (IRD), for gas chromatography, which, combined with a quadrupole mass spectrometer, can provide additional information especially for isomer mixtures [141]. Moreover, the cryogenic technique can also be successfully used for GC/FT-IR/MS coupling. The detection limit for naphthalene with using the "tracer" technique was given at 40 pg, while the detection limit with the mass spectrometer was even two orders of magnitude lower [142].

5.7.2 Thin-Layer Chromatography/IR Spectroscopy

For mixtures of poorly volatile or non-volatile substances, thin-layer chromatography (TLC) is just as an effective as well as inexpensive and uncomplicated separation method. The mass of the hereby-used substance fraction is sufficiently large with use of microtechniques, so that one can successfully obtain interpretable IR spectra from the separated fractions. Here, the substances found in the individual spots are eluted in the proper manner and transferred to KBr or into a solvent in order to analyze them as a micropellet by spectroscopy via diffuse reflectance or, for volatile substances, as a solution in a microcell [71, 143]. Making the spot visible, of course, may not occur in this case by spraying a reagent. When the spots cannot be recognized by fluorescence under an UV-lamp, the zones can be determined by spraying one of the parallel traces located on the same plate. Upon using strong polar solvent, one must take care that also coating fractions can be eluted, which disturb the spectrum.

Various methods have been described for transferring non-volatile substances form the carrier material into an IR-spectroscopically measurable form. The best method is the transferal in KBr. De Klien [144] recommends sprinkling the carrier material around the spot in question or extending the spot and to transfer the substance form the carrier material into the KBr by carefully adding a solvent. A few authors recommend the removal of the spot and elution of the substance by using a Wick-stick®, triangular KBr-tablets which are commercially available [145]. In any case, the substance is isolated by using a suitable solvent, is enriched and measured via transmittance or diffuse reflectance. The micromethods are limited not by the spectrometer technology but rather through the manual preparation, e.g. by contaminants or sample volatility.

In situ measurements on TLC plates are also possible with the diffuse reflection measurement technique. Figure 5-23 shows the DR-spectrum of a TLC-cellulose plate. Other materials are Al_2O_3, silica gel, and RP-substrates (reversed phase). One can see that in wide ranges, less energy is available for spectral measurement because of strong substrate absorption. This results in a clearly poorer detectability for the *in situ* measurement than for sample isolation and transfer into a non-absorbing substrate such as KBr. Another aspect is that, because of adsorption effects and interactions with the substrate, *in situ* spectra can differ from those of pure substances, like those present in spectral libraries [143, 146].

For background compensation, a subtraction must occur in KM-units (see Sect. 5.1.3), because $f(R/R_u)$ and $f(R)-f(R_u)$ differ,

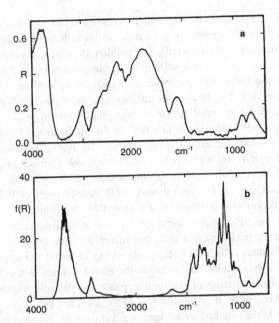

Fig. 5-23.
a) Diffuse reflectance of a DC-cellulose plate,
b) diffuse reflectance spectrum in Kubelka-Munk units.

whereby R and R_u represent the degree of reflection of the TLC-sample spot and of the neighboring TLC-substrate, respectively. The errors depending on R_u can be considerable [54]. In the experiments, the single-beam spectrum of a reflection standard is important besides the single-beam spectra of the TLC-fraction and of the pure substrate. Gilded sandpaper with a constant degree of reflection within the entire wavenumber range has been proven useful (see Sect. 5.1.3).

An attachment with an X-Y-platform for online measurement of TLC chromatograms was described [147]. With this, sought substances could be identified with using the HPTLC-UV/FT-IR-coupling [148]. As a whole, this technique allows the calculation of Gram-Schmidt traces and other presentation forms typical for GC/IR-coupling.

5.7.3 HPLC- and SFC/IR-Spectroscopy

The combination of various types of liquid chromatography-limited here to high-pressure liquid chromatography (HPLC) and supercritical fluid chromatography (SFC)-has the principle difficulty that the mobile phase itself exhibits strong IR absorption which limits the evaluable spectral region for detecting the eluate components. One must consider here the considerable excess of mobile phase. The eluate can be measured with corresponding flow-through cells, naturally with the limitations inherent in the use of the mobile phase. Transmission cells of various pathlengths have been recommended, whereby relatively small cell volumes must result (e.g. 3 mm pathlength, free diameter 3.5 mm, inner volume 30 µl). The pathlength is mostly a compromise in order to attain a sufficient sample absorbance in the mobile phase spectrum with sufficient transmission windows [149].

Especially for liquid chromatography with reversed phase (RP) column material, polar mobile phases mixed with water, among others, are used which require the application of compatible window materials. Their refraction index, however, is mostly relatively high so that disturbing sine-formed baseline modulations result from multi-interference on the windows (see also Sect. 3.6.5). For this reason, ATR flow-through cells have been recommended which do not show this disadvantage.

For the SFC, supercritical CO_2 is the most frequently used mobile phase. Because it is IR-transparent in wide regions, pressure-resistant cells with pathlengths of up to 5 mm are applicable. Xenon was also suggested as the mobile phase. In the fractions, the separation of the mobile phase is relatively unproblematic because of its volatility (see [150] for a review).

The latter coupling procedure is advantageous, because it has a greater detectability yet is also more difficult in handling. For concentrating, the fractions here are applied e.g. to KBr powder and then measured via diffuse reflectance. The advantage is that no spectral information is lost and the detection limits decrease due to the undisturbed IR spectra in contrast to the use of flow-through cells. Various accessories have described in the literature for eliminating the mobile phase, which allow online measurements of the chromatogram, among others, via transmission, diffuse reflection, or with IR microscopy. Also here, the aim is to keep the sample areas possibly small for low losses like in the use of IR-cryogenic methods in gas chromatography. Micro-HPLC is useful, because with this, as a whole only small amounts of mobile phase are present (see also reference data in [150]).

Other developments, e.g. the MAGIC-interface, lend from the LC/MS-coupling in which an aerosol generator is typically used. With this, non-polar mobile phases, but also water, can be completely eliminated under a vacuum [151] or increased temperature [142]. In the removal of buffered mobile phases, a part of the buffer remains on the substrate that can be compensated by corresponding reference spectra. However, the use of volatile buffers like ammonium acetate is recommended. In this respect, IR detection limits in the nanogram range and lower were reported by using an attachment with concentrated flow nebulizers [152]. Somsen et al. described an interesting study of steroids by means of LC-FT-IR-coupling by using a spray jet [153].

In the cited examples, the low volatility of the substances is decisively important. With higher vapor pressure of the components, other coupling techniques, such as obtained with HPLC/ GC-coupling [154], are necessary for separating the mobile phase. Nonetheless, the experimental expense is considerable for compatibly coupling liquid chromatography with IR spectroscopy, when online measurements with corresponding detection limits should be realized.

5.8 Vibrational Spectroscopic Imaging

Infrared imaging (IRI), based on mid-infrared (IR) and near-infrared (NIR), is a synergy of two traditionally distinct methods for studying the chemistry and morphology of a sample, infrared spectroscopy and optical microscopy. This technique is derived from the coupling of a focal plane array (FPA) detector with an infrared spectrometer. Infrared imaging analysis combines the capability of spectroscopy for molecular analysis with the ability to visualize the spatial distribution of a sample's constituents thus adding a third dimension to the optical image. Further advantages are the non-invasive, non-destructive character of the method as well as contrast enhancement compared to optical microscopy. This last advantage is especially important for biological samples, where the refractive indices of all materials present in the sample are so similar that there is only very little contrast in the visible image.

Further, with the immense amount of data implicit in the hyperspectral representation, statistical methods and artificial neural networks for testing the significance of observed spectral changes significantly enhance the ability to meaningfully interpret the spectral data.

In the mid-infrared spectral region, spectroscopic imaging has been successfully implemented by optically and electronically coupling an infrared focal-plane array detector (FPA) to a step-scan FT-IR bench [155]. The integration of a high performance infrared FPA, originally developed for military and surveillance applications, and a Fourier transform infrared microscope furnishes high fidelity, chemically specific images while simultaneously producing high-resolution infrared spectra of each pixel on the array. This is illustrated in Fig. 5-24. In step-scan interferometers (see also Sect. 3.4.2), the movable mirror is held stationary at each sampling position and then moved rapidly to the next retardation point. The interferogram signal is produced by integrating the detector output signal during the time interval that the mirror is held stationary. As exemplified shown in Fig. 5-24 for one pixel (n_i, m_j), at each sampling position of the interferometer, the signals of all $n \times m$ pixels are simulta-

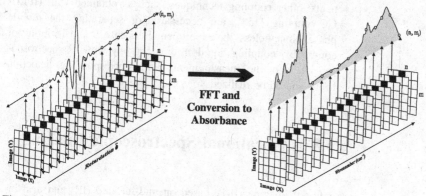

Fig. 5-24. Schematic representation of data generation in an infrared spectral imaging system.

neously detected by the FPA. The combination of all detected signals at each retardation point yields the full interferogram for each pixel, finally resulting in the corresponding spectrum after Fourier transform.

In particular, the use of an FPA detector affords a significant temporal advantage, as well as supplying superior image fidelity.

For the mid-infrared spectral region liquid nitrogen cooled Mercury-Cadmium-Telluride (MCT) focal-plane array detectors with 64×64 or 256×256 pixels are commercially available. With the latter, 65'536 interferograms can be collected in less than 15 minutes, and the sampling area varies between 30×30 m, 400×400 μm, or 4×4 mm. While the spectral resolution can be varied between 0.1 and 128 cm^{-1}, a spatial resolution up to 10 μm can be achieved. In this context, it should be mentioned that with a conventional IR microscope the signal deteriorates with apertures smaller than about 20 μm. Below this, signal-to-noise ratio falls drastically due to reduced throughput, and with apertures of 12 mm, spectra are dominated by noise. On the other side, with the same aperture of 12 μm excellent spectra are still obtained using synchrotron microscopy, since the highly focused synchrotron radiation has at least 1000-times the intensity of traditional IR sources.

After acquisition of thousands of infrared spectra within a few minutes, data analysis and presentation present unique challenges.

The most forward and most common method is so-called functional group mapping, where parameters obtained from a functional group of interest are plotted as a function of position within the map. The parameters that may be plotted depend upon the information sought, e.g. the frequency of an absorption maximum or peak resp. integrated intensity of an absorption band.

Functional group mapping, either as three-dimensional intensity plots or two-dimensional contour plots, is clearly a very powerful analytical tool. However, in order to understand the significance of such plots this methods requires absorption bands to be assigned to particular functional groups, which may not only be difficult but also highly subjective, even for the trained spectroscopist. To remove or reduce this subjectivity, various multivariate pattern recognition/classification techniques can be applied to spectroscopic data. In cluster analysis [156], such as Fuzzy C-means clustering spectra are grouped according some measure of similarity. Linear discriminant analysis (LDA) [156] and, increasingly emerging, artificial neural networks (ANN's) [157, 158] are further commonly applied techniques.

Although vibrational spectroscopic imaging methods are relative newcomers to the panoply of vibrational techniques, they inherit the versatility of the traditional single-point infrared approaches and, in combination with digital imaging technology, allow for new perspectives in the interpretation of the spectra. Imaging advantages are derived, in part, from the ability of an investigator to process and discern effectively a complex two-dimensional spatial representation of a sample in terms of a variety of spectrally related molecular parameters. In particular, a single spectroscopic image, or chemical map, is capable of summarizing and conveying a wealth of spatial, chemical, and molecular data of benefit to both specialists and non-specialists.

Vibrational spectroscopic imaging has been widely applied to diverse materials, including polymers, semiconductors, pharmaceuticals, cosmetics, consumer products, and biological materials. The strengths and adaptability of vibrational spectroscopic imaging rest not only on the ability to determine chemical compositions and component distribution within a sample, but also on being able to extract localized molecular information relevant to the sample architecture.

For example, vibrational spectra are sensitive to alterations in protein secondary and tertiary structures. Spectral shifts could therefore be used to image specifically the distributions of specific structural moieties within a biological sample as opposed to measuring simply an overall distribution of a general class of molecules. Multivariate approaches may also be used to generate a composite metric indicative of either disease progression or biological function. Although these metrics cannot be readily interpreted in terms of biochemical changes, they may often be statistically correlated with disease. In addition, by simultaneously recording data on a two-dimensional focal-plane array detector from a myriad of spatial locations within a biological matrix, pooled spectra may be treated statistically. In this manner, analysis of variance can be employed to more robustly test the significance of observed spectral changes appearing, for example, because of a perturbed or diseased state. Infrared imaging techniques as well as their biological applications are described in detail in [155].

References

[1] Crawford, B., Goplen T.G., Swanson, D., "The Measurement of Optical Constants in the Infrared by Attenuated Total Reflection", in: *Advances in Infrared and Raman Spectroscopy*, Vol. 4, Clark, R.J.H., Hester, R.E. (eds.), Heyden, London, **1978**, p. 47.

[2] Harrick, N.J., *Appl. Spectrosc.* **1977**, 31, 548

[3] Grosse, P., *Trends Anal. Chem.* **1989**, 8, 222

[4] Grosse, P., "Characterization of thin solid Films and Surfaces by Infrared Spectroscopy" in: *Festkörperprobleme/Advances in Solid State Physics*, Vol. 31, Rössler, U. (ed.), Vieweg, Braunschweig, **1991**, p. 77

[5] Weidner, M., Röseler, A., *Phys. Status Solidi A* **1992**, 130, 115

[6] Heise, H.M., *Fresenius J. Anal. Chem.* **1993**, 346, 604

[7] Ishino, H., Ishida, H., *Appl. Spectrosc.* **1992**, 46, 504

[8] Bortz, M.L., French, R.H., *Appl. Spectrosc.* **1989**, 43, 1498

[9] Yamamoto, K., Masui, A., Ishida, H., *Appl. Optics* **1994**, 33, 6285

[10] Korte, E.H., Staat, H., *Fresenius J. Anal. Chem.* **1993**, 347, 454

[11] Wurvey, C.J., Fairless, B.J., Kimball, H.E., *Appl. Spectrosc.* **1989**, 43, 1317

[12] Griffith, D.W.T., Schuster, G., *J. Atmos. Chem.* **1987**, 5, 58

[13] Allara, D.L., Swalen, J.D., *J. Phys. Chem.* **1982**, 86, 2700

[14] Francis, S.A., Ellison, A.H., *J. Opt. Soc. Am.* **1959**, 49, 130

[15] Greenler, R.G., *J. Chem. Phys.* **1966**, 44, 310

[16] Rabolt, J.R., Jurich, M., Swalen, J.D., *Appl. Spectrosc.* **1985**, 39, 281

[17] Herres, W., Zachmann, G., *Fresenius Z. Anal. Chem.* **1984**, 319, 701

[18] Yarwood, J., *Spectroscopy* **1990**, 5, 34

[19] Griffiths, P.R., Van Every, K.W., Wright, N.A., *Fresenius Z. Anal. Chem.* **1986**, 324, 571

[20] Hayden, B.E., "Reflection Absorption Infrared Spectroscopy" in: *Vibrational Spectroscopy of Molecules on Surfaces*, Yates, J.T. Jr., Madey, T.E. (eds.), Plenum Press, New York, **1987**, p. 267

[21] Porter, M.D., *Appl. Spectrosc.* **1984**, 38, 11

[22] Kawai, M., Onishi, T., Tamura, K., *Appl. Surf. Sci.* **1981**, 8, 361

[23] Urban, M.W., Koenig, J.L., "Recent Developments in Depth Profiling from Surfaces using FT-IR Spectroscopy" in: *Applications of FT-IR Spectroscopy. Vibrational Spectra and Structure*, Vol. 18, Durig, J.R. (ed.), Elsevier, Amsterdam, **1990**, p. 127

[24] Newton, I., *Optics*, New York: Dover, **1952**

[25] Harrick, N.J., *Internal Reflection Spectroscopy*. Ossining, New York: Harrick Scientific Corporation, Ossining, **1979**

[26] Fahrenfort, J., *Adv. Mol. Spectrosc. Proc. Int. Meet. 4th 1959*, **1962**, 2, 701

[27] Available from Harrick Scientific, 88 Broadway, Ossining, NY 10562, Home Page *http://www.harricksci.com*

[28] Harrick, N.J., Milosevic, M., Berets, S.L., *Appl. Spectrosc.* **1991**, 45, 944

[29] Compton, S., Stout, P., FTS 7/IR Notes no. 1, Bio-Rad, Digilab Division, Cambridge, MA 02139, **1991**

[30] Fringeli, U.P., in: Mirabella F.M. (ed.), *Internal Reflection Spectroscopy: Theory and Application*, New York: Marcel Dekker, **1992**, p. 255

[31] Fringeli, U.P., *Chimia* **1992**, 46, 200

[32] Available from SPECTRA-TECH, Shelton, CT 06484-0869

[33] Available from SPECAC, Orpington, Kent BR5 4HE, UK

[34] Doyle, W.M., *Appl. Spectrosc.* **1990**, 44, 50

[35] Miller, B.E., Danielson, N.D., Katon, J.E., *Appl. Spectrosc.* **1988**, 42, 401

[36] Available from Axiom Analytical, Inc., Irvine, CA 92614

[37] Doyle, W.M., *Process Control Quality* **1992**, 2, 11

[38] Heinrich, P., Wyzgol, R., Schrader, B., Hatzilazaru, A., Lübbers, D.W., *Appl. Spectrosc.* **1990**, 44, 1641

[39] Katz, M., Katzir, A., Schnitzer, I., Bornstein, A., *Appl. Optics* **1994**, 33, 5888

[40] Rosenberg, E., Kellner R., in: 9th International Conference on Fourier Transform Spectoscopy, Bertie, J.E., Wieser, H. (eds.), *Proc. Soc. Photo-Instrum. Eng.* **1994**, 2089, p. 228

[41] Kortüm, G., *"Reflexionsspektroskopie"*, Springer-Verlag, Berlin, **1969**

[42] Kubelka, P., Munk, F., *Z. Techn. Phys.* **1931**, 12, 593

[43] Otto, A., Korte, E.H., *Mikrochim. Acta [Wien]* II, **1988**, 141

[44] Olinger, J.M., Griffiths, P.R., *Anal. Chem.* **1988**, 60, 2427

[45] Ulbricht, R., *Elektrotechn. Z.* **1900**, 29, 595

[46] Richter, W., Erb, W., *Appl. Optics* **1987**, 26, 4620

[47] Stark, E., Luchter, H., Margoshes, M., *Appl. Spectrosc. Rev.* **1986**, 22, 335

[48] Murray, I., Cowe, I.A. (eds.), *"Making Light Work: Advances in Near Infrared Spectroscopy"*, VCH, Weinheim, **1992**

[49] Messerschmidt, R.G., *Appl. Spectrosc.* **1985**, 39, 737

[50] Korte, E.H., Otto, A., *Appl. Spectrosc.* **1988**, 42, 38

[51] Marbach, R., Heise, H.M., *Appl. Optics* **1995**, 34, 610

[52] Brimmer, P.J., Griffiths, P.R., *Appl. Spectrosc.* **1988**, 42, 242

[53] Yang, P.W., Mantsch, H.M., Baudais, F., *Appl. Spectrosc.* **1986**, 40, 974

[54] Otto, A.; "Infrarotspektroskopie mit diffus reflektierter Strahlung: in-situ Messungen an schwach streuenden Proben", *Fortschritt-Berichte VDI*, Reihe 8, Nr. 146, VDI-Verlag, Düsseldorf, **1987**

[55] Fraser, D.J.J., Norton, K.L., Griffiths, P.R., *Anal. Chem.* **1990**, 62, 308

[56] Fraser, D.J.J., Griffiths, P.R., *Appl. Spectrosc.* **1990**, 44, 193

[57] Krivacsy, Z., Hlavay, J., *Talanta* **1994**, 41, 1143

[58] TeVrucht, M.L.E., Griffiths, P.R., *Appl. Spectrosc.* **1989**, 43, 1492

[59] Krivacsy, Z., Hlavay, J., *Spectrochim. Acta* **1994**, 50A, 49

[60] Brimmer, P.J., Griffiths, P.R., *Anal. Chem.* **1986**, 58, 2179

[61] Christy, A.A., Tvedt, J.E., Karstand, T.V., Velapoldi, R.A., *Rev. Sci. Instrum.* **1988**, 59, 423

[62] Fuller, M.P., Griffiths, P.R., *Anal. Chem.* **1978**, 50, 1906

[63] Korte, E.H., "Infrarot-Spektroskopie diffus reflektierender Proben", in: *Analytiker-Taschenbuch*, Bd. 9, Günzler, H., Borsdorf R., Fresenius, W., Huber, W., Kelker, H., Lüderwald, I., Tölg, G., Wisser, H. (Hrsg.), Springer-Verlag, Berlin, **1990**, S. 91

[64] Gurka, D.F., Billets, S., Brasch, J.W., Riggle, C.J., *Anal. Chem.* **1985**, 57, 1975

[65] Brackett, J.M., Azarraga, L.V., Castles, M.A., Rogers, L.B., *Anal. Chem.* **1984**, 56, 2007

[66] Spragg, R.A., *Appl. Spectrosc.* **1984**, 38, 604

[67] Yang, P.W., Mantsch, H.H., *Appl. Optics* **1987**, 26, 326

[68] Brimmer, P.J., Griffiths, P.R., Harrick, N.J., *Appl. Spectrosc.* **1996**, 40, 258

[69] Brimmer, P.J., Griffiths, P.R., *Appl. Spectrosc.* **1987**, 41, 791

[70] Brimmer, P.J., Griffiths, P.R., *Appl. Spectrosc.* **1988**, 42, 242

[71] Yan, B., *Acc. Chem. Res.* **1998**, 31, 621

[72] Messerschmidt, R.G., Harthcock, M.A. (eds.), *Infrared Microspectroscopy: Theory and Applications,* New York: Marcel Dekker, **1988**

[73] Krishnan, K., Hill S.L. in: Ferraro, J.R., Krishnan, K. (eds.), *Practical Fourier Transform Spectroscopy,* San Diego: Academic Press, **1990**, p. 103

[74] Compton, S., Powell, J., FTS 7/IR Notes no. 2, Bio-Rad, Digilab Division, Cambridge, MA 02139, **1991**

[75] Harrick, N.J., Milosevic, M., Berets, S.L., *Appl. Spectrosc.* **1991**, 45, 944

[76] Naumann, D., Helm, D., Labischinski, H., *Nature* **1991**, 351, 81

[77] Peitscher, G.W., *Makromol. Symp.* **1986**, 5, 75

[78] Messerschmidt, R.G., Harthcock, M.A. (eds.), *"Infrared Microspectroscopy – Theory and Applications",* Marcel Dekker, New York, **1988**

[79] Schrader, B., "Infrarot- und Raman-Mikrospektroskopie", in: *Analytiker-Taschenbuch,* Bd. 13, Günzler, H., Bahadir, A.M., Borsdorf, R., Danzer, K., Fresenius, W., Huber, W., Lüderwald, I., Schwedt, G., Tölg, G., Wisser, H. (Hrsg.), Springer-Verlag, Berlin **1995**, S. 3

[80] Humecki, H.J. (ed.), *"Practical Guide to Infrared Microspectroscopy",* Marcel Dekker, New York, **1995**

[81] Sigrist, M.W., "Air Monitoring by Laser Photoacoustic Spectroscopy", in: *Air Monitoring by Spectroscopic Techniques,* Sigrist, M.W. (ed.), John Wiley & Sons, New York, **1994**, p. 163

[82] Oelichmann, J., *Fresenius Z. Anal. Chem.* **1989**, 333, 353

[83] Rosencwaig: *"Photoacoustics and Photoacoustic Spectroscopy",* John Wiley & Sons, New York, **1980**

[84] McClelland, J.F., *Anal. Chem.* **1983**, 55, 89A

[85] Krishnan, K., *Appl. Spectrosc.* **1981**, 35, 549

[86] Graham, J.A., Grim III, W.M., Fately, W.G., "Fourier Transform Infrared Photoacoustic Spectroscopy of Condensed-Phase Samples", in: *Fourier Transform Infrared Spectroscopy,* Vol. 4, Ferraro, J.R., Basilte, L.J. (eds.), Academic Press, Orlando, **1985**, p. 345

[87] Belton, P.S., Wilson, R.H., Saffa, A.M., *Anal. Chem.* **1987**, 59, 2378

[88] Smith, M.J., Manning, C.J., Palmer, R.A., Chao, J.L., *Appl. Spectrosc.* **1988**, 42, 546

[89] Sullivan, D.H., Conner, W.C., Harold, M.P., *Appl. Spectrosc.* **1992**, 46, 811

[90] Mink, J., Keresztury, G., *Appl. Spectrosc.* **1993**, 47, 1446

[91] Schäfer, K., Haus, R., Heland, J., Haak, A, *Ber. Bunsenges. Phys. Chem.* **1995**, 99, 405

[92] Kapff, S.F., *J. Chem. Phys.* **1948**, 16, 446

[93] Griffiths, P.R., *Appl. Spectrosc.* **1972**, 26, 73

[94] Ford, M.A., Spragg, R.A., *Appl. Spectrosc.* **1986**, 40, 715

[95] DeBlase, F.J., Compton, S., *Appl. Spectrosc.* **1991**, 45, 611

[96] Bates, J.B., "Infrared Emission Spectroscopy", in: *Fourier Transform Infrared Spectroscopy – Applications to Chemical Systems,* Vol. 4, Ferraro, J.R., Basile, L.J. (eds.), Academic Press, New York, **1978**, p. 99

[97] Horning, D.F., *Disc. Faraday Soc.* **1950**, 9, 115

[98] Hineno, M., Yoshinaga, H., *Spectrochim. Acta* **1972**, 28A, 2263

[99] Hermann, T.S., Harvey, S.R., *Appl. Spectrosc.* **1969**, 23, 435; Hermann, T.S., Harvey, S.R., Honts, C.N., *Appl. Spectrosc.* **1969**, 23, 451; Hermann, T.S., *Appl. Spectrosc.* **1963**, 23, 461, 473

[100] Whittle, E., Dows, D.A., Pimentel, G.C., *J. Chem. Phys.* **1954**, 22, 1943

[101] Barnes, A.B., Orville-Thomas, W.H., Garfres, R., Müller, A.: *"Matrix Isolation Spectroscopy"*, D. Reidel, Publ., Dordrecht, **1981**

[102] Andrews, L., Moskovits, M. (eds.), *"Chemistry and Physics of Matrix-Isolated Species"*, North-Holland, Amsterdam, **1989**

[103] Almond, M.J., Downs, A.J., "Spectroscopy of Matrix Isolated Species", *Advances in Spectroscopy*, Vol. 71, Clark, R.J.H., Hester, R.E. (eds.), J. Wiley & Sons, Chichester, **1989**

[104] Schnöckel, H., Willner, H., "Matrix-isolated-Molecules", in: *Infrared and Raman Spectroscopy – Methods and Applications*, Schrader, B. (ed.), VCH, Weinheim, **1995**, p. 297

[105] Chapman, O.L., McIntosh, C.L., Pacansky, J., *J. Am. Chem. Soc.* **1973**, 95, 614

[106] Ozin, G.A., "Matrix Isolation Laser Raman Spectroscopy", in: *Vibrational Spectroscopy of Trapped Species*, Hallam, H.E. (ed.), J. Wiley & Sons, Chichester, **1973**, p. 403

[107] Rochkind, M.M., *Anal. Chem.* **1967**, 39, 567

[108] Carr, B.R., Chadwick, B.M., Edwards, C.S., Long, D.A., Warton, F.C., *J. Mol. Struct.* **1980**, 62, 291

[109] Berger, E., Griffith, D.W.T., Schuster, G., Wilson, S.R., *Appl. Spectrosc.* **1989**, 43, 320

[110] Holloway, T.T., Failress, B.J., Freidline, C.E., Kimball, H.E., Kloepfer, R.D., Wurrey, C.J., Jonooby, L.A., Palmer, H.G., *Appl. Spectrosc.* **1988**, 42, 359

[111] Klawun, C., Sasaki, T.A., Wilkins, C.L., Carter, D., Dent, G., Jackson, P., Chalmers, J., *Appl. Spectrosc.* **1993**, 47, 957

[112] Bourne, S., Haefner, A.M., Norton, K.L., Griffiths, P.R., *Anal. Chem.* **1990**, 62, 2448

[113] Visser, T., Vredenbregt, M.J., *Vibrational Spectrosc.* **1990**, 1, 205

[114] Visser, T., Vredenbregt, M.J., Jackson, P., Dent, G., Carter, D., Schofield, D., Chalmers, J., in: *9th International Conference on Fourier Transform Spectroscopy*, Bertie, J.E., Wieser, H. (eds.), *Proc. Soc. Photo-Opt. Instrum. Eng.* **1993**, 2089, 184

[115] Leipertz, M., Spiekermann, M., "Low- and high-temperature techniques, spectrometric determination of sample temperature", in: *Infrared and Raman Spectroscopy*, Schrader, B. (ed.), VCH, Weinheim, **1995**, p. 658

[116] Lam, C.K.Y., Tilotta, D.C., Busch, K.W., Busch, M.A., *Appl. Spectrosc.* **1990**, 44, 318

[117] Hunter, G.B., Allemand, C.D., Eagar, T.W., *Opt. Engineering* **1985**, 24, 1081

[118] Bauer, H., Richter, W., Schley, U., *PTB Mitteilungen Forschen + Prüfen* **1977**, 87, 217

[119] Nerheim, A.G., "Applications of Spectral Techniques to Thermal Analysis", in: *Fourier Transform Infrared Spectroscopy*, Vol., 4, Ferraro, J.R. Basile, L.J. (eds.), Academic Press, Orlando, **1985**, p. 147

[120] Ferraro, J.R., Basile, L.J., *Appl. Spectrosc.* **1974**, 28, 505

[121] Lauer, J.L., "High-Pressure Infrared Interferometry", in: *Fourier Transform Infrared Spectroscopy*, Vol. 1, Ferraro, J.R., Basile, L.J. (eds.), Academic Press, New York, **1978**, p. 169

[122] Ferraro, J.R., *"Vibrational Spectroscopy at High External Pressures – The Diamant Anvil Cell"*, Academic Press, New York, **1984**

[123] Buback, M., "Applications of high-pressure techniques", in: *Infrared and Raman Spectroscopy*, Schrader, B. (ed.), VCH, Weinheim, **1995**, p. 640

[124] Challener, W. A., Thompson, J. D., *Appl. Spectrosc.* **1986**, 40, 298

[125] Jordanov, B., Korte, E. H., Schrader, B., *J. Mol. Struct.* **1988**, 174, 147

[126] Michl, J., Thulstrup, E. W., *"Spectroscopy with Polarized Light"*, 2nd Ed., VCH, Weinheim, **1995**

[127] Noda, I., Dowrey, A. E., Marcott, C., *Appl. Spectrosc.* **1993**, 47, 1317

[128] Heise, H. M., Kolev, D., *Appl. Spectrosc.* **1988**, 42, 878

[129] Nafie, L. A., *J. Mol. Struct.* **1995**, 347, 83

[130] Purdie, N., Brittain, H. G. (eds.), *"Analytical Applications of Circular Dichorism"*, Elsevier Science Publ., Amsterdam, **1994**

[131] White, R., *"Chromatography/Fourier Transform Infrared Spectroscopy and its Applications"*, Marcel Dekker, New York, **1990**

[132] Beyermann, K., *Fortschr. Chem. Forsch.* **1969**, 11, 484

[133] Edwards, R. A., Fagerson, J. S., *Anal. Chem.* **1965**, 37, 1630

[134] Witte, K., Dissinger, O., *Fresenius Z. Anal. Chem.* **1969**, 236, 119

[135] de Haseth, J. A., Isenhour, T. L., *Anal. Chem.* **1977**, 49, 1977

[136] Herres, W., *"HRGC-FTIR: Capillary Gas Chromatography – Fourier Transform Infrared Spectroscopy"*, Dr. Alfred Hüthig Verlag, Heidelberg, **1987**

[137] Welti, D., *"Infrared Vapour Spectra"*, Heyden & Son, London, **1970**

[138] Nyquist, R. A., *"The Interpretation of Vapor-Phase Infrared Spectra"*, Vols. 1, 2. Stadtler-Heyden, Philadelphia, London, **1984**

[139] Wurrey, C. J., Gurka, D. F., "Environmental Applications of Gas Chromatography/Fourier Transform Infrared Spectroscopy (GC/FT-IR)", in: *Vibrational Spectra and Structure – Applications of FT-IR Spectroscopy*, Vol. 18, Durig, J. R., Elsevier, Amsterdam, **1990**, p. 1

[140] Kuckuk, R., Hill, W., Burba, P., Davies, A. N., *Fresenius J. Anal. Chem.* **1994**, 350, 528

[141] Leibrand, R. J. (ed.), *"Basics of GC/IRD and GC/IRD/MS"*, Hewlett-Packard, Palo Alto, **1993**

[142] Smyrl, N. R., Hembree, D. M. Jr., Davis, W. E., Williams, D. M., Vance, J. C., *Appl. Spectrosc.* **1992**, 46, 277

[143] Otto, A., Bode, U., Heise, H. M., *Fresenius Z. Anal. Chem.* **1988**, 331, 376

[144] de Klein, W. J., *Anal. Chem.* **1969**, 41, 667

[145] Krohmer, P., Kemmner, G., *Fresenius Z. Anal. Chem.* **1968**, 243, 80

[146] Beauchemin, B. T. Jr., Brown, P. R., *Anal. Chem.* **1989**, 61, 615

[147] Glauninger, G., Kovar, K.-A., Hoffmann, V., *Fresenius J. Anal. Chem.* **1990**, 338, 710

[148] Kovar, K.-A., Dinkelacker, J., Pfeifer, A. M., Pisternick, W., Wössner, A., *GIT Spezial · Chromatographie* **1995**, 1, 19

[149] Vidrine, D. W., "Liquid Chromatography Detection using FTIR", in: *Fourier Transform Infrared Spectroscopy*, Vol. 2, Ferraro, J. R., Basile, L. J. (eds.), Academic Press, New York, **1979**, p. 129

[150] Fujimoto, C., Jinno, K., *Anal. Chem.* **1992**, 64, 476A

[151] Robertson, R. M., de Haseth, J. A., Browner, R. F., *Appl. Spectrosc.* **1990**, 44, 8

[152] Lange, A. J., Griffiths, P. R., *Appl. Spectrosc.* **1993**, 47, 403

[153] Somsen, G. W., Gooijer, C., Brinkman, U. A. Th., Velthorst, N. H., Visser, T., *Appl. Spectrosc.* **1992**, 46, 1514

[154] Grob, K., *"Online Coupled LC-GC"*, Hüthig Verlag, Heidelberg, **1991**

[155] Schaeberle, M.D., Levin, I.W., Lewis, E.N., in: Gremlich, H.-U., Yang, B. (eds.), *Infrared and Raman Spectroscopy of Biological Materials,* New York: Marcel Dekker, **2001**, 231

[156] Nauman, D., in: Gremlich, H.-U., Yang, B. (eds.), *Infrared and Raman Spectroscopy of Biological Materials,* New York: Marcel Dekker, **2001**, p. 323

[157] Schmitt, J., Udelhoven, T., in: Gremlich, H.-U., Yang, B. (eds.), *Infrared and Raman Spectroscopy of Biological Materials,* New York: Marcel Dekker, **2001**, 379

[158] Udelhoven, T., Naumann, D., Schmitt, J., *Appl. Spectrosc.,* **2000**, 54, 1471

6 Qualitative Spectral Interpretation

6.1 Fundamentals

As was discussed in the previous sections, IR radiation is taken up for exciting certain atomic movements within a molecule. The energy required for such vibrational motions is definite for a molecular species but differs from substance to substance. Hence, the IR spectrum is also characteristic for every substance and can serve for its identification (spectral comparison). The task of interpretative IR spectroscopy, however, is more extensive. One aims to assign the observed spectrum to the molecular design. Of course, the spectra are just as varied as the chemical substances. In other words, the IR spectrum incorporates the entire information about the molecular design of the compound. Consequently, a total spectral analysis is very complicated and succeeds only after extensive calculation for relatively simply designed compounds. In most cases, one has to be satisfied with a partial interpretation and bases conclusions on empirical rules. Starting from an existing spectrum, in this chapter we will handle the fundamentals, the rules, and the way to proceed in interpreting spectra.

6.1.1 IR Spectrum

The spectrum can be presented in two ways: As percent transmittance (to be exact: degree of pure spectral transmittance) vs. wavenumber or absorbance vs. wavenumber. While transmittance is often used in the interpretation of chemical groups present in the sample, absorbance, due to its linear dependence on concentration (see Sect. 1.2.3), provides one with the ability to make direct quantitative analytical measurements.

First, we will be acquainted with the general presentation of a spectrum and elucidate some terms. Figure 6-1 will be used as an example here.

1) Baseline.
2) Area between baseline and transmittance 1: background absorption or reflection losses on e.g. cell windows.
3) Absorption maximum, i.e. transmittance minimum and absorbance maximum, respectively, of the vibrational bands: all of these synonymous terms serve for labeling the position of

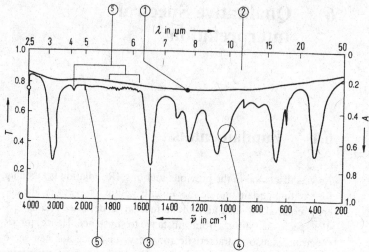

Fig. 6-1. Model spectrum for explaining fundamental terms (see text).

maximum uptake of radiation energy in the spectrum by the molecule.

4) Shoulder: two non-separated bands.
5) Disturbances by absorption of CO_2 and H_2O of the air.

Important data for each spectrum are:
- Sample characterization (name or code name)
- Origin (source, journal etc.)
- Preparation technique (film, solution, pellet, matrix)
- Purity (GC, distillation fraction)
- Date and name of the operator
- Filename.

Furthermore important for collections that are more comprehensive is the following: Spectral number and instrument; data are desirable about the purpose of the spectral acquisition and about reference spectra of the same substance (IR, MS, GC, NMR, UV/VIS).

6.2 Initial Inspection of Spectra

Before a spectrum is interpreted, it should first be examined for errors, disturbances and anomalies. Already at first glance of a spectrum, it can be seen if too little or too much substance was used in the sample preparation (Fig. 6-2). With insufficient amounts, only the most intense bands are recognized.

If too much sample material is used "the transmittance will hang on the ground", so to speak. It cannot be interpreted with certainty, because the many bands are confusing.

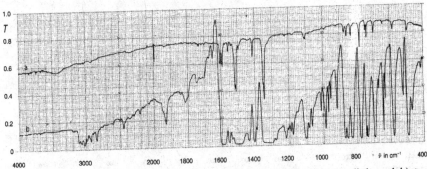

Fig. 6-2. Example of spectral acquisition with a) too little and b) too much substance in the sample cell.

The band shape allows conclusions to be drawn on the quality of sample preparation: an increase of the transmittance of the background absorption towards smaller wavenumbers in pellet spectra infers large crystals proving that the sample had been poorly pulverized (Fig. 6-3). A distortion of the band shape for a single band such that the transmittance on the short-wave flank is greater than on the long-wave flank indicates embedding processes (Christiansen effect, see Sect. 4.2.1.6).

For spectra of dissolved substances without compensation, bands of the solvent often dominate the image (Fig. 6-4): very wide and intense absorptions overly the substance spectrum. As it is not easy to compensate for the solvent over the entire spectrum, one often then finds locations at which the bands point in the wrong direction.

The substance type under question can often be indicated from the general image (Fig. 6-5 a–d). At good resolution, organic materials generally show 30 to 40 bands in the region between 2000 and 400 cm^{-1} with a concentration in the fingerprint region (see Sect. 6.3.1). Inorganic substances are indicated by a few wide and intense bands. Polymeric material is concluded from very strong and low-structured bands. An undifferentiated spectral image with very many bands of similar intensity infers a mixture.

Despite often fewer vibrational bands, gases assume very wide wavenumber regions for one rotational-vibrational band with their P- and R-branches because of the rotational-vibrational fine structure. The gaseous aggregate state of smaller molecules is clearly recognizable from this.

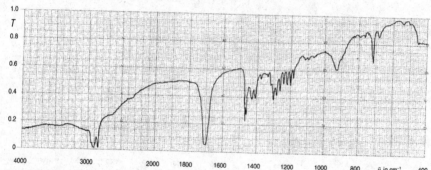

Fig. 6-3. Example of crystallites in a pellet spectrum that are too big: the scattering losses are mainly so large in the short-wave part of the spectrum that the background absorption strongly increases.

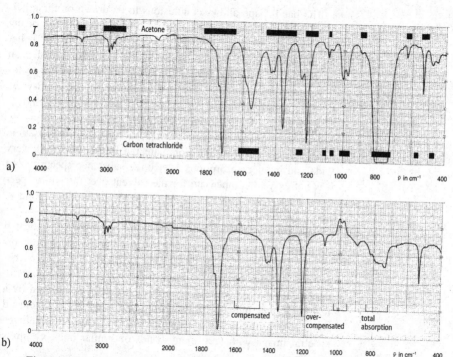

Fig. 6-4. Spectra of acetone in carbon tetrachloride.

a) Non-compensated spectrum; the bands of acetone (upper row) as well as of CCl_4 (lower row) appear. The bands of CCl_4 may disturb the spectral interpretation.

b) Compensated spectrum. The bands belonging to carbon tetrachloride no longer occur, e.g. see the region around 1500 cm^{-1}. At about 1000 cm^{-1}, it is overcompensated. The radiation is completely absorbed between 820 cm^{-1} and 770 cm^{-1}. Hence, the detector gets no energy, and a purely arbitrary curve appears in the spectrum. Often, the noise amplitude in these intervals is noticeably large.

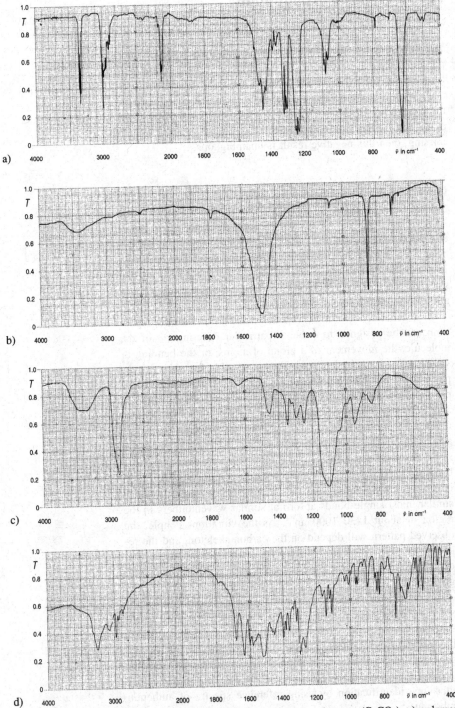

Fig. 6-5. Example spectra: a) gas (1-butyne), b) inorganic substance (CaCO$_3$), c) polymer (polyethylene oxide) and d) mixture (distillation residue).

6.3   General Assignments

The intense bands are primarily conclusive for the spectral interpretation. Every absorption is based on a certain vibrational
form of the molecule. Thus, a weak band just as clearly indicates a certain atomic grouping as intense signals. For the latter,
the change of the dipole moment is especially large through the
deflection of the atoms concerned. If a known intense band is
missing, then this negative result is unambiguous.

6.3.1   The Fingerprint Region

At wavenumbers with values greater than 1500 cm^{-1} it is generally possible to assign each absorption band in an infrared spectrum. This is not the case for most absorptions observed below
1500 cm^{-1}. This region is referred to as the *fingerprint region*,
since quite similar molecules give different absorption patterns
at these frequencies.

It is generally assumed that each band in an infrared spectrum can be assigned to either a particular deformation of the
molecule, the movement of a group of atoms, or the bending or
stretching of a particular bond. This is possible for many bands,
particularly for the 'well behaved' stretching vibrations of multiple bonds. However, many vibrations are not so well behaved
and may vary by hundreds of wavenumbers, even for similar
molecules. This applies to most bending and skeletal vibrations,
absorbing in the 1500–650 cm^{-1} region and for which small
steric or electronic effects in the molecule may result in large
band shifts. In addition, because organic compounds are based
on C-, N- and O-atoms, and the strength of the single bonds is
similar, most single bonds absorb at similar wavenumbers in the
region of about 1430–1000 cm^{-1}. As the vibrations couple, the
observed pattern will depend on the carbon skeleton, and the resulting bands will originate from the oscillations of large parts
of the skeleton, or the skeleton and the attached functional
groups.

A spectrum of a molecule may have a hundred or more absorption bands present in this region, but there is no need to assign the vast majority, and only the most intense bands serve as
indicators (esters, ethers, alcohols). The spectrum can be re-
garded as a *fingerprint* of the molecule. Consequently, the fingerprint region between 1500 and 650 cm^{-1} is thus preferred for
substance identification by using reference spectra of authentic
substances.

6.3.2   Group Frequencies

It has to be asked to which extent valid conclusions can be made from the band positions regarding assigning a narrowly limited partial structure [1], e.g. assignment to ester-, aryl- or nitro-groups. From the previous statements, this applies only when vibrational coupling is insignificant.

Only such molecular fractions can interact strongly with one another, the mass and force constant of which have similar sizes. With only small mutual influence, however, the absorption maxima of the partial structures are bound to quite narrow and characteristic regions. A sufficient valid assignment can be made with empirical rules. One then speaks of group frequencies:

The absorption of the fixed $C \equiv C$ triple bond lies at higher wavenumbers than that of the $C=C$ bond, and this, in turn, has a higher frequency that the band of the C–C single bond. This is also expressed by a corresponding numerical value of the force constants, because the masses of the vibrating C-atoms stay the same (Table 6-1).

The band of the C–X single bond shifts towards smaller wavenumbers when the mass of X increases; in our example, the energy of the C–X bond remains about the same (Tab. 6-1). Using this information, one can already try to initially classify the spectrum (Fig. 6-6). Nevertheless, based on the group frequencies, the various band positions, and their intensities, several various organic substance classes can be identified with certainty already at first glance. This includes the following:

The areas of multiple bonds are clearly defined. Moreover, the C–H, C–X (with X = C, O, N) and C-halogen regions are sepa-

Table 6-1. Dependency of the absorption wavenumber on the bond strength ΔE and on the atomic mass; k is the respective bond force constant. *

Bond	$\tilde{v}(C \cdots C)$ cm^{-1}	ΔE kJmol^{-1}	k mdyn Å^{-1}	Bond	$\tilde{v}(C \cdots C)$ cm^{-1}	ΔE kJmol^{-1}	k mdyn Å^{-1}	Atomic Mass of X (relative)
$-C \equiv C-$	2000	890	15.6	$-C-H$	3000	420	4.8	1
$\rangle C=C\langle$	1600	680	9.6	$-C-C$	1000	370	4.5	12
$-C-C-$	1000	370	4.5	$-C-Cl$	700	335	3.6	35

* Conversion factor to the SI-units are: 1 mdyn/Å = 10^2 N/m, 1 mdyn/Å $\cdot 10^{-18}$ J.

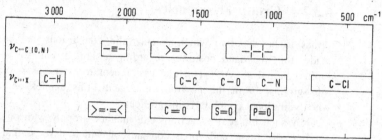

Fig. 6-6. Initial regional classification of the IR spectrum (see text).

rated from each other. The third line in Fig. 6-6 already infers one of the interpretation difficulties. For complicatedly built molecules, the absorption regions of various molecular partial structures overlap; for instance, the C=O region with the C=C double bond region. Nonetheless, based on the group frequencies, various band positions and their intensities can be clearly identified, often already at first glance, to various organic substance classes which include: alkanes, alkenes, aromatics, alcohols, ethers, esters, ketones, acids, aldehydes, anhydrides, amines, nitro-compounds, amides, nitriles and compounds with cumulated double bonds. The aromatic and pure aliphatic representatives of these groups also can be differentiated to the most part. We will get to know the spectra of the individual compounds in this order. As the specified band positions are mostly taken from standard works of the IR literature (e.g. [2–10], among others) reference will be cited only in a few special cases.

6.4 IR Spectra of Individual Compound Classes [11]

Using alkane chains as an example, we will first discuss the general fundamentals and then functional groups on the C-skeleton and observe how the spectrum changes. Only in special cases will we consider the behavior of several functional groups on the same basic skeleton.

6.4.1 Alkanes

The very simply designed paraffins also show a corresponding simple IR spectrum. According to Eq. (2-31), Sect. 2.2.2.1, *n*-butane should exhibit 36 bands, but only a few intense absorption regions are found. This finding is extremely important: by no means will all possible vibrational modes be clearly observable as bands in the routine spectrum. The various reasons for this are as follows:

- some bands lie beyond the region recorded,
- many bands fall together,
- a few bands are weak, i.e. show low intensity,
- many bands are "symmetry forbidden".

Methylene Group

For the stretching vibration of the methylene group, a motion of the atomic nuclei in the bond direction, only two modes of movement come into question (Fig. 6-7 a, b).

Fig. 6-7. Fundamental vibrations of the CH$_2$-fragment. Large sphere: C-atom; small spheres: H-atoms. Here, as in all corresponding figures of this type, the arrows indicate the direction of deflection of the atomic nuclei.

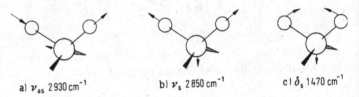

a) ν_{as} 2930 cm^{-1} b) ν_s 2850 cm^{-1} c) δ_s 1470 cm^{-1}

Both vibrations differ from each other by 80 cm^{-1} and are well-separated and observable as intense bands for cycloalkanes. For the deformation vibration, a movement of the atoms in which their bond angle is altered, scissoring is obvious (Fig. 6-7 c). Other possibilities are the rotating motions of the CH$_2$-fragments around the Cartesian coordinates (Fig. 6-8 a–c).

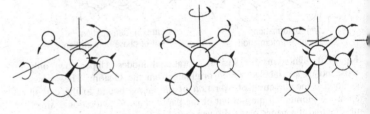

Fig. 6-8. Deformation vibrational modes of the CH$_2$-fragment in alkanes.

a) rocking: 720 cm^{-1} b) twisting: $\approx$ 1300 cm^{-1} c) wagging: 1305 cm^{-1}

The atomic displacements correspond to the rotational possibilities of the non-linear triatomic model (e.g. H$_2$O, see Fig. 2-15). The corresponding IR band, however, is attributed to a fundamental vibration and not to a rotational motion of the entire molecule. For this reason, the deflection of the H-atoms must be compensated by a corresponding counter motion of the C-atom, because, otherwise, components of the rotation and translation for the entire molecule would occur. However, this counter vibration of the C-atom is again coupled with motions of the neighboring C-atom.

The rocking vibration yields a characteristic absorption at 720 cm^{-1}, when more than three CH$_2$-groups are connected together in a chain (Fig. 6-9). For crystalline polymethylene compounds, this band is split into two components.

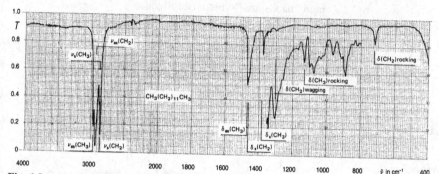

Fig. 6-9. *n*-Tridecane, film, insert: 200 μm pathlength.

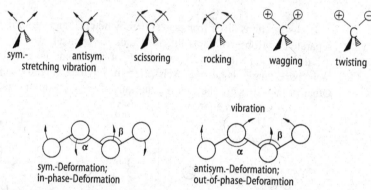

Fig. 6-10. Synonyms for various vibrational modes. Upper row: only those bonds are labeled which originate from the C-atom. The arrows symbolize the direction of vibration of the atoms bound to the carbon atom. ⊕ vibrational motion out of the paper plane; ⊖ vibrational direction behind the paper plane. Lower row: for explaining the symmetrical and the antisymmetrical deformation vibrational mode for linear multi-atomic molecules (accord. to [2]).

In solution, the wagging form is recognizable as a weak band mostly only with a large pathlength. For the twisting mode, the dipole moment also changes only to a slight degree, analogous to the wagging vibrational mode. Because both have their absorption maxima at around 1300 cm^{-1}, an assignment is difficult to make. In crystalline carboxylic acids, the wagging vibrations cause a characteristic multi-band pattern (see Sect. 6.4.11.3).

Table 6-2. Abbreviations for various vibrational modes.

Symbol	Characterization of the Vibrational Mode
a	i.p. deformation vibration
β	i.p. deformation vibration
Γ	o.o.p. deformation vibration of skeleton atoms
γ	o.o.p. deformation vibration
Δ	i.p. deformation vibration of skeleton atoms
δ	i.p. deformation vibration of a X–H bond
δ_s	symmetric deformation vibration (bending)
δ_{as}	antisymmetric deformation vibration (bending)
δ'	deformation vibration (twisting, rocking)
κ	o.o.p. wagging vibration of a XH_2-group $(X \neq C)$
r	rocking vibration
r_β	i.p. rocking vibration
r_γ	o.o.p. rocking vibration
ρ	i.p. rocking vibration of a XH_2-group $(X \neq C)$
v	stretching vibration of a X–H bond
v_s	symmetric stretching vibration
v_{as}	antisymmetric stretching vibration
v_β	i.p. stretching vibration
v_γ	o.o.p. stretching vibration
t	twisting vibrations
τ	torsion, twisting vibration of a XH_2-group $(X \neq C)$
Φ	o.o.p. ring deformation vibration
$\overline{\omega}$	wagging vibration
$\overline{\underline{\omega}}$	stretching vibration of skeleton atoms without H-bond

They are clearly observable there and applicable for an assignment.

The various vibrational modes are differentiated in the literature by different symbols: we were already introduced to v for the stretching vibration and δ for a deformation vibration, and we will continue to use them in this general meaning. Abbreviations with special meaning are compiled in Table 6-2, because one occasionally finds them in the literature. The abbreviations *i.p.* for in-plane and *o.o.p.* for out-of-plane are generally common (Tab. 6-2, see Fig. 6-15 and 6-30). Furthermore, the terms "asymmetric" and, better, "antisymmetric" are synonymous. Besides these, the synonyms cited in Fig. 6-10 appear for typical vibrational modes.

Methyl Group

An IR spectrum mostly has more than two strong bands at around 2800–3000 cm^{-1}. These additional maxima result from C–H stretching vibrations of the methyl group. This group is already composed of four atoms and thus causes more compli-

cated vibrational motions than are possible for the CH_2-group.
The stretching vibrational modes are illustrated in Fig. 6-11.

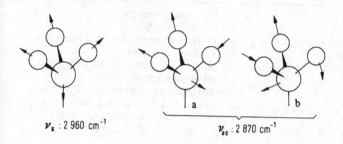

ν_s : 2 960 cm^{-1} a   b ν_{as} : 2 870 cm^{-1}

Fig. 6-11. Stretching vibrational modes of the methyl group.

   The symmetric C–H stretching vibration is totally analogous
to the motion form of the CH_2-group (see Fig. 6-7). There are
two antisymmetric stretching vibrations that have very similar
energies, whereby a two-fold degenerate vibration can occur
with corresponding molecular symmetry (see Sect. 2.2.2.2). The
CH_3-deformation vibrations are composed of the fundamental
vibrations illustrated in Fig. 6-12.

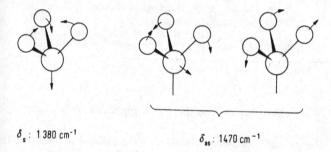

δ_s : 1 380 cm^{-1} δ_{as} : 1 470 cm^{-1}

Fig. 6-12. Deformation modes of the H-atoms in the methyl group.

   The symmetric form is again analogous to $\delta(CH_2)$; the antisymmetric vibration with two similar-energy possibilities is
shown in Fig. 6-12. As for the CH_2-group, the rocking and
twisting motions are rotations of the CH_3-fragments around the
Cartesian coordinates. With isopropyl or tertiary butyl groups,
the 1380 cm^{-1} band splits into a doublet.

Methine Group
A C–H stretching vibrational band is assigned to it at 2890 cm^{-1}
as well as a C–C–H deformation band at 1340 cm^{-1}. This assignment, however, does not suffice as a correlation criterion.

Carbon Skeleton

The non-polar C–C chain does not result in strong absorption of electromagnetic radiation. Nevertheless, with a larger sample thickness, one can determine a band-rich region between 1350 cm^{-1} and 750 cm^{-1} that permits certain substance identifications (see Fig. 6-9). In spectra of cyclic hydrocarbons, more intense bands that originate from ring-deformation motions generally occur in the fingerprint region.

Identification of Alkanes

These bands occur in almost all spectra of organic substances: C–H stretching vibrational bands $2940–2855 \text{ cm}^{-1}$, deformation vibrational bands 1470 cm^{-1}; the symmetric CH_2-deformation bands at 1380 cm^{-1} which is split for isopropyl and tertiary-butyl derivatives, as well as the CH_2-rocking absorption at 720 cm^{-1}. Chain branching leads to further bands in the region of $1250–910 \text{ cm}^{-1}$.

6.4.2   Halogen Compounds

In general, very large regions are specified for the occurrence of C-halogen bands. Nevertheless, the corresponding band can be recognized and assigned among the others by its high intensity (Tab. 6-3).

One can notice that the aryl-halogen wavenumbers all lie in a narrower range than the corresponding alkyl-halogen compounds. The reason for the wider ranges generally lies in the mechanical vibrational coupling. This important phenomenon in the IR spectroscopy will be discussed by using an example [12]: alkyl chlorides exhibit two C-halogen bands (Fig. 6-13). These can be assigned to two trans- and gauche conformers.

In the v(C–Cl)-stretching vibration, the a-C-atom is strongly deflected from its resting position. For the *trans* conformation, this in turn causes a change of the C_1–C_2–C_3-valence angle (Fig. 6-14). In this case, the mechanical coupling takes place with the angle deformation vibration of the C-skeleton. In con-

Table 6-3. Absorption regions of organic halogen compounds (in cm^{-1}).

Group	Alkyl Halogenide	Aryl Halogenide
C–F	1365–1120	1270–1100
C–Cl	830– 560	1096–1034
C–Br	680– 515	1073–1028
C–I	610– 485	1061–1057*

* *para*-substituted iodine compounds

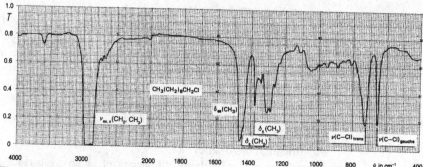

Fig. 6-13. *n*-Decylchloride, film 50 μm.

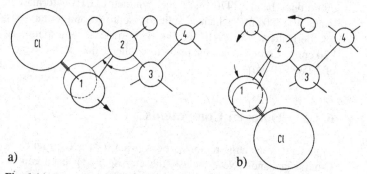

a)   b)

Fig. 6-14. For explaining the conformation dependency of coupling occurrences:
a) *trans*-Cl–C_3 conformation: in the C_1–Cl stretching vibration, the C_1–C_2–C_3 angle is deformed by the movement of C_1 in the plane of the $C_2C_3C_4$ atoms (the Cl-atom experiences hardly any deflection because of its large mass). Thus: coupling of ν(C–Cl) and δ(CCC) without an effect on the (CH$_2$)-rocking vibration, ν(C–Cl)=726 cm^{-1}.
b) *gauche*-Cl–C_3 conformation: in the C_1–Cl stretching vibration, the C_1–C_2–C_3 angle is not changed in the $C_2C_3C_4$ plane. The C_1 atom vibrates perpendicular to this plane. Thereby, the C_2-atom is "pulled along" effecting a counter action (CH$_2$)-rocking vibration. Thus: coupling of ν(C–Cl) and δ(CH$_2$)-rocking, ν(C–Cl)=645 cm^{-1}. The thinly drawn circle represents the deflected C_1-atom.

trast, if the molecule exists in the *gauche* conformation, the deflection of the C_1-atoms does not result in any large change of the C-valence angle at C_2. However, the C_1-atom drawn perpendicular out of the plane of the C_2–C_3–C_4-chain produces a corresponding counter motion of the CH$_2$-rocking vibration. In all cases of coupled vibrations, the deflections of the atoms involved have at least one vibrational direction (vibrational coordinate) in common. The ν(C–Cl) and the δ(CH$_2$)-rocking modes move in trans conformers in perpendicular directions and thus cannot influence each other, even when the wavenumbers of the coupled vibrations are numerically similar (trans conformers: ν(C–Cl) ≈ 726 cm^{-1}, δ(CH$_2$)-rocking ≈ 720 cm^{-1}). Hence, this explains the narrower range in Tab. 6-3 for ν(C–Br) and ν(C–I) as well as for ν(C$_{aryl}$–halogen).

Identification of Halogen Alkanes

The C-halogen bands lie within relatively narrow limits (± 10 cm^{-1}), providing that only terminally bound halogen on higher alkyl chains are taken into account: R–Cl: 726 cm^{-1} and 645 cm^{-1}; R–Br: 625 cm^{-1}. For the interpretation of middle-positioned halogenated and multi-halogenated alkanes, only reference spectra can help.

6.4.3   Alkenes

Alkenes are characterized by their C=C double bond and the resulting planar design around both sp^2-centers. This creates several substitution possibilities, likewise expressed in the IR spectrum.

C–H-Vibrations

The vibrational modes of the olefins are compiled in Figures 6-15 to 6-18. Because the heavy substituents (R) do not strongly couple with the movements of the H-atoms, the vibrations of the substituents is not accounted for in the figures. The C–H stretching vibrations show their absorption maximum in the region of 3125–3010 cm^{-1}. Compared to the CH-band of saturated aliphatic compounds (a(CH) in –CH$_2$–: 18500 m·mol^{-1}), their intensity is lower (a(CH) in =CH$_2$: 2800 m·mol^{-1} [14]). However, because an organic molecule generally has many aliphatic H-atoms rather than vinyl H-atoms, the alkyl-CH band mostly predominates in the 3000 cm^{-1} region.

More intense bands but which are less suitable for characterization are caused by such in-plane deformation vibrations of the H-atoms stretched by the σ-bonds. These are the so-called in-plane vibrations (abbreviated i.p.).

The strongest bands, however, originated from those deflections of the H-atoms directed out of the plane defined above, i.e. the so-called out-of-plane modes (abbreviated o.o.p.).

Fig. 6-15. Modes of motion of the H-atoms in olefins:
a) C–H stretching vibration,
b) in-plane C–C–H deformation vibration and
c) out-of-plane C–C–H deformation vibration.

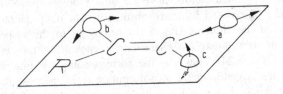

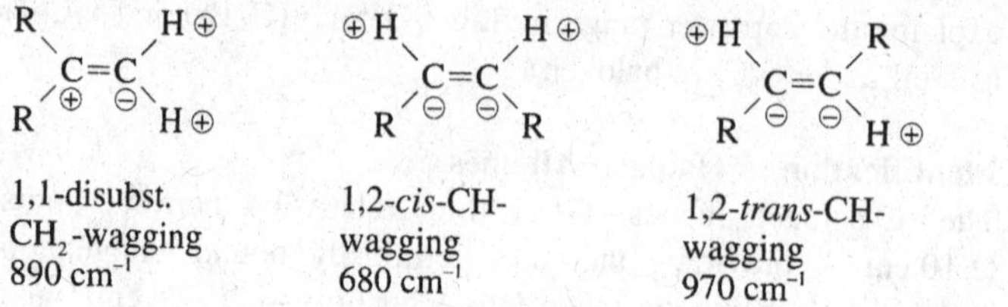

1,1-disubst.
CH$_2$-wagging
890 cm^{-1}

1,2-*cis*-CH-
wagging
680 cm^{-1}

1,2-*trans*-CH-
wagging
970 cm^{-1}

Fig. 6-16. Out-of-plane deformation vibrations of di-substituted olefins.
⊕ Direction of vibration out of the paper plane.
⊖ Direction of vibration behind the paper plane.

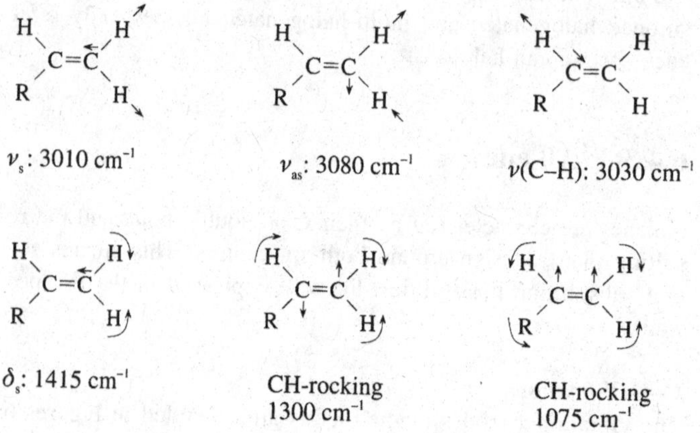

ν_s: 3010 cm^{-1}

ν_{as}: 3080 cm^{-1}

ν(C–H): 3030 cm^{-1}

δ_s: 1415 cm^{-1}

CH-rocking
1300 cm^{-1}

CH-rocking
1075 cm^{-1}

Fig. 6-17. In-plane deformation and C–H stretching vibrations for mono-substituted olefins.

CH$_2$-wagging
910 cm^{-1}

trans-CH-wagging
990 cm^{-1}

cis-CH-wagging
630 cm^{-1}

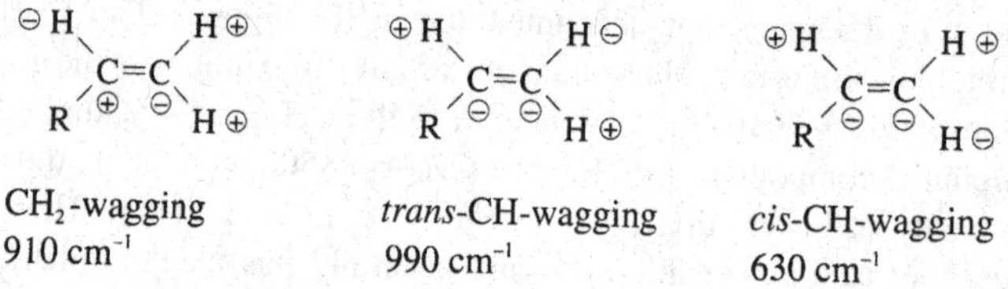

Fig. 6-18. Out-of-plane deformation vibrations for mono-substituted olefins.

Tri-substituted Olefins

Here, there is only one o.o.p. mode. Because it is found at ≈ 825 cm^{-1} having moderate intensity, it does not distinguish itself as a clear criterion besides other bands in this region (see Fig. 6-19). The corresponding i.p. vibrational mode is not diagnostic.

Di-substituted Olefins

For di-substituted olefins, there are three isomeric possibilities: vicinal, 1,2-*cis*- and 1,2-*trans*-substitution. The o.o.p. vibrational bands of the H-atoms (Fig. 6-16) belonging to the respective type of substitution are characteristic enough to use as key bands (Fig. 6-24). However, the corresponding twisting vibrations are insignificant for all substitution types, because the dipole moment does not change hereby (Figures 6-20 to 6-22).

Mono-substituted Olefins (Vinyl Compounds)

Vibrational modes similar to those of the di-substituted olefins can be specified for this compound class (see Fig. 6-17 and 6-18). Accordingly, their bands lie in similar regions (Fig. 6-23 and 6-24).

910 cm^{-1} – corresponding to the vibrational mode of the 890 cm^{-1} band of 1,1-di-substituted olefins,

990 cm^{-1} – corresponding to the 970 cm^{-1} band of *trans*-1,2-di-substituted olefins,

630 cm^{-1} – corresponding to the 680 cm^{-1} band of *cis*-1,2-di-substituted olefins.

One assignment is definitely possible due to the number and the position of the bands (Fig. 6-24). Examples of spectra of all olefin isomers are shown in Figures 6-19 to 6-23.

Monosubstituted olefins are already clearly characterized with the bands of the –C–H-stretching vibrations and the various =C–H-deformation vibrations. Low-intensity control bands,

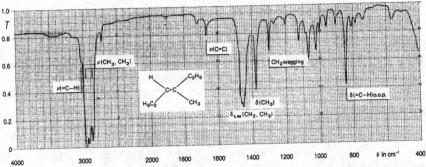

Fig. 6-19. 3-Methyl-*trans*-3-hexene, film, 20 μm. Example for a 1,1,2-tri-substituted-ethylene.

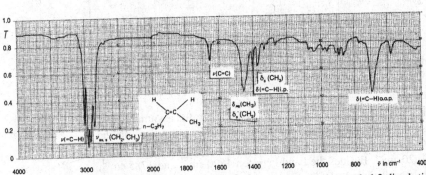

Fig. 6-20. *cis*-2-hexene, film, 20 μm. Example for a *cis*-1,2-di-substituted-ethylene.

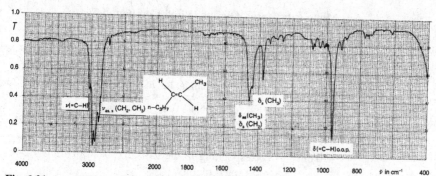

Fig. 6-21. *trans*-2-hexene, film, 20 μm. Example for a *trans*-1,2-di-substituted ethylene.

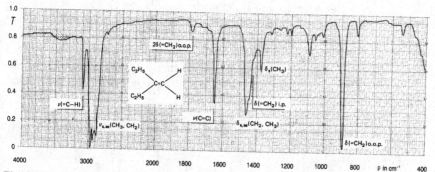

Fig. 6-22. 2-Ethyl-1-butene, film, 20 μm. Example for a 1,1-di-substituted ethylene (region at about 3400 cm^{-1}: OH-contamination).

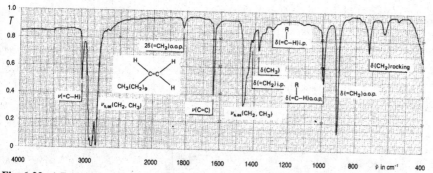

Fig. 6-23. 1-Dodecene, film, 20 μm. Example for a mono-substituted ethylene.

however, can further solidify the assignment: for asymmetrically substituted olefins, one can recognize the absorption of the C=C-valence vibration as a weak band in the region of 1680–1620 cm^{-1}. Besides the band between 1820 cm^{-1} and 1785 cm^{-1}, monosubstituted olefins have a relatively strong overtone band of the δ-o.o.p.-fundamental vibration (at ca. 910 cm^{-1}).

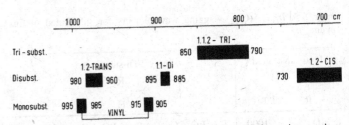

Fig. 6-24. Characteristic frequencies of the o.o.p. deformation motions for olefins.

Table 6-4. Geminal $\delta(CH_2)$-wagging wavenumbers for substituted olefins (in cm^{-1}).

Monosubstituted		1,1-Di-substituted	
R = RO–CO–	961	R = NC–	895
NC–	960	CH$_3$–	887
CH$_3$–	910	Br–	877
R–CO–O–	870	Cl–	867
R–CO–NH–	840	RO–	711
R–O–	813		
D–	943		
Si–	950		
Zn–	952		

The following should be noted for influencing the –CH=CH$_2$-wagging band position: in general, group frequencies occur in a relatively narrow region interval up to around 100 cm^{-1}. However, a more precise examination shows that various substituents cause slight shifts of the band position. Such maximum shifts can often be correlated to resonance, inductive or other effects. These phenomena will be discussed later in Sect. 6.5.

Geminal wagging frequencies of substituted olefins are insensitive to mass effects, i.e. the conversion of e.g. propylene ($\delta(CH_2) = 910$ cm^{-1}) to vinyl bromide ($\delta(CH_2) = 924$ cm^{-1}) hardly affects the frequency of =CH$_2$-wagging (Tab. 6-4). In contrast, its position is much more dependent on resonance effects:

$$\left\{ \begin{array}{c} R-O-\overset{\overset{\textstyle O}{\|}}{C}-CH=CH_2 \\ \updownarrow \\ R-O-\overset{\ominus}{\underset{|\bar{O}|}{C}}=CH-\overset{\oplus}{CH_2} \end{array} \right. \qquad \left\{ \begin{array}{c} R-\bar{O}-CH=CH_2 \\ \updownarrow \\ R-\overset{\oplus}{O}=CH-\overset{\ominus}{\bar{C}H_2} \end{array} \right.$$

$\delta(CH2)$-wagging: 961 cm^{-1} $\delta(CH_2)$-wagging: 813 cm^{-1}

Table 6-5. trans-δ(CH$_2$)-wagging wavenumbers for substituted olefins (in cm^{-1}).

Mono-substituted $\oplus$H H / R H$\oplus$		trans-1,2-Di-substituted $\oplus$H R$_2$ / R$_1$ H$\oplus$		
R = D–	1000	R$_1$= RO–CO–	R$_2$ = RO–CO–	976
Si–	1009	CH$_3$–	RO–CO–	968
Sn–	1000	CH$_3$–	CH$_3$–	964
Zn–	1008	CH$_3$–	Cl–	926
F–	925	Cl–	Cl–	893
Cl–	938	C$_6$H$_5$–	C$_6$H$_5$–	958
Br–	936			
I–	943			

For the trans-=CH$_2$-wagging vibration of the vinyl group (Fig. 6-18) and in the case of the trans-1,2-di-substitution (Fig. 6-16), however, resonance and mass-related effects exert only a small influence. Nonetheless, the electronegative groups here cause a band shift towards lower wavenumbers (Tab. 6-5).

For the other olefin bands, the substituent effects are no so clearly recognizable. They partially oppose each other and, thus, conclusions can only be made to a limited extent. A few examples elucidate this effect.

Cyclic Compounds with C=C-Groups

Double bonds located on a ring show different reactivity than compared to open-chain olefins. This behavior is generally explained by ring-tension effects. When we consider the position of the ν(C=C)-band as a function of the ring size, a correlation is likewise found (Tab. 6-6).

While in chemical reactions, steric interactions of the H-atoms (Pizer tension) or angular changes in the transition state

Table 6-6. Dependency of ν(C=C)-wavenumbers on the ring size (in cm^{-1}).

n	H H / (CH$_2$)$_n$		(CH$_2$)$_n$ C=CH$_2$	
0	acetylene	(1974)	ethylene	(1621)
1	cyclopropene	1641	allene	(1980)
2	cyclobutene	1566	methylene cyclopropane	1781
3	cyclopentene	1611	methylene cyclobutane	1678
4	cyclohexene	1646	methylene cyclopentane	1657
5			methylene cyclohexane	1651

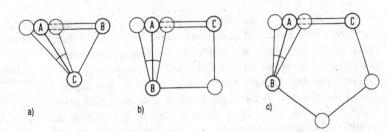

Fig. 6-25. For explaining the vibrational coupling for endocyclic olefins:
a) $v(C=C)$ and $v(C–C)$ influence one another, because the atomic distance d_{AB} and d_{AC} simultaneously change (1640 cm^{-1}).
b) $v(C=C)$ has no effect on $v(C–C)$, because d_{AC} is practically independent of d_{AB} (1565 cm^{-1}).
c) $v(C=C)$ again has effects on $v(C–C)$, because d_{AB} changes with d_{AC} (1610 cm^{-1}). Thinly drawn: deflected position of the C-atoms.

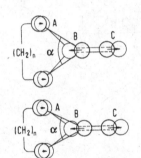

Fig. 6-26. For explaining the vibrational coupling in exocyclic double bonds. Boldface: starting position of the C-atoms; lightface: deflected position of the C-atoms.

of the reaction – often connected with resonance effects – are decisive for the way in which cyclic olefins react, in IR spectroscopy we must totally disregard the geometry of the molecules in the transition state or electron excitation states and may only discuss the first excited vibrational state. Then, it is apparent that we can explain the (C=C)-band position in Tab. 6-6 as a function of a vibrational coupling (Fig. 6-25).

The change of the C=C-distance also affects the position of the neighboring C-atoms and their mutual distance. This again has repercussions on the C=C-wavenumber. For cyclobutene, this interaction is minimal, because no movements of the single bond C-atoms are induced.

The analogous applies likewise for exocyclic double bonds. In addition, here, the wavenumber of the C=C stretching vibration depends on the ring size (Fig. 6-26). Upon changing d_{BC}, the distance d_{AB} is simultaneously changed. This change is greater the smaller the angle a is.

Identification of Alkenes
The short-wave vibration $v(C–H)$-band at 3125–3040 cm^{-1} is mostly present and likewise the $v(C=C)$-band of variable intensity between 1680 and 1620 cm^{-1}. Depending on the substitution type, deformation bands are characteristic in the region between 1000 and 905 cm^{-1}, near 890 cm^{-1} and 730–665 cm^{-1} and 850–790 cm^{-1}. The overtone of the vinyl group appears at 1820–1785 cm^{-1} (see Fig. 6-24).

6.4.4 Molecules with Triple Bonds

Because of the stronger bonding, the absorption of the triple bond vibration occurs in the region between 2300–2000 cm^{-1} far removed from that of the single- and double bond. Because of the small change in the dipole moment during the vibration, however, monosubstituted acetylenes show only a weak band here, which may also be completely missing for substituted acetylenes. The regions cited in Table 6-7 generally apply.

The band may be very intense when a heteroatom replaces one of the C-atoms in the triple bond or is located next to the triple bond (Fig. 6-27). The assignments in this region are mostly unambiguous, because only few groupings come into question. The examples are compiled in Table 6-8.

Table 6-7. Band positions for acetylene compounds.

Vibration	Wavenumber Region
$\nu(\equiv C-H)$	3340–3267 cm^{-1}
$\nu(-C\equiv H-)$	2260–2190 cm^{-1}
$\delta(\equiv C-H)$	700– 610 cm^{-1}

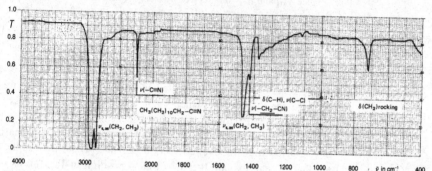

Fig. 6-27. Tridecanoic nitrile, film, 20 μm.

Table 6-8. Absorption regions of $C\equiv X$-stretching vibrational bands in cm^{-1}.

Molecular Fragment	Substance Group	Wavenumber Region
$-C\equiv C-H$	mono-substituted acetylenes	2140–2100
$-C\equiv C-$	di-substituted acetylenes	2260–2190
$-CH_2-C\equiv N$	nitriles	2260–2240
$-C=C-C\equiv N$	acrylonitriles	2235–2215
$Aryl-C\equiv N$	benzonitriles	2240–2220
$-C\equiv N \rightarrow O$	nitriloxides	2304–2288
$-S-C\equiv N$	thiocyanates	2170–2135
$>N-C\equiv N$	aminonitriles	2225–2175
$Aryl-\overset{\oplus}{N}\equiv N$	diazonium salts	2309–2136
$-\overset{\oplus}{N}\equiv\overset{\ominus}{C}|$	isonitriles	2165–2110
$[C\equiv N]^{\ominus}$	cyanides	2200–2070
$Me \leftarrow |\overset{\ominus}{C}\equiv\overset{\oplus}{O}|$	carbonyls	2170–1900

6.4.5   Aromatic Compounds

Like olefins, aromatic compounds have the sp^2-configuration of the carbon atoms and the planar molecular design attributed to this. Otherwise, both compound classes chemically differ so strongly from one another that, from this viewpoint, they cannot be handled at all in one section together.

Moreover, the electron spectra of the olefins differ so strongly from those of the aromatic compounds that both groups must be treated under very different aspects. Because in IR spectroscopy, we are concerned only with the state of the molecule upon transfer from the ground state into the first excited vibrational state, we again find many fundamental ideas from the IR spectra of the olefins in the spectra of the aromatics. The differences between the IR spectra of the olefins and of the aromatics are based primarily only on the interactions of the C–C- and C–H-fragments. Nevertheless, one subdivides for current reasons the vibrational modes of the aromatic ring into those mainly originating from C–C-bonds and those originating from C–H-bonds (Fig. 6-28).

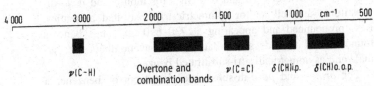

Fig. 6-28. Absorption regions of aromatic compounds.

C–C-Vibrations

All spectra of aromatic compounds display a few especially characteristic bands attributed to C–C-vibrational modes, namely at $1600–1585$ cm^{-1}, $1500–1430$ cm^{-1} and 700 cm^{-1}. They are relatively little affected by ring substituents. The responsible fundamental vibrations for benzene are presented in Fig. 6-29.

$1600–1585$ cm^{-1} region: both C–C-stretching vibrational modes, responsible for the absorptions according to Fig. 6-29, mostly fall together into one band. Conjugation with π-systems such as C=O, C=N, C=C or NO_2 cause a splitting just like heavy elements (Cl, S, P, Si) without considerable band shifting. Nevertheless, the higher frequency bands may also be mistaken for olefinic C=C-bands.

$1500–1430$ cm^{-1} region: this vibrational mode also has two components but which are generally further split than the double band in the 1600 cm^{-1} region. One of the bands is typically found above 1470 cm^{-1}, and the other lies between 1465 and

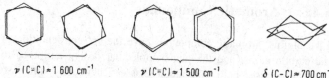

$\nu\,(C=C)\approx 1\,600\ cm^{-1}$ $\nu\,(C=C)\approx 1\,500\ cm^{-1}$ $\delta\,(C-C)\approx 700\ cm^{-}$

Fig. 6-29. Changes of the benzene ring out of the regular hexagonal symmetry that are responsible for the characteristic aromatic bands (without accounting for substituents). The corners of the bold-drawn hexagon indicate the deflection of the C-atoms before the reverse motion. The other extreme mode is respectively the return motion of the C-atoms beyond the points of the regular hexagonal symmetry.

$1430\ cm^{-1}$. *Para*-substituted representatives show this latter band mostly at wavenumbers $10-20\ cm^{-1}$ higher than other substituted aromatics. The intensity of the $1500\ cm^{-1}$ band allows conclusions to be drawn about the substitution: electron-releasing groups increase its intensity and, without these groups, this band may even be missing. In contrast, the intensity of the second component at $1450\ cm^{-1}$ is only little affected by substituents.

Region around $700\ cm^{-1}$ region: this band appears with high intensity at between 710 and $675\ cm^{-1}$ for mono- and *meta*-disubstituted as well as for symmetric-tri-substituted aromatics. Less pronounced and appearing at $730-690\ cm^{-1}$, this band is found shifted towards somewhat higher wavenumbers for vicinal- and asymmetrically-tri-substituted benzenes.

For *ortho*- and *para*-di-substitution products of benzene, a weak absorption likewise occurs in the same area but only when both substituents vary. With identical residues, this band is IR-inactive.

C–H-Vibrations

Analogous to the alkenes, the various possible C–H-vibrations are differentiated according to the deflection of the H-atoms with respect to the plane defined by the 6 C-atoms (Fig. 6-30). The C–H-stretching vibrations defined in this plane appear, like the vinyl-bound H-atoms, at between 3100 and $3000\ cm^{-1}$. Deformation vibrations in the ring plane lead to bands within the fingerprint region of $1300-1000\ cm^{-1}$.

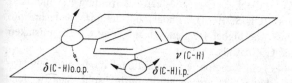

$\delta(C-H)$o.o.p. $\nu\,(C-H)$ $\delta(C-H)$i.p.

Fig. 6-30. For defining in-plane and out-of-plane vibrational modes on the aromatic ring.

The bands attributed to deformation vibrations of the H-atoms perpendicular to the plane are found in the 1000-675 cm^{-1} region:

- 3100–3000 cm^{-1} region: in this region, all vinyl-bound H-atoms absorb with weak intensity, thus, also olefins. Hence, they are not specific to aromatics. Because of H-bond absorptions, these bands are often hard to find.
- 1300–1000 cm^{-1} region: the in-plane rocking bands of the C–H-fragments fall right in the middle of the fingerprint region. Because of changing intensities, the coupling with C–C-vibrations and the general complication of this spectral region, an interpretation is appropriate only by taking extreme precaution.
- 910–660 cm^{-1} region: the C–H-deformation motions out of the ring plane are mutually influenced. The coupling of neighboring C–H-groups is strong, but is interrupted by substituents. In this way, the position of the absorption maxima of these vibrational modes depends on the position of the ring substituents (Tab. 6-9).

Table 6-9. Band positions of the δ(C–H) o.o.p. vibrations of substituted benzene derivatives (see Sect. 6.4.14).

Substitution Type	Region (cm^{-1})	Configuration	Ring Deformation (cm^{-1})
mono-	751±15	5 neighboring H-atoms	697±11
ortho-	751± 7	4 neighboring H-atoms	
meta-	782± 9	3 neighboring H-atoms	690±15
para-	817±13	2 neighboring H-atoms	

These correlations via the number of neighboring hydrogen atoms on an aromatic ring apply also for condensed ring systems and heterocycles (see Sect. 6.4.15), e.g.:

4 neigh-
boring
H-atoms:
740 cm^{-1}

R

3 neigh-
boring
H-atoms:
789 cm^{-1}

R

3 neigh-
boring
H-atoms:
788 cm^{-1}

On the other hand, some ring substituents such as $-NO_2$, $-COOH$, $-COO^-$, $-CONR_2$ cause complications so that in such cases, the data of Tab. 6-9 are not applicable. What only helps then is the comparison with spectra of similarly substituted compounds.

$2000-1600 \text{ cm}^{-1}$ region: in this region, the $\delta(C-H)$ o.o.p. overtones fall together with a few combination bands. Through this, a pattern that is characteristic for the substitution mode of the aromatic ring occurs, which is quite helpful for determining the degree of substitution and the position of the substituents. This illustrates the fact that for recognizing a compound class, the general picture is much more decisive than the exact band position.

Fig. 6-31 gives an example for every substitution possibility on the benzene ring, respectively. For easier interpretation, one acquires this region best using a larger pathlength. Carbonyl groups in the molecule naturally mask the absorptions by the much stronger C=O-stretching vibration.

Identification of Aromatic Rings

The absorptions by vibrations of aromatic ring are so character-istic that a clear assignment is possible by the following re-gions: the substitution bands in the $910-660 \text{ cm}^{-1}$ region, the C=C-stretching vibrational bands at around 1600 cm^{-1} and around 1515 and 1450 cm^{-1}, the overtone bands and combina-tion bands at between 2000 and 1600 cm^{-1}, as well as the $\nu(C-H)$-bands in the range of 3100 to 3000 cm^{-1}.

6.4.6   Ethers

If a CH_2-group in an aliphatic chain is replaced by an oxygen, then one finds a new band in the IR spectrum of the ether at around 1100 cm^{-1} besides the pattern of the aliphatic chain (Fig. 6-32).

According to the frequency position and intensity, it is ob-vious that this band at around 1100 cm^{-1} is caused by the anti-symmetric C–O–C-stretching vibration. The band is very wide and the only characteristic absorption of simple ethers. The band of the symmetric C–O–C-stretching vibrational mode $\nu_s(COC)$ is weak, because the dipole moment here changes much less pronouncedly than $\nu_{as}(COC)$. Moreover, it is found in a wider range, so that it is unsuitable for characterizing ethers.

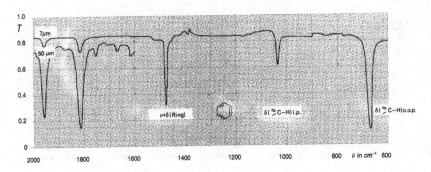

a) Benzene (pathlengths 7 μm and 50 μm)

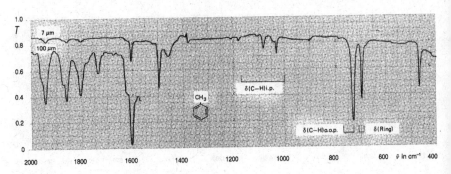

b) Toluene (pathlengths 7 μm and 100 μm)

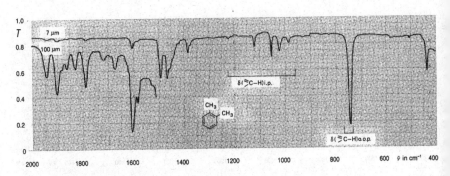

c) *o*-Xylene (pathlengths 7 μm and 100 μm)

Fig. 6-31 a–m. Substitution pattern bands of various benzene derivatives.

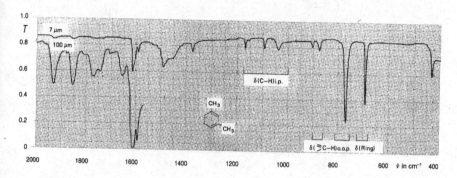

d) *m*-Xylene (pathlengths 7 μm and 100 μm)

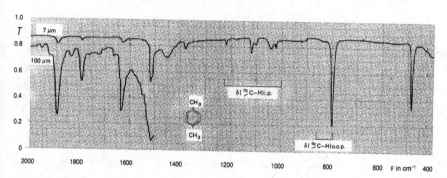

e) *p*-Xylene (pathlengths 7 μm and 100 μm)

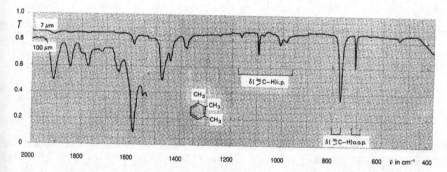

f) 1,2,3-Trimethylbenzene (pathlengths 7 μm and 100 μm)

Fig. 6-31. (continued)

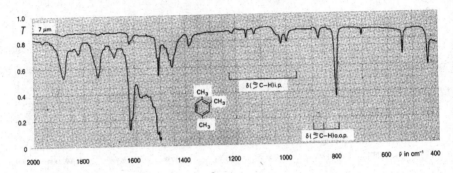

g) 1,2,4-Trimethylbenzene (pathlengths 7 μm and 100 μm)

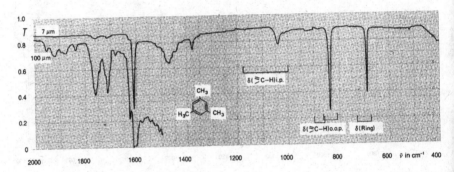

h) 1,3,5-Trimethylbenzene (pathlengths 7 μm and 100 μm)

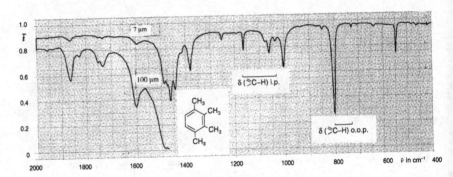

i) 1,2,3,4-Tetramethylbenzene (pathlengths 7 μm and 100 μm)

Fig. 6-31. (continued)

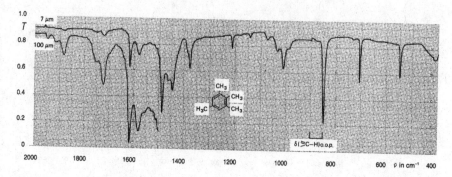

k) 1,2,3,5-Tetramethylbenzene (pathlengths 7 μm and 100 μm)

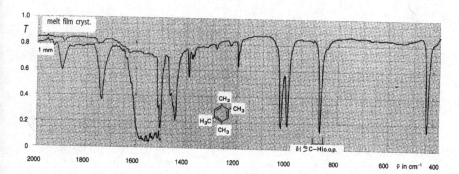

l)1,2,4,5-Tetramethylbenzene (pathlengths 7 μm and 100 μm)

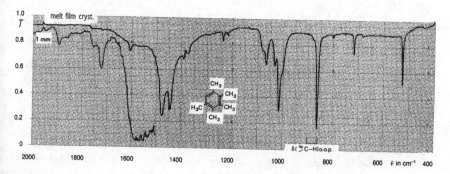

m) Pentamethylbenzene (film, cryst., insert: 0.59% (m/m) in CCl₄, pathlength 1 mm)

Fig. 6-31. (continued)

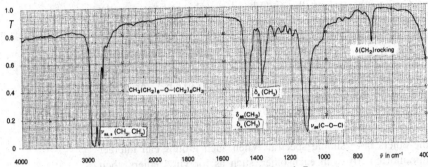

Fig. 6-32. Di-*n*-heptylether, film, pathlength 7 μm

The C–O-bond in vinyl ethers is more fixed, so that the corresponding IR band lies at a higher frequency than the corresponding saturated ether, namely at 1200 cm^{-1}. Of course, the ether oxygen exerts its effect on the double bond. Hence, for vinyl ethers the δ(C–H) o.o.p. bands shift from 990 cm^{-1} and 910 cm^{-1} to 960 cm^{-1} and 820 cm^{-1}, and the v(C=C)-band reaches intensities that approach those of the C=N-group. Therefore, there is especially a big danger of confusing them with each other here.

Spectra of purely aromatic ethers have their strong band near 1250 cm^{-1}. This can be again attributed to the C–O-stretching vibration. The shift towards shorter wavelengths for vinyl ethers as well as for aryl ethers can be explained by resonance effects.

For aryl alkyl ethers, a second strong band also occurs at about 1040 cm^{-1} besides the 1250 cm^{-1} absorption. Because this falls in the region of aliphatic ethers and is missing for di-aryl ethers, it can be classified to the alkyl-O-stretching vibration. Naturally, we do observe any pure group vibration here. The movements of the oxygen atom and the a-C-atom are coupled with the other atoms of the molecule. This coupling is clearly observable for cyclic compounds (Tab. 6-10).

Table 6-10. Band positions of cyclic ethers in cm^{-1}.

n	Substance	v_{as}(C–O–C)	v_s(C–O–C)	Ring Size
2	ethylene oxide	839	1270	3-ring
3	oxetane	983	1028	4-ring
4	tetrahydrofuran	1071	913	5-ring
5	tetrahydropyran	1098	813	6-ring

Cyclic ethers show the C–O-band distributed over a wider range (1250–910 cm^{-1}) than aliphatic ethers. Because of the ring skeleton vibrational absorptions likewise occurring in this region, a rapid interpretation is often hindered and reference spectra must be used. Mainly characteristic is the epoxy band at about 860 cm^{-1} (890 cm^{-1} for *trans*- and 830 cm^{-1} for *cis*-disubstituted epoxy compounds).

Note: Viewed alone the structural elements like the alkenes or the ethers have simple spectra. When such groups can interact intramolecularly with each other, such as for vinyl ethers, then the spectrum becomes more complicated and shows new bands, maximum shifts, and intensity changes. Hence, we will discuss the individual functional groups alone when these are particularly important. The mutual influence of the substituents of the aryl nucleus and their expression in the spectrum will be handled in Sect. 6.4.14.

6.4.7   Acetals and Ketone Acetals

Because several C–O-single bonds exist in the molecule, acetals and ketone acetals absorb accordingly intensely in the region from 1190 cm^{-1} to 1035 cm^{-1} with generally four C–O-stretching vibrational bands which belong to the following fundamental vibrations [15]:

1060–1035 cm^{-1}	$\begin{array}{c} R \quad \overrightarrow{O}-\overrightarrow{C} \\ \diagdown C \diagdown \\ R \quad O-C \end{array}$	This band generally belongs to the symmetric vibrational mode.
1100–1060 cm^{-1}	$\begin{array}{c} R \quad \overrightarrow{O}-\overleftarrow{C} \\ \diagdown C \diagdown \\ R \quad O-C \end{array}$	The most intense band belongs to the vibration in which the dipole moment change is greatest.
1145–1120 cm^{-1}	$\begin{array}{c} R \quad \overrightarrow{O}-\overleftarrow{C} \\ \diagdown C \diagdown \\ R \quad O-C \end{array}$	This band is greatly affected by the residue *R*.
1190–1160 cm^{-1}	$\begin{array}{c} R \quad \overleftarrow{O}-\overleftarrow{C} \\ \diagdown C \diagdown \\ R \quad O-C \end{array}$	This remaining vibrational mode can be assigned to the band in this region.

Ortho-esters likewise fall in this category. Hence, they may be easily mistaken for acetals and ketone acetals. They mostly also have several bands in the C–O-single-bond region, the positions of which strongly depend on coupling phenomena.

In addition, other band positions are affected by the ether group:

R–O–CH$_3$: ν_{as}(C–H)   2992–2955 cm^{-1}
              ν_s(C–H)   2897–2867 cm^{-1}
              $\delta_{s, as}$(C–H)   1470–1440 cm^{-1}

R–O–CH$_2$-: ν_{as}(C–H)   2955–2922 cm^{-1}
              ν_s(C–H)   2878–2835 cm^{-1}
              δ(C–H)   1475–1447 cm^{-1}

Identification of Ethers and Acetals

Simple ethers can only be recognized by their wide intense band at around 1100 cm^{-1}. The absence of other characteristic bands supports this assignment. Acetals, ketone acetals, and ortho-esters show several strong absorptions in the region from 1175 cm^{-1} to 1065 cm^{-1}.

6.4.8   Alcohols

Alcohols have the C–O-single bond in common with ethers. Thus, they likewise show a strong band in the region of 1210–1000 cm^{-1}. In addition, absorptions occur by O–H-stretching and deformation vibrations that differ from the alkanes and ethers. In Sect. 6.4.9, we will discuss the hydrogen-bond effects and their influences in the spectrum.

Vibrations of the O–H-Group

In the liquid and solid state, alcohols and phenols, the OH-group of which is not sterically shielded, have a very wide and intense absorption region from ≈ 3500–2800 cm^{-1}; the maximum lies at around 3300 cm^{-1} (Fig. 6-33). In non-polar solvents, however, one finds in diluted solution a sharp OH-band between 3650 and 3590 cm^{-1}, the position of which is constant to a few wavenumbers for the various forms of OH-groups (primary, secondary, tertiary, axial, equatorial, phenolic (Tab. 6-11) [16].

In the associated state (see Sect. 6.4.9), the atoms –O–H···O– lie in one plane. The deflection of the hydrogen atom out of this plane corresponds to an o.o.p. deformation vibration. The corresponding band is thus sensibly interpretable only in the H-bond state as a diffuse absorption at ≈ 650 cm^{-1}. In the non-associated state, this vibration corresponds to the H-torsion around the C–O-bond axis. The C–O–H i.p. deformation vibration yields an undefined absorption at 1400–1300 cm^{-1} only in concentrated solution or in the liquid phase. Because it is overlapped by δ(CH), it mostly is insignificant for the interpretation. All OH-deformation vibrations are strongly coupled with CH-wagging and the C–O-stretching vibration.

Vibrations of the C–O-Group

Analogous to ethers, alcohols and phenols show the strong C–O-single-bond band. However, alcohols and ethers can be clearly differentiated through the OH-absorption. The coupling of the C–O-vibration with the neighboring C-atoms is so large here due to similar heavy masses and about the same force constants, that a-C-substituents still affect the C–O-band position. This suffices for differentiating the strongly branched alcohols (Tab. 6-11, Fig. 6-33 to Fig. 6-36). Phenols can be clearly distinguished from primary and secondary aliphatic alcohols. The short-wave shift of the (C–O)-band for phenols can be explained by resonance effects (see Sect. 6.4.6).

Table 6-11. v(C–O) and δ(O–H) band positions for alcohols in cm^{-1}.

	v(C–O)	v(O–H)
Primary alcohols	1075–1000	3640
Secondary alcohols	1120–1090	3630
Tertiary-alcohols	1210–1100	3620
Phenols	1260–1180	3615–3590
Cyclohexanol		
• equatorial	1068	
• axial	973	
Triterpenes		
• equatorial		3628–3630
• axial		3635–3638
δ(O–H) i.p.	region: 1400–1300	assignment unclear
δ(O–H) o.o.p.	750– 650	without diagnostic value

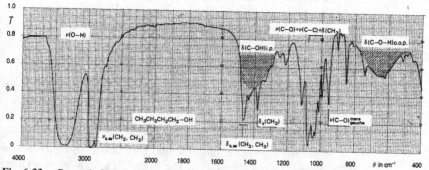

Fig. 6-33. *n*-Butanol, film, pathlength 20 μm. Example for a primary alcohol.

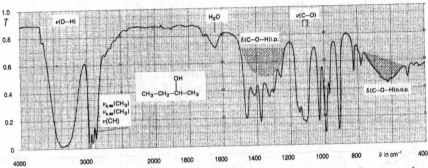

Fig. 6-34. 2-Butanol, film, pathlength 20 µm. Example for a secondary alcohol.

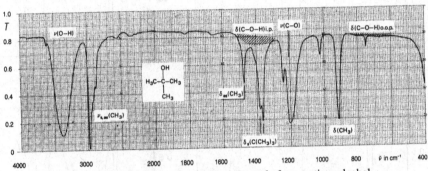

Fig. 6-35. tert.-Butanol, film. Example for a tertiary alcohol.

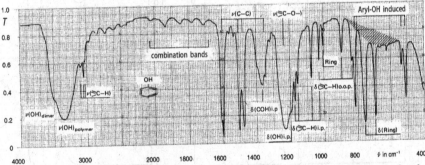

Fig. 6-36. Phenol, film. In the δ(CH) o.o.p. region, induced new bands, the so-called "X-sensitive bands", occur by the aryl substituents.

Identification of Alcohols

Alcohols are surely identifiable by the strong OH-band at ≈ 3330 cm^{-1} and an intense and wide absorption between 1250 and 1000 cm^{-1}. The COH-deformation bands are mostly marked by the broad background absorption of the δ(OH)-deformation vibration of the alcohol.

6.4.9   Hydrogen Bonding

Alcohols are special with respect to many of their properties. This is linked to the individual association of the molecules via hydrogen bonding. In the IR spectrum, particular evidence of such molecular interactions is found [17, 18]. In the case of alcohols, for example, we can determine, only in much diluted solution, the sharp undisturbed OH-valence vibration of the alcohols at

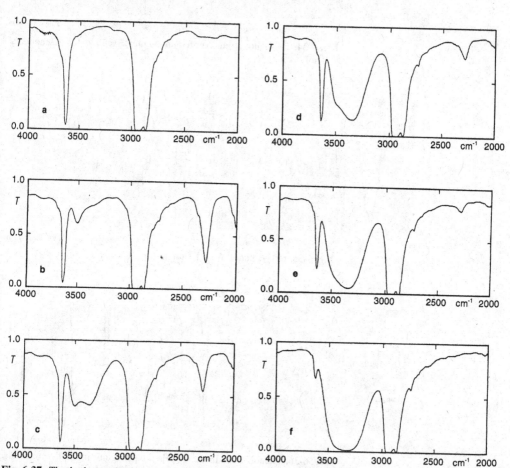

Fig. 6-37. The hydrogen bond and its manifestation in the IR spectrum: *n*-butanol in carbon tetrachloride. The following concentrations and pathlengths were used:

Spectrum	*n*-Butanol (mol/l)	Pathlength (mm)
a	0.00490	50.00
b	0.0499	5.00
c	0.126	2.00
d	0.247	1.00
e	0.510	0.50
f	4.22	0.05

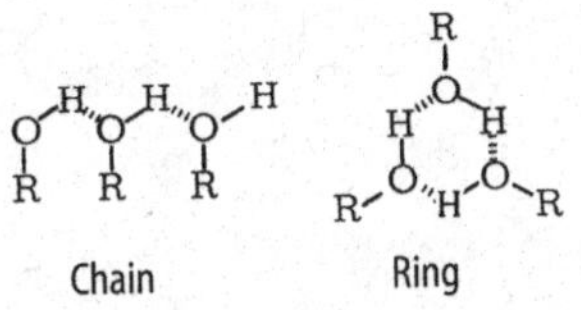

Fig. 6-38. Various forms of H-bond complexes for alcohols.

≈ 3600 cm^{-1}. With increasing concentration, the intensity of this band decreases, however, and a new, very strong and broad band is formed in the region of 3400–3200 cm^{-1} (Fig. 6-37).

Essential for the formation of hydrogen bonds is a polar RO–H-bond and the acceptor HO–R-group with the free electron pair on the oxygen atom (Fig. 6-38).

Upon placing the molecules next to each other according to the scheme X–H$\cdots$Y, the X–H-bond is stretched and thus more strongly polarized. The shift of the absorption maximum towards lower wavenumbers is attributed to the easing of the bond. The increased bond polarization results in a large change of the dipole moment during vibration. This explains the high absorbance of the shifted band. For interpreting the large half-width of the band, we must account for the fact that the aforementioned complexes are ideal boundary cases. In reality, all feasible cluster formations occur simultaneously to which somewhat various force constants belong, respectively. The same conceptions apply also for the δ(COH)-bands.

Of course, the position and intensity of the H-bond bands depend also on the solvent, because in protic solvents, their molecules compete with the dissolved substance for the H-bonds. The sharp undisturbed OH-absorption is hence found only in non-polar solvents with a low concentration of OH-compound. With increasing concentration, the band for the dimer occurs at first and is later joined by absorptions for trimers and higher complexes (see Fig. 6-37). Nowadays, there is a tendency to assume that dimeric alcohols are linear but that trimers are cyclic [19].

One occasionally finds that the H-bond band is maintained in its form and position upon diluting the solution, if the solvent is aprotic (see Sect. 4.3.2.5). In these cases, an *intramolecular* hydrogen bond exists, which naturally cannot be affected by concentration changes (Fig. 6-39):

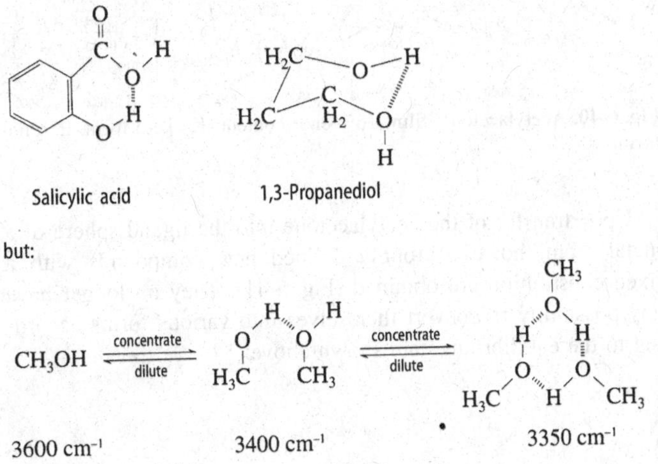

Salicylic acid   1,3-Propanediol

but:

Fig. 6-39. Examples for intra- and intermolecular hydrogen-bond formation.

3600 cm^{-1} 3400 cm^{-1} 3350 cm^{-1}

The manifestation of H-bonds in the IR spectrum, exemplified by alcohols, is found in many cases in other compound classes, too, despite being found to a lesser degree: acids, phenols, amides and amines are the best known examples of this. Besides, studies in mixtures produce valuable conclusions. Especially with regard to solvent influences, acetylene, haloforms, ethers and carbonyl compounds are cited here as examples of this.

6.4.10 Chelates

In the previous section, we have seen that band shifts, band widening and intensity changes are characteristic for H-bonds. The IR spectroscopy hereby covers simultaneously all feasible molecular configurations. The spectrum provides information about all individual molecules. An example is acetylacetone (Fig. 6-40).

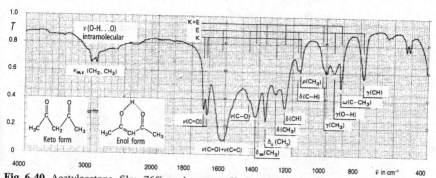

Fig. 6-40. Acetylacetone, film, 76% enol content: K=keto form, E=enol form.

Upon transfer of the acetylacetone into the ligand sphere of a metal atom, however, totally defined new compounds with a fixed constitution are obtained (Fig. 6-41). They no longer have any possibility to convert themselves into various forms according to the equilibrium state shown above:

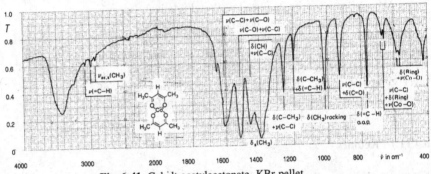

H bonds, intramolecular   Chelate, ridig
tautomerism possible.   geometry
Asymmetric design,
various conformers are
possible

Fig. 6-41. Cobalt acetylacetonate, KBr pellet.

We will see how the chelate formation is manifested in the IR spectrum, and we want to accentuate hereby the differences to the H-bond:

We can ascertain band shifts for hydrogen bonds only for the γ(OH)- and δ(COH)-bands. This is caused by the various energy quantities of the vibration (mass and force constant). For coordination compounds, considerably more bands will change their position, because pronounced vibrational coupling takes place with similarly large masses and force constants of that atoms involved in a fundamental vibration.

Half-width changes through concentration shifts are found in chelates only to a small degree, if at all, because all molecules exist in defined atomic distances and are saturated intramolecularly, and intermolecularly, tend towards aggregation only via van der Waals forces.

Affecting the intensity of the ligand bands by complex formation is only possible, insofar as dipole moment changes can be introduced into the molecule by the metal atom. This applies principally for the bonds originating from the ligand atoms.

For H-bonds, new bands appear for the $-H \cdots Y--$bond in the far-IR region below 200 cm^{-1}. Chelates show the metal-ligand bond likewise in the long-wave region. In particular, chelates hav-

ing various central atoms but the same ligands are often differentiable only in the region below 800 cm^{-1} (e.g. heteropolyacids).

6.4.11 Carbonyl Compounds

The high-intensity absorption band of the C=O-stretching vibration lies in the region between 1800 and 1650 cm^{-1} that is only
sparsely inhabited by other groups. The carbonyl group is especially interesting for IR spectroscopy because of its tendency towards intra- and intermolecular interactions. In addition, in view
of the molecular design, it can be likewise influenced in many
ways. Many effects can be analyzed well and relatively specifically here. For comparison, the behavior of the carbonyl group
is generally based on the ketones with saturated linear alkyl
groups. We will be acquainted with the various effects through
suitable examples and will again handle them briefly from another viewpoint in Sect. 6.5. Data on the absorption regions of
the various carbonyl compounds are summarized in Table 6-12.

6.4.11.1 Ketones

Ketones with linear alkyl groups represent the basic type of all
carbonyl spectra. For long-chain representatives of this compound class, the pattern of the alkane spectra can be recognized
as the basic scheme. New and obvious is a very strong band at
1715 cm^{-1} (Fig. 6-42), which stems from the stretching vibration of the C=O-group.

For lower alkyl groups on the C=O-residue, a few bands appear more strongly than for the n-alkanes. These are induced by
the C=O-fraction of the molecule and concern the vibrations of
the methylene- and methyl group neighboring the carbonyl
group. The band regions are compiled in Tab. 6-13.

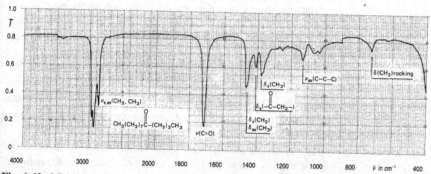

Fig. 6-42. 5-Tridecanone, film.

Table 6-12. Absorption regions of carbonyl compounds.

Compound Class	Structural Element	Carbonyl Band Region in cm^{-1}	Remarks
Ketones	$-CH_2-\overset{O}{\underset{}{C}}-CH_2-$	1725–1700	no control band
	$-CH=CH-\overset{O}{\underset{}{C}}-CH_2-$	1695–1660	additionally bands of the C=C-double bond or of the aryl group
	$Aryl-\overset{O}{\underset{}{C}}-CH_2-$	1700–1680	
	$-\overset{O}{\underset{}{C}}-\overset{O}{\underset{}{C}}-$	1730–1710	possibly 2 bands for asymmetric substitution
	$-\overset{O\cdots H\cdots O}{\underset{CH}{C\diagdown\diagup C}}-$	1640–1535	as they are not ketones in the actual sense anymore, also ketone-atypical spectral properties
	$O=\langle\rangle=O$	1690–1655	
Aldehydes	$-CH_2-\overset{O}{\underset{}{C}}-H$	1740–1720	double band at 2850 cm^{-1} and 2750 cm^{-1} by Fermi resonance
	$-CH=CH-\overset{O}{\underset{}{C}}-H$	1705–1685	
	$Aryl-\overset{O}{\underset{}{C}}-H$	1715–1695	
Acids	$-CH_2-\overset{O}{\underset{}{C}}-OH$	1725–1700	v(OH)-bond band, δ(OH)-absorption, v(C–O)-band aside from possibly C=C- or aryl-bands
	$-CH=CH-\overset{O}{\underset{}{C}}-OH$	1715–1680	
	$Aryl-\overset{O}{\underset{}{C}}-OH$	1700–1680	
	$\overset{Hal}{\underset{}{-CH}}-\overset{O}{\underset{}{C}}-OH$	1740–1715	possibly band splitting of the v(C=O)

Table 6-12. (continued)

Compound Class	Structural Element	Carbonyl Band Region in cm^{-1}	Remarks
Esters	$-CH_2-\overset{O}{\overset{\|}{C}}-O-CH_2-$	1750–1735	ester C–O-band in 1200 cm^{-1} region next to C=C- or aryl-bands
	$\left.\begin{array}{l}-CH=CH-\\ Aryl-\end{array}\right\} -\overset{O}{\overset{\|}{C}}-O-CH_2-$	1730–1715	
	$-CH_2-\overset{O}{\overset{\|}{C}}-O-\left\{\begin{array}{l}-CH=CH-\\ -Aryl\end{array}\right.$	1800–1770	
	γ-lactones	1780–1760	
	δ-lactones	1750–1735	
Amides	$-CH_2-\overset{O}{\overset{\|}{C}}-NH_2$	1650	additional amide-II-band, N–H-stretching and deformation bands, insofar as NH exists. For cyclic secondary amides up to 6-ring, no amide II-band
	$-CH_2-\overset{O}{\overset{\|}{C}}-NH-CH_2-$	1680–1630	
	$-CH_2-\overset{O}{\overset{\|}{C}}-N\overset{CH_2-}{\underset{CH_2-}{}}$	1670–1630	
	γ-lactams	1700	
	δ-lactams	1670	
	$-NH-\overset{O}{\overset{\|}{C}}-NH-$	1660	
	$-CH_2-O-\overset{O}{\overset{\|}{C}}-N{<}$	1735–1700	
Acid chlorides	$-CH_2-\overset{O}{\overset{\|}{C}}-Hal$	1800	no definite control bands
	$\left.\begin{array}{l}-CH=CH-\\ Aryl-\end{array}\right\} -\overset{O}{\overset{\|}{C}}-Hal$	1780–1750	
Anhydrides	$-CH_2-\overset{O}{\overset{\|}{C}}-O-\overset{O}{\overset{\|}{C}}-CH_2-$	1820 and 1760	intensity ratio of the band cleavage gives further information
	$\left[\begin{array}{l}-CH=CH-\\ Aryl-\end{array}\right\}-\overset{O}{\overset{\|}{C}}-\Bigg]_2 O$	1785-1725	

Table 6-13. Band positions for ketones.

Group	Band Position in cm^{-1}	Assignment	Remarks
$>$C=O	1725–1705	ν(C=O)	
$\overset{O}{\underset{\parallel}{-C}}$–CH$_3$	1370–1350	δ_s(CH$_3$)	lower frequency and more intense than alkyl-CH$_3$
$\overset{O}{\underset{\parallel}{-C}}$–CH$_2$–	1440–1405	δ(CH$_2$)	strong
–CH$_2$–$\overset{O}{\underset{\parallel}{C}}$–CH$_2$–	1230–1100	ν_{as}(C–C)	
Alkyl–$\overset{O}{\underset{\parallel}{C}}$–CH$_3$	1170		

Using ketones as an example, we find the effect of polar substituents on the spectrum, again, to be clearly pronounced. The polar group itself results in a very intense and characteristic band. Absorptions of groups in the immediate neighborhood of the dipole alter its position and mostly become more intense (compare the spectrum of 5-tridecanone, Fig. 6-42 with that of n-tridecane, Fig. 6-9). A prediction of the shifting direction is often possible with empirical rules. Intensity changes are very difficult to specify and generally presume an analysis of the effect of the substituents on the change of the dipole moment of the fundamental vibration responsible for the band.

The comparison of various ketone spectra indicates the applicability of the relative intensity expression "strong-moderate-weak". Because the spectra are recorded such that the most intense band lies at $T \approx 0.05$, the alkane chain in higher ketones can, by all means, dominate the spectral picture. Then, the intensity of the C=O-band, known to be high, falls behind that of these bands. On the other hand, for lower ketones, the typical C–H-bands occur only weakly so that the "alkane type" is hardly evident. Therefore, upon making qualitative statements about intensity, one may never oversee the fact that the band strength is a consequence of the number of absorbing groups pro volume unit (extreme case: anoxidized polyethylene and acetone).

The band position of saturated ketones is affected intermolecularly only by association effects and the aggregate state or by the solvent. Intramolecularly, polar substituents, π-electron systems, coupling effects and conformation influences shift the absorption maximum (see Fig. 6-43). For identifying ketones, also see the end of Sect. 6.4.11.

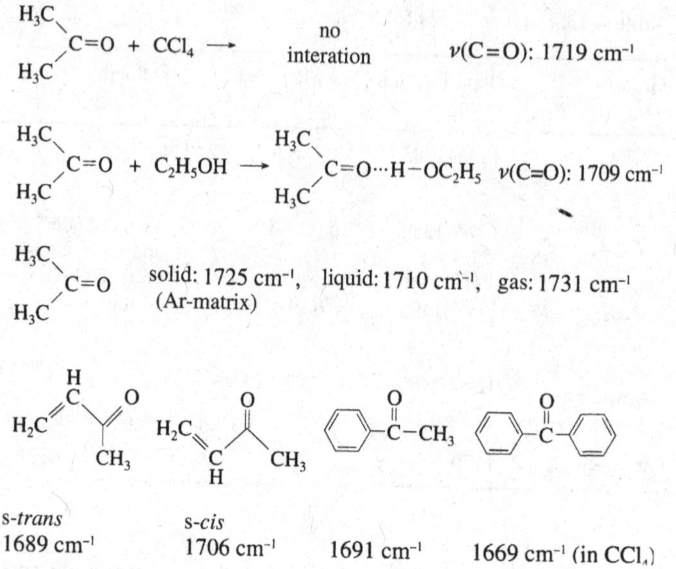

$$\begin{array}{c} H_3C \\ \diagdown \\ \diagup \\ H_3C \end{array} C{=}O + CCl_4 \longrightarrow \quad \begin{array}{c} \text{no} \\ \text{interation} \end{array} \quad \nu(C{=}O){:}\ 1719\ cm^{-1}$$

$$\begin{array}{c} H_3C \\ \diagdown \\ \diagup \\ H_3C \end{array} C{=}O + C_2H_5OH \longrightarrow \begin{array}{c} H_3C \\ \diagdown \\ \diagup \\ H_3C \end{array} C{=}O{\cdots}H{-}OC_2H_5 \quad \nu(C{=}O){:}\ 1709\ cm^{-1}$$

$$\begin{array}{c} H_3C \\ \diagdown \\ \diagup \\ H_3C \end{array} C{=}O \quad \begin{array}{l} \text{solid: } 1725\ cm^{-1}, \quad \text{liquid: } 1710\ cm^{-1}, \quad \text{gas: } 1731\ cm^{-1} \\ \text{(Ar-matrix)} \end{array}$$

s-trans	*s-cis*		
1689 cm^{-1}	1706 cm^{-1}	1691 cm^{-1}	1669 cm^{-1} (in CCl$_4$)

Fig. 6-43. Wavenumber regions of ketone absorption bands.

6.4.11.2 Aldehydes

An aldehyde is formed by substituting one of the alkyl chains of a ketone with an H-atom. Because of lower mass effects, the C=O absorption maximum lies at a higher frequency, namely at 1740–1720 cm^{-1}, than the ketone band. Due to resonance effects, a conjugation of the C=O-group with an a,β-olefin bond shifts the C=O-band to 1705–1685 cm^{-1}, with an aromatic ring to 1715–1695 cm^{-1}, and for polyene conjugations down to 1680–1660 cm^{-1}. Electronegative groups act in the opposite direction: CCl$_3$CHO: 1768 cm^{-1}; CHCl$_2$CHO: 1748 cm^{-1}; CH$_2$ClCHO and H$_3$CCHO: 1730 cm^{-1}, respectively (all measured in CCl$_4$, see Sect. 6.5.5). Aldehydes cannot always be surely distinguished from the ketones from the CO-band position alone. However, help for differentiating them is found in the C–H-stretching vibrational region. Here, aldehydes show a characteristic double band at 2900–2800 cm^{-1} and 2775–2695 cm^{-1} (Fig. 6-44). This originates from an interaction of the aliphatic C–H-stretching with the overtone of the likewise characteristic H–C=O-bending vibration at 1410–1380 cm^{-1} (Fermi resonance, see Sect. 2.2.2.4).

For aldehydes without any double bond, the 1390 cm^{-1}-band is shifted so far that no coincidence of the 2δ(CHO) bending vibration occurs anymore with ν(CH) and the interaction of both vibrational modes ceases, as is the case for CCl$_3$CHO (trichloroacetaldehyde), for example.

Of course, only free aldehydes display the double band. Oligomerized aldehydes, (paraldehyde, trioxane derivatives) have a totally different spectrum (see Sect. 6.4.7). An overview for identifying carbonyl compounds is found at the end of Sect. 6.4.11.

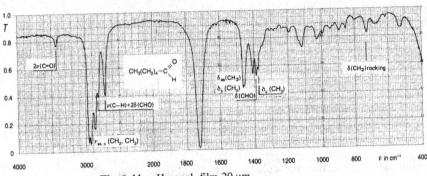

Fig. 6-44. *n*-Hexanal, film 20 μm.

6.4.11.3 Carboxylic Acids

Besides the C=O-group, organic carboxylic acids also carry the OH-function. This polar group creates new possibilities of molecular interactions, and further bands can contribute towards characterization of this substance class. Especially their tendency towards forming hydrogen-bond aggregates distinguishes them from other substances.

Carbonyl Group

Despite the hydroxy substituent, the IR absorption band of the carbonyl (COOH)-group (≈ 1705 cm^{-1}) is surprisingly close to that of the ketones at 1715 cm^{-1}. The reason for this is the opposing action of two different effects: on the one hand, the strongly electronegative O–H-substituent increases the double-bond character of the C–O-group and, thus, its stretching frequency (structures I, II (compare with acid chlorides R–CO–X). On the other hand, resonance structures must be applied for carboxylic acids for describing their properties (structures III, IV):

$$\left\{ \underset{\text{I}}{R-C\overset{\displaystyle O}{\underset{\displaystyle O-H}{\Big\langle}}} \longleftrightarrow \underset{\text{II}}{R-C\overset{\displaystyle \overset{\delta\oplus}{O}}{\underset{\displaystyle \underset{\delta\ominus}{O-H}}{\Big\langle}}} \longleftrightarrow \underset{\text{III}}{R-C\overset{\displaystyle \overset{\ominus}{O}}{\underset{\displaystyle \underset{\oplus}{O-H}}{\Big\langle}}} \longleftrightarrow \underset{\text{IV}}{R-C\overset{\displaystyle \overset{\ominus}{O}}{\underset{\displaystyle O}{\Big\langle}}} \overset{\oplus}{H} \right\}$$

It can be seen from this that the C=O-double-bond character is weakened and that the absorption frequency shifts to a longer wavelength region.

Acids have the aforementioned ability to exert a great influence on the position of the C=O-band and to form hydrogen bonds. Even in much diluted solutions, carboxylic acids exist mostly in dimeric form, then through the dissociation tendency of the proton and by the simultaneous basicity increase of the carbonyl-O-atom, the otherwise weak bond is quite strengthened by retrocoupling.

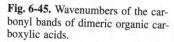

For this reason, the molecule must be treated as a dimer from a spectroscopic viewpoint. Therefore, we can determine two vibrational modes (Fig. 6-45): an IR-active form for the antisymmetric C=O-stretching vibration and an IR-inactive, but Raman-active, symmetric C=O-stretching vibrational mode (also see Sect. 8.3).

If the dimer formation is prevented by offering the acid e.g. intra- or intermolecular alcohol- or ether groups, then the C=O-frequencies shift to shorter wavelength regions, since then the bonding compensation is no longer so complete (Fig. 6-46).

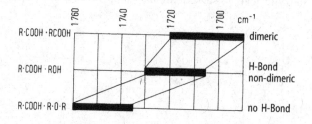

asymmetric vibrational mode;
IR-active 1720–1680 cm^{-1},
Raman-inactive

symmetric vibrational mode;
IR-inactive
Raman-active
1680–1640 cm^{-1}

Fig. 6-45. Wavenumbers of the carbonyl bands of dimeric organic carboxylic acids.

Fig. 6-46. Effect of the solvent on the carbonyl wavenumber of organic carboxylic acids.

Hydroxy Groups

A pronouncedly wide and moderately strong H-bridge band at around 3000 cm^{-1} is very characteristic for dimeric carboxylic acids (Fig. 6-47). The CH-stretching vibrations overlap this band. Weak but characteristic shoulders, originating from overtones and combination bands, lie beside it at around 2500–2700 cm^{-1}. A rather sure indicator for dimeric carboxylic acids

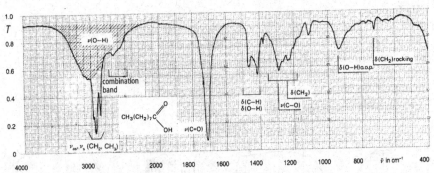

Fig. 6-47. *n*-Nonanoic acid, film.

is the o.o.p.-deformation vibrational band $\delta(-OH \cdots O)$ of the bridge complex at 960–875 cm^{-1}, recognizable from its larger half-width compared to other bands in this region.

C–O-Group with a Single Bond

One may not consider the C–O-stretching vibration as being isolated, because it is coupled with the δ(COH)-deformation vibration. Conversely, this applies also for the δ(COH)-vibration, which partially contains the C–O-stretching vibrational energy (also see Fig. 6-48).

The simultaneous motion leads to a like-phase and counter-phase vibrational mode, respectively. The band at 1315–1200 cm^{-1} (1325–1280 cm^{-1}) is assigned mainly to the C–O-stretching vibration, analogous to the ether band. The band at 1440–1396 cm^{-1} is interpreted as the COH-deformation band, analogous to the δ(CH$_2$)-band. For spectra of dimeric carboxylic acids, we find the following absorption regions attributable to the –COOH-group (Fig. 6-49).

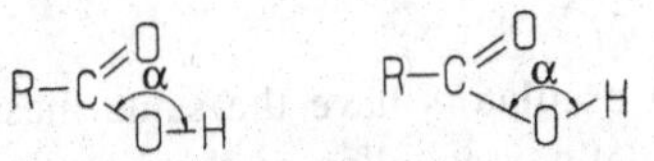

$$R-C\overset{O}{\underset{O-H}{\big\backslash}} \qquad R-C\overset{O}{\underset{O}{\big\backslash}}{}_H$$

Fig. 6-48. Deformation vibration of the C–O–H-fragment of carboxylic acids. Coupling of v(C–O) with δ(COH): upon stretching the C–O-bond, the bonding angle C–O–H is simultaneously affected.

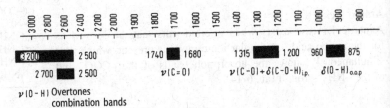

Fig. 6-49. Absorption regions of organic carboxylic acids.

CH$_2$-Groups

In the crystalline state, long-chain fatty acids and their esters with a chain of more than 12 C-atoms have several equidistant bands between 1350 and 1180 cm^{-1}, the so-called progression bands (Fig. 6-50). These stem from the twisting and rocking vibrations of the *trans*-oriented CH$_2$-groups. The number of these bands depends on the chain length.

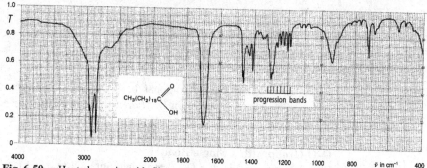

Fig. 6-50. *n*-Heptadecanoic acid, film.

Carboxylic Acid Salts

If an organic carboxylic acid is converted to its salt, this will result in an ideal bonding equilibrium between the C=O-double- and the C–O-single bond due to the complete dissociation of the anion and cation:

Because both oscillators have the same mass and force constant, there is a defined coupling of the vibrations. The band for the antisymmetric vibrational mode lies at 1650–1550 cm^{-1}, and for the symmetric mode, at 1440–1360 cm^{-1}. For anhydrous salts, all absorption bands attributable to the OH-group in the acid are then missing, of course.

Amino Acids

A particular case exists when internal salts may be formed as for the amino acids. A dipolar ionic mode is present, which thus simultaneously shows the absorption bands of the –COO$^-$ and of the –NH$_3^+$-group (Tab. 6-14).

Table 6-14. Band positions of α-amino acids.

Group	Band Position in cm^{-1}
ν_{as} (–COO$^-$)	1605–1555
ν_s(–COO$^-$)	1425–1393
ν(–NH$_3^+$)	3100–2600
δ_{as}(–NH$_3^+$)	
$+\tau$(–NH$_3^+$)	2200–2000
δ_{as}(–NH$_3^+$)	1665–1585
δ_s(–NH$_3^+$)	1530–1490

6.4.11.4 Acid Chlorides

Carboxylic acid halogenides are formed by substituting the alde-hyde-H-atom with a halogen atom. The functional group R–CO–X is so simple that a band assignment is easily possible (Fig. 6-51).

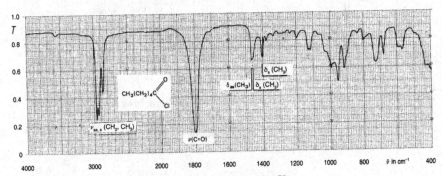

Fig. 6-51. *n*-Hexanoic acid chloride, film.

The band at 1815–1770 cm^{-1} corresponds to the C=O-vibration. Its strong shifting to the short-wave range, in comparison to the ketones, is a result of the high electronegativity of the halo-gen group. This infers that in carboxylic acid chlorides, the degree of C=O-double bonding is higher than in the esters, for example. Thus, the R–CO–Cl-carbonyl band lies, for example, near the upper limit in the double-bond region. The same thinking applies also for the remaining halogens, so that the carboxylic acid halo-genides have their C=O-band in the 1800 cm^{-1}-region.

$$\text{Alkyl} - \overset{\displaystyle O}{\underset{\displaystyle X}{C}} \qquad
\begin{array}{ll}
X = F: & \nu(C = O): \; 1800 \text{ cm}^{-1} \\
\quad Cl: & 1790 \text{ cm}^{-1} \\
\quad Br: & 1800 \text{ cm}^{-1} \\
\quad I: & 1785 \text{ cm}^{-1}
\end{array}$$

6.4.11.5 Esters

Like other carbonyl compounds, esters are primarily recognized by their strong carbonyl band (1740 cm^{-1}) and characterized by the control band of the C–O-single-bond vibrations at 1200 cm^{-1}.

Carbonyl Group

The COOR-fragment is so similar to the carboxyl group of the carboxylic acids that it is first surprising to find the C=O-band of the ester lying at a relatively high frequency, namely at

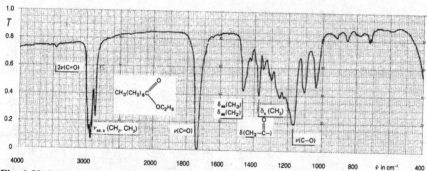

Fig. 6-52. Ethyl octanoate, film, 7 μm.

1750–1735 cm^{-1} (Fig. 6-52). Recalling that carboxylic acids can ease the C=O-double bond by dimerization via H-bridges, the short-wave shift for the esters gets understandable.

As in all previous cases, an α, β-double bond proximal to the

$$\left\{ R-CH=CH-\overset{\text{IOI}}{\underset{}{C}}-O-R' \longleftrightarrow R-\overset{\oplus}{CH}-CH=\overset{\overset{\ominus}{\text{IOI}}}{\underset{}{C}}-O-R' \right\}$$

$$1740–1715 \text{ cm}^{-1}$$

C=O-group shifts the absorption into the long-wave region:

Unsaturated groups on the ether-like oxygen, however, shifts the bands towards shorter wavelengths, because the central oxygen atom exists in various resonance forms:

$$\left\{ R-\overset{O}{\underset{}{C}}-O-CH=CH-R' \longleftrightarrow R-\overset{\overset{\ominus}{\text{IOI}}}{\underset{}{C}}=\overset{\oplus}{O}-CH=CH-R' \longleftrightarrow R-\overset{O}{\underset{}{C}}-\overset{\oplus}{O}=CH-\overset{\ominus}{CH}-R' \right\}$$

$$1800–1770 \text{ cm}^{-1}$$

C–O-Group with a Single Bond

An ester cannot be surely recognized from the position of the C=O-band alone. Identification is only possible by the strong control band at around 1200 cm^{-1}. This is the antisymmetric stretching vibration of the C–O–C-fragment with additional energy contributions from the C–C-motions. The high wavenumber position of this band with respect to ethers (1125 cm^{-1}) is caused by ester-resonance. Due to the pronounced vibrational interaction, various esters can be differentiated by IR spectroscopy (Tab. 6-15).

Table 6-15. Band positions for differentiating various esters.

Substance Group	ν(C–O) in cm^{-1}
Formiates	1200–1180
Acetates	1260–1230
Higher aliphatic esters	1210–1160
Acrylates	1300–1200
Benzoates	1310–1250 and 1150–1100
Acetates with primary alcohols	1260–1230 and 1060-1035
Acetates with secondary alcohols	1260–1230 and 1100
Acetates with phenols	≈ 1205

In general, the band around 1250 cm^{-1} is assigned to the ν_{as}(C–COO–)-mode and the band around 1050 cm^{-1} to the ν_{as}(O-CH$_2$–C)-mode.

Absorptions of the Neighboring Groups

For the ketones, the α-positioned fragments relative to the carbonyl group have shifted and mostly more intense IR bands. The same effect is also found for esters (Tab. 6-16).

Table 6-16. Effect of the ester group on the absorption bands of α-positioned methyl- and methylene groups (in cm^{-1}). The band positions for paraffin hydrocarbons are in parentheses.

$$\underset{\text{CH}_3-\overset{\overset{\text{O}}{\|}}{\text{C}}-\text{O}-}{} \qquad \underset{-\text{CH}_2-\overset{\overset{\text{O}}{\|}}{\text{C}}-\text{O}-}{}$$

δ_s: 1374	δ_{as}: 1430	δ: 1430
(1370)	(1460)	(1460)
	ν_{as}: 2990	
	(2962)	

$$\underset{-\overset{\overset{\text{O}}{\|}}{\text{C}}-\text{O}-\text{CH}_3}{} \qquad \underset{-\overset{\overset{\text{O}}{\|}}{\text{C}}-\text{O}-\text{CH}_2-}{}$$

δ_{as}: 1430	δ: 1475
(1460)	(1420)
ν_{as}: 2990	δ_{wag}: 1400
(2962)	(1305)

Lactones

In IR spectroscopy, tension-free lactones behave like the corresponding open-chain esters. However, the C–C–O-angle is under tension in small rings. Consequently, the C=O-stretching vibrational frequency (in cm^{-1}, respectively) increases as shown in the following series:

| 1727 | 1750 | 1774 | 1841 | unstable (2350) |

6.4.11.6 Anhydrides

Here, we deal with a substance class, which is already clearly characterized by its spectrum in the 1800 cm⁻¹ region. For all anhydrides, two C=O-bands are found [20]: one located at a very high frequency around 1820 cm⁻¹ and the other, at around 1750 cm⁻¹ (Fig. 6-53).

This applies for open-chain, non-conjugated derivatives. Although the resonance structures II and III

$$\left\{ \underset{\text{I}}{\overset{\text{O}\ \ \ \text{O}}{\underset{\text{R}\ \ \ \text{O}\ \ \ \text{R}}{\text{C}\ \ \ \text{C}}}} \longleftrightarrow \underset{\text{II}}{\overset{\ominus|\overline{\text{O}}|\ \ \ \text{O}}{\underset{\text{R}\ \ \ \overset{\oplus}{\text{O}}\ \ \ \text{R}}{\text{C}\ \ \ \text{C}}}} \longleftrightarrow \underset{\text{III}}{\overset{\text{O}\ \ \ |\overline{\text{O}}|^{\ominus}}{\underset{\text{R}\ \ \ \overset{\oplus}{\text{O}}\ \ \ \text{R}}{\text{C}\ \ \ \text{C}}}} \right\}$$

contribute little towards stabilizing the molecule; they are nevertheless responsible for the planar design of the molecule around this –C(O)–O–C(O)–arrangement. Because of this, band splitting into a symmetric and an antisymmetric vibrational mode of the coupled oscillators occurs in the IR spectrum. Anhydrides are one of the few cases in which the ν_{as}-vibration has lower energy than the ν_s-vibration. The intensity of the C=O-band depends on the molecular structure, i.e. on the mutual orientation of the C=O-oscillators. Open-chain anhydrides exist mainly in the conformation IV (or V).

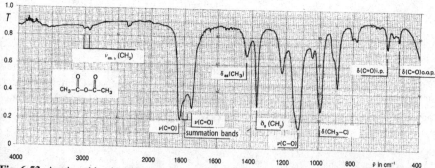

Fig. 6-53. Acetic acid anhydride, film.

symmetric streching vibration　　　　antisymmetric streching vibration

In the symmetric stretching vibration, there is a large change in the dipole moment and, thus, also in the intensity (specified are the wavenumber intervals occurring for the bands):

ν_s: 1825–1815 cm^{-1} strong　and　ν_{as}: 1755–1745 cm^{-1} weak

For cyclic anhydrides, the case is reversed:

symmetric streching vibration　　　　antisymmetric streching vibration
ν_s: 1860 cm^{-1}　　　　　　　　ν_{as}: 1785 cm^{-1}

In the vibrational mode I, the almost parallel dipoles lying in the opposite direction compensate one another. Thus, the stronger band is produced by the long-wave ν_{as}-vibration:

ν_s: 1870–1845 cm^{-1} weak　and　ν_{as}: 1800–1715 cm^{-1} strong

Additional strong bands for the single bond C–O-groups lie at 1050–1040 cm^{-1} for aliphatic open-chain anhydrides, and at 955–895 cm^{-1} and 1300–1180 cm^{-1} for alicyclic anhydrides.

6.4.11.7 Amides

The amide group shows complicated vibrational modes, so that the spectra are correspondingly band-rich. The possibility of forming cis-trans-isomers in monosubstituted amides and the ability of primary and secondary amides to form multiple H-bridge systems adds to further complication in the spectral interpretation. Especially spectral images of the same substance may be quite different depending on the sample preparation. The aggregate state and the concentration of the amide are hereby the most important influences.

For the band assignment, the absorptions that are characteristic for the amide group were first numbered with Roman numerals. Thus, the name 'amide-II band' is only significant as a name. The assignment appears in Tab. 6-17. The intense bands are assigned to vibrations of the C=O-, C–N- and the NH$_2$-group.

Table 6-17. Absorption regions, name and assignment of the amide bands.

Name	Range in cm^{-1}		Assignment
	Solid	Solution	
Amide I	≈ 1650	≈ 1690	prim. amides ν(C=O)
	1680–1630	1700–1670	sec. amides ν(C=O)
	1670–1630		tert. amides ν(C=O)
Amide II	1650–1620	1620–1590	prim. amides ν(C–N)+δ(C–N–H)
	1570–1515	1550–1510	sec. amides ν(C–N)+δ(C–N–H)
Amide III	1330–1200		sec. amides [δ(NH)+δ(OCN)]*
Amide IV	≈ 620		prim. amides
	630–600		sec. amides ν(C–C)+δ(O–C–N)*
Amide V	≈ 720		sec. amides δ(N–H$\cdots$bridge)
Amide VI	1440–1400		prim. amides δ(skeleton)

* Assignment uncertain.

Primary Amides

In resonance systems, the NH_2-group is considered one of the strongest resonance-electron donor groups. Hence, it is easy to explain that the amide C=O-band appears at a very low frequency, namely 1670-1620 cm^{-1} (Fig. 6-54).

This band is very intense and is generally accompanied by a second band, that of the δ(NH_2)-vibration at between 1650 and 1620 cm^{-1}. Both these bands often coincide, whereby the absorption range of primary amides becomes very wide. The difference to carboxylic acid salts is recognizable from the double NH_2-stretching vibrational band (3350 cm^{-1} and 3180 cm^{-1}). The wavenumber region applies for the solid. In solution, the band positions shift by the following amounts (Tab. 6-18):

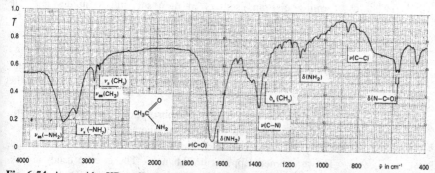

Fig. 6-54. Acetamide, KBr pellet, example for a primary amide.

Table 6-18. Effect of the aggregate state on the band positions of primary amides.

Vibration	Solid	Solution	$\approx (v_{sol.} - v_{solid})$
$v_{as}(NH_2)$	3350 cm^{-1}	3520 cm^{-1}	$+170$ cm^{-1}
$v_s(NH_2)$	3180 cm^{-1}	3400 cm^{-1}	$+220$ cm^{-1}
$v(C=O)$	≈ 1650 cm^{-1}	≈ 1690 cm^{-1}	$+40$ cm^{-1}
$+(N-H)$	$1650-1620$ cm^{-1}	$1620-1590$ cm^{-1}	-30 cm^{-1}

Secondary Amides

1. When monosubstituted amides are not forced into the *cis*-conformation because of the molecular design, then they exist in the *trans*-conformation:

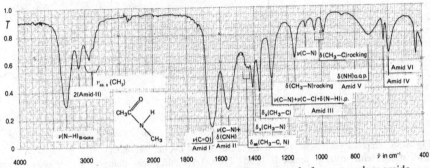

trans-conformation   cis-conformation

In the solid state, the C=O-band is found between 1680 and 1630 cm^{-1} with the v(NH)-control band as a single band at 3300 cm^{-1} (see Fig. 6-55). The weak accompanying band at 3100 cm^{-1} is an overtone of the characteristic and strong 1550 cm^{-1} band that corresponds to an interaction of the v(C–N)- and the δ(CNH)-vibration. Likewise, a moderately strong band at 1250 cm^{-1} is characteristic for secondary amides. In addition, there is a very wide δ(NH) wagging band at around 700 cm^{-1}. In solution, band shifts are again observed (Tab. 6-19).

2. *cis*-configured secondary amides are found practically only in form of lactams, where the *trans*-configuration is impossible because of the high ring tension. However, the bands lie essentially in the same regions as for the *trans*-configured derivatives.

Fig. 6-55. *N*-Methylacetamide, film, example for a secondary amide.

Table 6-19. Effect of the aggregate state on the band positions of secondary amides.

Vibration	Solid	Solution	$\approx (\nu_{sol.} - \nu_{solid})$
$\nu(N–H)$	3330 cm^{-1}	3430 cm^{-1}	$+140 \text{ cm}^{-1}$
$\nu_s(N–H)$	3160 cm^{-1}	3430 cm^{-1}	$+270 \text{ cm}^{-1}$
$\nu(C=O)$	$1680–1630 \text{ cm}^{-1}$	$1700–1670 \text{ cm}^{-1}$	$+30 \text{ cm}^{-1}$
$\delta(C–N)$	$1570–1515 \text{ cm}^{-1}$	$1550–1510 \text{ cm}^{-1}$	-10 cm^{-1}

3200 cm^{-1}: $\nu(N–H)$ $1650–1750 \text{ cm}^{-1}$: $\nu(C=O)$
3100 cm^{-1}: $\nu(C=O)+\delta(N–H)$ $1490–1440 \text{ cm}^{-1}$: $\delta(N–H)$
                    (combination band) $1350–1310 \text{ cm}^{-1}$: $\nu(C–N)$

Similar as for the lactones, the ring size of the lactams has a considerable effect on the position of the $\nu(C=O)$ band (in cm^{-1}):

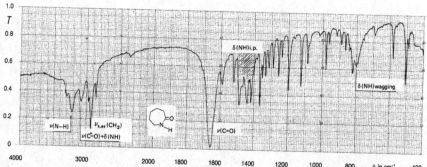

| 1669 | 1673 | 1717 | 1750 | 1850 | (2336) (in CCl$_4$) |

Inherent for the *cis*-configuration is that $\delta(CNH)$ can no longer couple with $\nu(C=O)$, thereby causing one of the bands at 1550 cm^{-1} of the *trans*-form to be missing (Fig. 6-56). In contrast, the *cis*-$\delta(NH)$ appears at a lower frequency at $1490–1440 \text{ cm}^{-1}$ and the $\nu(C–N)$ at $1350–1310 \text{ cm}^{-1}$. $\delta(NH)$ wagging appears as a wide band at about 800 cm^{-1}.

Tertiary Amides (*N, N*-di-substituted)
The low position of the carbonyl band at $1680–1630 \text{ cm}^{-1}$ (solid and solution) is sufficient for identifying tertiary amides (Fig. 6-57).

Fig. 6-56. Caprolactam, KBr pellet. Example for a cyclic, secondary acid amide with missing amide-II band. Typical for alicycles: many sharp bands in the fingerprint region.

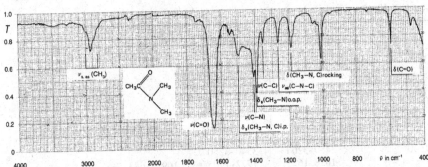

Fig. 6-57. *N,N*-Dimethylacetamide, film. Example for a tertiary amide.

The other frequencies (v(NH), δ(CNH)) are missing. The difference between them and the carboxylic acids can be recognized from the missing OH-band.

6.4.11.8 Carboxylic Acid Derivatives

Having discussed the amide group, we have already handled a relatively complicated molecular building block. The degree of the vibrational coupling included all atoms and, therefore, resulted in the characteristic frequencies of the amides. Now, it will be even more complicated when still further structural elements are found on the molecule. The spectra thereby become very substance-specific. The carboxylic acid derivatives exemplify this. The band positions of frequently occurring structural groups are compiled in Table 6-20.

Overview for Identifying Carbonyl Compounds
Simple carbonyl compounds can be surely distinguished by their very intense C=O-absorption in the region of between 1800 and 1650 cm^{-1}. The finer differentiation takes place by using control bands.

Ketones: They exist when other C=O-derivatives can be excluded; v(C=O) at around 1715 cm^{-1}.

Aldehydes: Besides the C=O-band at 1740–1720 cm^{-1}, these reveal two C–H bands with about the same intensity at around 2850 cm^{-1} and 2750 cm^{-1}.

Carboxylic acids: Besides being recognized by v(C=O) at 1705 cm^{-1}, they are clearly discerned by the wide H-bridge band in the 3000 cm^{-1} region, which is overlapped by v(C–H) and an intensity-weak, but characteristic, on-

Table 6-20. Characteristic bands of various amide derivatives in cm^{-1}.

$$\underset{\text{Ureas}}{\overset{\displaystyle R\diagdown\;\;\overset{\textstyle O}{\overset{\|}{C}}\;\;\diagup R}{\underset{R\diagup\text{N}\qquad\text{N}\diagdown R}{}}}$$

$$\underset{\text{Urethanes}}{R-O-\overset{\textstyle O}{\overset{\|}{C}}-\overset{\displaystyle R}{\underset{\displaystyle R}{N}}}$$

Ureas	Urethanes		
	primary	secondary	tertiary
1640	3450–3200	3340–3250	–
	1628–1622	1722–1705	1691–1683
	1540–1530	1620	

$$R-\overset{\textstyle O}{\overset{\|}{C}}-NH-OH \qquad R-\overset{\textstyle O}{\overset{\|}{C}}-NH-NH_2-$$

Hydroxamic acids   Hydrazides

Hydroxamic acids	Hydrazides
3300	3320–3180
2800	1700–1640
1640	1633–1652
1550	1542–1502
1440–1360	1150–1050
900	

$$R-\overset{\textstyle O}{\overset{\|}{C}}-NH-\overset{\textstyle O}{\overset{\|}{C}}-R \qquad R-\overset{\textstyle O-}{\overset{\|}{C}}-NH-NH-\overset{\textstyle O}{\overset{\|}{C}}-R-$$

Imides   Diacylhydrazides

Imides		Diacylhydrazides	
trans-trans	*cis-trans*		
3280–3200	3245–3190	3380–3280	3210–3100
1737–1733	1700		3060–3020
1505–1503	1734	1742–1700	
1236–1167	1650	1707–1683	1623–1580
739–732	836–816		1506–1480
			1260–1200

hanger at 2500 cm^{-1}. The δ(O–H)-control band at 900 cm^{-1} and the ν(C–O)-single bond absorption (1325–1200 cm^{-1}) as well as δ(CHO) (1440–1395 cm^{-1}) support the assignment.

Acid chlorides: Here, the high-frequency 1800 cm^{-1} band is diagnostic.

Esters: A somewhat less intense band at around 1200 cm^{-1} belongs to the 1750–1753 cm^{-1} band.

Anhydrides: They have a clear double band with a very high-frequency component above 1800 cm^{-1}.

Amides, primary: Aside from a very intense and wide band for ν(C=O) at 1670–1620 cm^{-1}, they also show two bands for ν_{as}(NH$_2$) and ν_s(NH$_2$) and a wide δ(NH$_2$)-deformation vibrational band at around 1620 cm^{-1}.

Amides, secondary: They have two strong bands at 1680–1620 cm^{-1} and 1550 cm^{-1} with a sharp ν(N–H)-stretching vibrational band at 3300 cm^{-1} and a very wide δ(NH) absorption in the 700 cm^{-1} region.

Amides, tertiary: The very low wavenumber of the intense carbonyl band at between 1680 and 1630 cm^{-1} suffices for identification.

Carboxylic acid derivatives: Here, the spectra are often nebulous and, without preliminary information about the substance being investigated (origin, synthesis), no certain conclusions can be drawn about the assignment.

6.4.12 Nitrogen Compounds

Regarding the nitrogen compounds, one can often draw upon knowledge about the corresponding oxygen derivatives and alkanes. The C=N-double bond lies in the same region as the C=C-oscillator; the effect of electrostatic and resonance effects on the band position is similar to that for other unsaturated systems. The intensity of the ν(C=N)-vibration lies between that of ν(C=O) and ν(C=C). We have already been acquainted with some N-containing compound classes in previous sections (nitriles, amides). Here, the assignment of the C–N-single bond was rather difficult to do, and this again will be confirmed this. Moreover, the ν(NH)-bands served more for control; alone, they infer little about the substance type. Therefore, it is possible to differentiate the various substance classes only by the position of other group frequencies.

6.4.12.1 Amines

1. N–H-stretching vibrational bands: the bands of the ν(N–H) are found at somewhat lower frequencies than the corresponding –OH-band. Because symmetric and antisymmetric fundamental vibrations occur in primary amines, they have two well-separated bands near 3335 cm^{-1}. They are about 70 cm^{-1} apart. Inherently, secondary amines have only one possible band. Compared to the –OH-group, the NH-bands

are much more sharper (Fig. 6-58 and 6-60). N–H-stretching vibration bands of aromatic amines are shifted still further into the short-wave region and lie at 3450–3339 cm^{-1} (see Sect. 6.4.14).

2. Deformation bands: for primary amines, the band of the NH$_2'$-i.p. deformation vibration (scissoring) lies at higher wavenumbers than the corresponding CH$_2$-frequency, namely at 1650–1590 cm^{-1}. The C–N–H-deformation vibration of secondary amines appears in the same region at 1650–1550 cm^{-1} but with much weaker intensity. The wagging and twisting vibrations for primary amines fall together in a very wide and strong band in the region from 850–750 cm^{-1}. Secondary amines show the corresponding band near 715 cm^{-1}.

3. C–N-stretching vibration: analogous to the alcohols and ethers, likewise strong, mass-dependent CN-vibrational bands should occur for the amines. Although one can find corresponding, but only moderately strong absorptions for the simple amines (Tab. 6-21), an interpretation is not easy without reference spectra. On the other hand, exactly here, one can differentiate between aliphatic and aromatic amines. Tertiary amines are recognized easiest as hydrochlorides.

Table 6-21. Absorption regions of the v(C–N)-vibration.

Amine v(C–N) aromatic		Amine v(C–N) aliphatic	
prim.	1340–1250 cm^{-1}	prim.	
sec.	1350–1280 cm^{-1}	sec.	1220–1020 cm^{-1}
tert.	1360–1310 cm^{-1}	tert.	

4. Ammonium salts: they are all characterized by a wide, intense absorption region between 3300 and 2000 cm^{-1} which is overlapped by a complicated fine structure. This cannot be solely attributed to the N–H-stretching vibrations but is also caused by combination bands of the NH$_4^+$-group (Fig. 6-61 to 6-63).

Identification of Amines

a) Primary amines (see Fig. 6-58 and 6-61) show two sharp bands at around 3335 cm^{-1} and a wide band at 1615 cm^{-1}, both having about the same intensity but different half-widths. An additional very wide, intense absorption in the region of 850–750 cm^{-1} (δ(NH$_2$)-wagging) supports the assignment. Protonation shifts the 3335 cm^{-1} band to 3030–2630 cm^{-1} with widening. Characteristic is a weak band between 2220 and 1820 cm^{-1}.

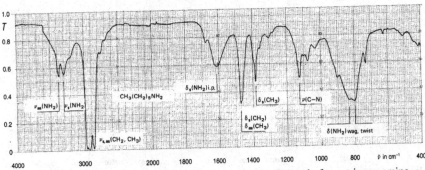

Fig. 6-58. n-Hexylamine, film, 20 μm. Example for a primary amine.

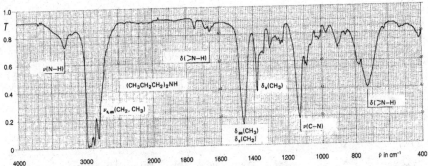

Fig. 6-59. Di-n-propylamine, film, 20 μm. Example for a secondary amine.

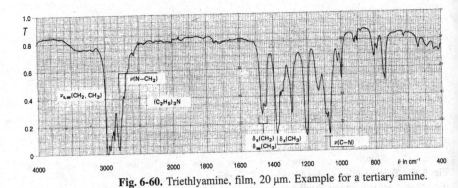

Fig. 6-60. Triethlyamine, film, 20 μm. Example for a tertiary amine.

b) Secondary amines (Fig. 6-59 and 6-62) have only one sharp band at 3335 cm^{-1}, and around 1615 cm^{-1}, any absorption is found in contrast to the primary amines. The NH-wagging band is shifted to 715 cm^{-1}. Protonation again shifts the 3335 cm^{-1} band to 3030–2630 cm^{-1}. Additional absorptions appear between 2500 and 2220 cm^{-1}.

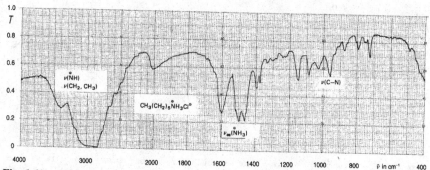

Fig. 6-61. *n*-Hexylamine-hydrochloride, KCl pellet 1:300. Example for the salt of a primary amine.

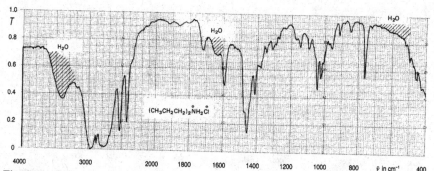

Fig. 6-62. Di-*n*-propylamine-hydrochloride, KCl pellet 1:300. Example for the salt of a secondary amine. Amine-hydrochlorides are hygroscopic; thus, additional water bands occur.

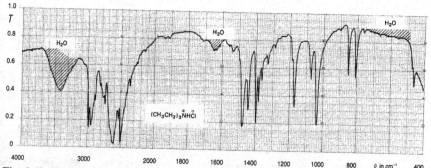

Fig. 6-63. Triethylamine-hydrochloride, KCl pellet 1:300. Example for the salt of a tertiary amine. Amine-hydrochlorides are hygroscopic; thus, additional water bands occur.

c) Tertiary amines (Fig. 6-60 and 6-63) are recognized by the long-wave N–CH$_2$-band at 2780 cm^{-1}. The regions around 3335 cm^{-1}, 1650 cm^{-1} and 850–750 cm^{-1} are free from wide bands. The protonated form shows a characteristic band at 2500 cm^{-1}.

6.4.12.2 C=N-Double Bond

The position of the C=N-band is partially the result of the opposing effects of resonance and inductive influences and of coupling effects. Protonation of the basic nitrogen decisively affects the spectral behavior of the compounds and may thus be additionally used for characterization. From Table 6-22, we can note the effects of the substituents by using azomethone as an example [21].

Table 6-22. Dependency of the v(C=N)-band position in cm^{-1} on the substituents (liquid, measured in $CHCl_3$ or CCl_4).

R_1 $>$C=N–R_3 R_2		N-Substituent R_3		
R_1	R_2	H	Alkyl-	Aryl-
H	H	1643*	1642–1657	
	Alkyl		1674–1665	
	Aryl		1665–1629	1637–1625
Alkyl-	Alkyl-	1646–1640	1662–1649	1670–1658
	Aryl-	1633–1620	1650–1640	1640–1627
Aryl-	Aryl-	$\approx$ 1603	1630–1615	1615–1610

* Gas

For open-chain C=N-derivatives, a sharp band can still be definitely assigned at 1690–1640 cm^{-1}. Aryl conjugation causes a shift towards longer wavelengths, regardless of whether the substituent is located on N or C. The greatest uncertainty for an assignment for olefin-substituted azomethines stems from the strong interaction between the C=C- and C=N-vibrations. An assignment to a particular oscillator is no longer sensible.

Heterosubstituents on the N- or C-fraction of the C=N-bond affect the absorption frequency by changing the N–X- or C–X-force constants, so that other substance classes can be recognized here (Tab. 6-23).

6.4.12.3 Nitro-Group

At first glance, this planar trimass oscillator is related to the carboxylate group; i.e. the two intense bands of the antisymmetric (1550 cm^{-1}) and symmetric (1360 cm^{-1}) stretching vibration occur (Fig. 6-64).

Table 6-23. Position of the ν(C=N)-band for various substance in cm^{-1} (R=alkyl, φ=aryl).

Compound Class	Substituent				ν(C=N)	
Oximes R_1 $\quad$C=N—OH R_2 ν(O–H): 3300–3150 ν(N-O): $\approx$ 930	R_1	R_2				
	R	R			1684–1652	
	R	φ			1640–1620	
	R	H			1652–1673	
	φ	H			1614–1645	
Imidoesters $\quad$OR$_3$ $\quad$ | R_1—C=N—R$_3$	R_1	R_2	R_3			
	φ	R	φ		1666	CCl$_4$
	R	R	R		1670	CCl$_4$
Carboxylic acid amides R_1O $\quad$C=N—R$_3$ R_2O ν(N–H): 3300 ν(C–O): 1225–1100	R_1	R_2	R_3			
	R	R	H		1690–1645	
Isoureas R_1 $\quad$N $R_2\quad$C=N—R$_4$ R_3—O	R_1	R_2	R_3	R_4		
	φ	H	R	φ	1655	dioxan
	φ	H	R	H	1672 and 1655	dioxan
	H	H	R	φ	1632	liquid
Amidines R_1 $\quad$N $R_2\quad$C=N—R$_4$ $\quad R_3$	R_1	R_2	R_3	R_4		
	R	R	H	H	1654 and 1634	solid
	R	H	R	H	1680 and 1593	solid
	R	H	R	R	1641	solid
	φ	H	φ	φ	1637	solid
Hydrazones R_1 $\quad$ R_3 $\quad$C=N—N R_2 $\quad$ R_4	R_1	R_2	R_3	R_4		
	H	R	H	R	1612–1604	
	R	R	H	R	1635–1625	
	H	R	R	R	1647–1634	
	R	R	R	R	1652–1625	

The dipole caused by the semipolar $N \rightarrow O$-bond makes the band position especially sensitive for polar influences. Therefore, for example, the bands of the antisymmetric stretching vibration for *p*-substituted nitrophenols can be correlated with the electronic-donating or electron-accepting property of the substituents (Hammett relationship). Such a correlation does not apply for the symmetric NO$_2$-vibration.

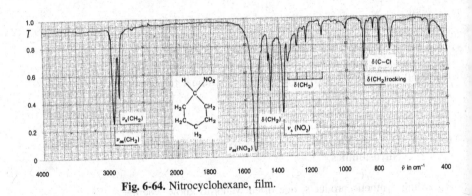

Fig. 6-64. Nitrocyclohexane, film.

The applicability of the Hammet $\sigma\rho$-relationship is probably purely arbitrary, because there is absolutely no theoretical foundation for a linear free enthalpy relationship of the IR vibrational frequencies [22].

Table 6-24. Effect of substituents on the position of the NO_2-bands (in cm^{-1}).

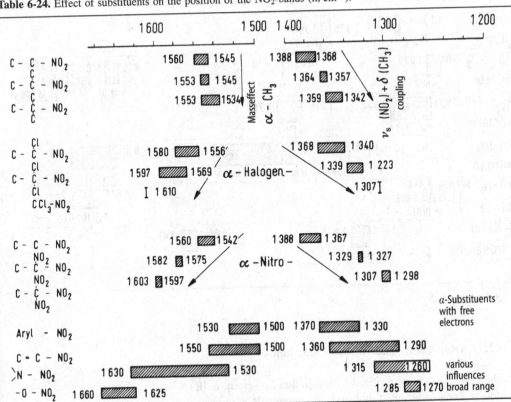

As the C–N- and N–O-vibrational frequencies in the nitro-compounds lie closer to each other than those of the C–C and C–O in carboxylates, we now have to additionally account for the vibrational coupling and may no longer consider the individual oscillators alone. The NO_2-stretching vibrational frequencies are compiled in Tab. 6-24.

6.4.12.4 Organic Nitrates, Nitramines, Nitrites, Nitrosamines

Regarding synthetic products, one often wants to know whether the desired compound was obtained or, for example, isomers (nitro-nitrite) or oxidation products (nitrite-nitrate, nitrosamine-nitramine) are formed. In the case of the cited compounds, one can already answer this question on hand of the IR spectrum (Tab. 6-25).

Table 6-25. Absorption regions of organic nitro- and nitro-derivatives (in cm^{-1}).

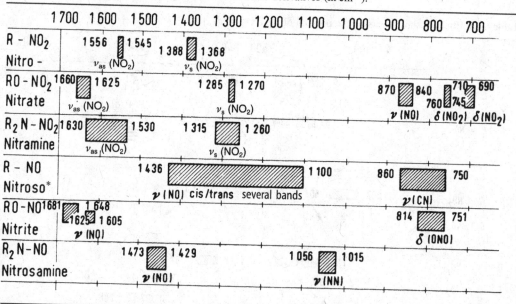

* See Sect. 6.4.12.5

6.4.12.5 Azo-, Azoxy- and Nitroso-Compounds

The band of the symmetric *trans*-di-substituted azo-group is IR-forbidden. For 4-substituted diphenylazo compounds with asymmetric substitution, the N=N-frequency appears weakly at 1410–

1420 cm^{-1}. Oxidation of the azo compounds to their azoxy-derivatives lifts the symmetry, and we again find absorption at:

1480–1450 cm^{-1}: δ_{as} (NNO) (mainly v(N=N))
1335–1315 cm^{-1}: δ_s (NNO) (mainly v(N → O))

$$\overset{O}{\overset{\uparrow}{}}$$

1530–1495 cm^{-1}: $v(-N=N-)$

Monomeric nitroso-compounds are recognized by very strong bands (the narrower wavenumber region is given in parentheses) at:

1621–1540 cm^{-1} (1555 ± 12 cm^{-1}): monomeric aliphatic
                                          C-nitroso-compounds,
1513-1488 cm^{-1}:   monomeric aromatic
                                          C-nitroso-compounds.

Control bands are found for v(C–N) and δ(CNO) at 1100 cm^{-1} and 460–400 cm^{-1}, respectively.

In solids, however, nitroso-compounds exist in dimeric form, whereby *cis*- and *trans*-configured molecules have different IR spectra (Tab. 6-26).

Table 6-26. Absorption regions of dimeric nitroso-compounds.

		v_{as}(N → O)	v_s(N → O)
aliphatic	trans	1300–1160	IR-inactive
	cis	1426–1387	1350–1323
aromatic	trans	1299–1253	IR-active
	cis	1409–1389	1409–1397

6.4.13 Compounds with Cumulated Double Bonds

In most of the cases discussed yet, the vibrational coupling of the atoms always played a large role. We have seen that the effect depends on the mass, the force constants and on the geometric arrangement of the atoms. The ideal case of coupling is met with compounds having cumulated double bonds. The masses of the vibrating atoms are equal or similarly large and the geometric arrangement is ideal for an interaction. A good example is the CO_2-molecule. There is no absorption band in the normal range of the C=O-vibration, but a strong absorption occurs at 2350 cm^{-1} for the antisymmetric stretching vibration. The symmetric mode (1336 cm^{-1}) is IR-forbidden. The mechanical coupling of the molecular vibrations shifts the antisymmetric stretching vibrational band of cumulated systems into the characteristic region of 2300–2000 cm^{-1}. Thus, they are easily and surely identifiable. The symmetric stretching vibration for linear X=Y=Z-molecules lies in the region of 1400 to 1300 cm^{-1} (excluding sulfur-containing compounds). However, the intensity is mostly too weak to pick them out clearly and immediately from the other bands. The absorption positions of important compound classes are found in Table 6-27.

Table 6-27. Band positions of compounds with cumulated double bonds in cm^{-1}.

Compound Class	Group	ν_{as}	ν_s	Others
Carbon dioxide	O=C=O	2349	*	δ: 667
Carbon oxide sulfide	O=C=S	2062	859	δ: 520
Carbon disulfide	S=C=S	1523	*	δ: 397
Allenes	$\diagdown$C=C=C$\diagup$	2000–1900	1665	δ(CH$_2$): 850
Ketenimines	$\diagdown$C=C=N–	2050–2000		
Carbodiimides	–N=C=N–	2150–2100	*	
Isocyanates	–N=C=O	2275–2263	1395–1375	
Isothiocyanates	–N=C=S	2150–2050	700–650 (alkyl) 945–925 (aryl)	
Ketenes	$\diagdown$C=C=O	2155–2130	1175	
Azides	–N=N=$\underline{N}$| $^{\oplus}$ $^{\ominus}$	2170–2080	1343–1177	
Aliphatic diazo-compounds	$\diagdown$C=N=$\underline{N}$| $^{\oplus}$ $^{\ominus}$	2135–2010		
Cyanates	[N=C=O]$^{\ominus}$	2220–2130	1334–1292	
Thiocyanates	[N=C=S]$^{\ominus}$	2090–2020	700	

* IR-forbidden.

6.4.14 Aromatic Compounds: Substituent Effects

In Section 6.4.5, we had learned about the characteristic bands of the basic skeleton. In this section, the interactions of the aryl nucleus with functional groups and their effects on the IR spectrum will be handled.

6.4.14.1 Aromatic Hydrocarbons

The spectra show only the absorptions in the five aromatic-characteristic regions; however, the substitution pattern can be interpreted from the o.o.p. CH-bands (see Sect. 6.4.5):

$3100-3000$ cm^{-1} Mostly several weak bands (ν(C–H)).

$2000-1650$ cm^{-1} Substitution-characteristic overtones and combination bands (summation bands) having low intensity.

$1620-1460$ cm^{-1} The bands between 1620 and 1460 cm^{-1} are ring modes involving C–C partial double bonds of the aromatic ring. The band at 1500 cm^{-1} is usually the strongest of these bands, which generally show medium to strong infrared intensities.

$1300-1000$ cm^{-1} Little suited for characterizing but valuable for identification purposes.

$910-660$ cm^{-1} Intense bands in the 910–660 cm^{-1} region, due to out-of-plane CH wagging and out-of-plane sextant ring bending vibrations, are indicative of the type of substitution.

Number of H-atoms between the Substituents	Substitution Type	Band Region in cm^{-1}
1	1,3-di-* 1,2,4-tri- 1,3,5-tri* 1,2,3,5-tetra- 1,2,4,5-tetra- 1,2,3,4,5-penta-	910–835
2	1,4-di- 1,2,4-tri-* 1,2,3,4-tetra-	885–800
3	1,3-di- 1,2,3-tri*	810–750
4	1,2-di-	770–735
5	mono-*	770–730
*Ring	mono- 1,3-di- 1,2,3-tri- 1,2,4-tri- 1,3,5-tri-	730–675

6.4.14.2 Aromatic Halogen Compounds

Besides the aromatic bands, the bands of the aryl-halogen bond occur at:

$$
\begin{array}{ll}
\text{Aryl-F} & 1270\text{--}1100 \text{ cm}^{-1} \\
\text{Aryl-Cl} & 1096\text{--}1034 \text{ cm}^{-1} \\
\text{Aryl-Br} & 1073\text{--}1028 \text{ cm}^{-1} \\
\text{Aryl-I} & 1061\text{--}1057 \text{ cm}^{-1}.
\end{array}
$$

The absorptions all lie in the fingerprint region and are thus frequently overlapped. The position of the substituent bands of the aromatic ring is mostly retained.

6.4.14.3 Aromatic Ethers

This substance class has already been handled in Sect. 6.4.7. The aryl-O-ether band at 1250 cm^{-1} is clearly removed from the O-alkyl absorption (1040 cm^{-1}). Noteworthy is the $-O-CH_3$-band for anisoles at 2835 cm^{-1}.

6.4.14.4 Phenols

The phenolic $-OH$-group has its strong and wide absorption maximum at between 3705 and 3125 cm^{-1}. A sharp line at about 3705 cm^{-1} is found for big *o*-substituents, which sterically prevent an H-bridge bond.

The $-OH$-deformation band lies at $1390\text{--}1315 \text{ cm}^{-1}$ and the aryl--OH-vibration at between 1335 and 1165 cm^{-1}, mostly split in a number of various maxima.

The positions of the substituent bands are unaffected by the OH-group: however, additional new bands also occur in this region (see Fig. 6-36), the so-called "X-sensitive bands". Such bands appear when certain in-plane ring-deformation vibrations of the benzene ring are coupled with the C–X-stretching vibration of the substituents [11].

6.4.14.5 Aromatic Amino Compounds

Primary aromatic amines have their N–H-stretching vibrational band with $3450\text{--}3390 \text{ cm}^{-1}$ at somewhat higher frequencies than primary aliphatic amines and, in addition, a weak accompanying band at 3225 cm^{-1}. Secondary amines show a much more sharper N–H-band at 3450 cm^{-1} than aliphatic secondary amines. Otherwise, the aromatic absorptions are not disturbed by the amino function. The C–N-band was already discussed in Sect.

6.4.12.1. We essentially have a sum spectrum of bands of the amino group and of the aromatic ring.

6.4.14.6 Aromatic Nitro Compounds

A resonance effect of the phenyl ring is observed here:

$$v_{as}(NO_2): 1540 \text{ cm}^{-1} \quad\quad v_s(NO_2): 1350 \text{ cm}^{-1}$$
$$(\text{aliphatic: } \approx 1550 \text{ cm}^{-1}) \quad (\text{aliphatic: } \approx 1378 \text{ cm}^{-1})$$

The substitution pattern between 910 and 660 cm^{-1} is strongly affected by the NO_2-group.

6.4.14.7 Aromatic Carbonyl Compounds

Ketones
The aromatic ring here generally acts as an electron donor and causes a shift of the $v(C=O)$-band toward lower frequencies compared with aliphatic ketones.

R–C–R   R–C–Aryl   Aryl–C–Aryl

1725 cm^{-1} 1664 cm^{-1} 1615 cm^{-1}

The carbonyl group affects the band group in the 910–660 cm^{-1} region to a lesser degree than the NO_2-substituent.

Aldehydes
Aromatic aldehydes likewise absorb at smaller wavenumbers than the aliphatic aldehydes. The double band of the $v(CH)/2\delta$-(CHO)-Fermi resonance is retained (2750 and 2850 cm^{-1}). The substitution bands are changed analogous to the ketones.

Carboxylic Acids
Again, the phenyl residue shifts the C=O-band to 1695–1665 cm^{-1}, i.e. to smaller wavenumbers. The other bands of the dimeric carboxylic acids remain unchanged: $v(OH)$: 3335–2500 cm^{-1}; $\delta(OH)$: 1430 cm^{-1}; $\delta(C-O)$: 1250 cm^{-1}; $\delta(OH)$: 950–870 cm^{-1}. On the contrary, the effect of the COOH-substi-

tuent on the phenyl ring is so great that substitution interpretations are very uncertain in this case.

Acid Halogenides

Analogous to the aliphatic derivatives, the electronegative halogen causes a strong short-wave shift (1770 cm^{-1}), but which is again attenuated by the aryl ring and is often accompanied by band splitting. In addition, one again finds the disturbance of the $910–660 \text{ cm}^{-1}$-region by the carbonyl conjugation.

Halogen	F	Cl	Br	I
ν(C=O)/cm^{-1}	1821	1783	1785	1752

Esters

The carbonyl band lies very low at $1725–1695 \text{ cm}^{-1}$. The C–O-single bonds are found in the fingerprint region at $1300–1100 \text{ cm}^{-1}$. Phenol esters, however, have a C=O-band shifted to a higher frequency at 1755 cm^{-1}, as was also seen for the vinyl esters (Sect. 6.4.11.5). In ring-conjugated lactones, both effects are cancelled, and the carbonyl absorption lies at $1740–1720 \text{ cm}^{-1}$, like for the corresponding aliphatic lactones.

1724 cm^{-1} 1754 cm^{-1} $1740–1720 \text{ cm}^{-1}$

The COOR-group on the aromatic ring also disturbs the interpretation of the substitution bands.

Anhydrides

They are clearly identifiable by the mostly definite, double band at between 1885 and 1725 cm^{-1} but not with regard to the substitution type. The reason for this is that here, too, the position of the substitution bands are shifted by the electronegative substituents.

Amides

This substance group has almost the same band positions as the aliphatic amides, because the aryl conjugation has little influence on the amide-I band. The phenyl-N-band lies between 1335 and 1250 cm^{-1}. The effect of the amide group on the substitution bands is not so great as that of the other carbonyl groups.

6.4.14.8 Aromatic Conjugated Triple Bonds and Cumulated Double Bonds

In essence, the absorption regions of the corresponding aliphatic compounds are retained:

Aryl–C$\equiv$N	2245 cm^{-1}	shifted somewhat towards higher wavenumbers; dependent on other substituents. The substitution bands are interpretable analogous to the alkyl-substituted derivatives.
Aryl–N=C=S Aryl–S=C=N Aryl–N=C=O Aryl–N=C=N–aryl Aryl–N=N=N	2175–2130 cm^{-1}	

6.4.14.9 Aromatic Oximes, Azomethines and Azo Compounds

The intense OH-stretching vibrational band of the oximes lies in the expected region at 3450–3030 cm^{-1}. The band of the C=N-group appears at between 1615 and 1585 cm^{-1} and, depending on other aryl substituents, mostly at somewhat lower frequency that the aliphatic oximes (1695–1585 cm^{-1}). Because the aromatic ring vibrations occur in the same region, the C=N-band is mostly not clearly recognizable.

The N–O-stretching vibrational band lies in the region of 1055 cm^{-1} and 870 cm^{-1} and is often split in several maxima (rotational isomers).

In addition, the C=N-band of Schiff's bases is mostly overlapped from the aromatic bands and therefore difficult to assign.

The azo group shows only weak bands due to induced change of dipole moment. Only *para*-substituted azobenzenes give strong absorptions near 1370 and 1150 cm^{-1}. Because very many bands generally occur for this compound type at between 1615 and 1000 cm^{-1}, definite conclusions are seldom possible to make.

The description of the aromatic spectra infers that in these complicated systems, several effects are at play, the individual contributions of which are not clearly differentiable. Simple explanations can only be given when one effect predominates (Tab. 6-28).

Table 6-28. Examples for substituent effects on the band position of group vibrations (in cm^{-1}).

Resonance Effects

R	$\nu(C=O)$ in CCl_4	
–NH$_2$	1677	electron-donating
–H	1691	substituents lower
–Cl	1692	$\nu(C=O)$.
–NO$_2$	1700	

CH$_3$–CH$_2$–CH$_2$–COOH $\nu(C=O)$ in CCl_4: 1721

CH$_3$–CH=CH-COOH   1709

⬡—COOH   1704

Inductive Influences

R	$\nu(C=O)$ in CCl_4	
–NH$_2$	1689	electronegative sub-
–H	1691	stituents increase
–Cl	1692	$\nu(C=O)$.
–NO$_2$	1701	

Steric Interactions

R$_1$	R$_2$	R$_3$	$\nu_{as}(NO_2)$	$\nu_s(NO_2)$ in CHCl$_3$ solution
H	H	H	1527	1351
H	H	I	1531	1352
I	H	H	1531	1352
I	I	I	1541	1368

6.4.15 Heterocycles

In order to be able to interpret heterocycles, at best one compares the spectra of similarly built compounds with each other [23, 24]. Among the individual aspects of interpretation, finding out the basic type is first important. Then, one orients oneself with respect to the substitution type and position of the substituents. Finally, one compares the spectrum under consideration with spectra of derivatives of the supposed compound.

a) *Basic type*: determining the skeleton, e.g. pyran, furan, pyrazine etc. at best occurs by methods other than IR spectroscopy (UV). Because of the low number of well-studied examples, the certain diagnosis remains limited to few substance classes.

b) *Substituents*: the nature of the substituents can often be recognized from the position of key bands. The rules specified for the phenyl type apply here. Upon first glance, the δ(CH) o.o.p. bands in the 1000–600 cm^{-1} region provide information about the position of substituents on the ring. Thus, pyridine behaves similarly to a mono-substituted aromatic ring. Decisive is the number of neighboring H-atoms. For alkyl-substituted derivatives, one can observe a likewise characteristic band pattern at 2000–1600 cm^{-1}, as we got to know for benzene derivatives (Sect. 6.4.5).

c) *Comparison with spectra of similar derivatives*: To discuss here are the band shifts primarily in relationship to the bands having a constant position. In all such cases, the IR spectral analysis can only give tips. One may not expect valid conclusions with this type of comparative interpretation.

6.4.16 Boron, Silicon, Sulfur and Phosphorus Compounds

The partially very extensive compound classes bring only few new aspects for the basic spectral interpretation, so that only the most important frequency positions are listed here. An assignment of the spectral bands generally requires a study of relevant special spectral libraries. However, then valid conclusions may likewise be made about the molecular design, as it is possible for the simple organic compounds. In the following overviews, Sections 6.4.16.1 thru 6.4.16.4, the bond type is given in the first column. The second column gives a more precise specification, and the third column cites the region in cm^{-1}, where the band assigned to the partial structure in column 2 is expected.

6.4.16.1 Boron Compounds [25, 26]

B–H		2612–2494
		2137–1543
		≈ 2240
	BH_4^+	
B–C		1185–1100
		844–770
	$B–(C_{aliphat.})_3$	1237–1142
		718–620
	$B–(C_{aromat.})_3$	1450–1075
B–N	$(-\underset{R}{B}-\underset{R'}{N}-)_3$	1517–1374

	R_2B-NR_2	1550–1330
	Hal_3B-NR_2	776–681
B–O	$B-(OC)_3$	1350–1310
	$(-B-N-)_3$	1330–1318
	OR R	
	$(-B-O-)_3$	1378–1335
	OR	
B–Hal	B–F	1497–844
	B–Cl	1086–297
	B–Br	1076–240

6.4.16.2 Silicon Compounds [6]

Si–H	R_2Si-H	2157–2117
	$-OSi-H$	2230–2120
	$\diagdown SiH_2$	965–923
		920–843
Si–C	$C-SiCH_3$	760–620
	$OSi-CH_3$	800–770
Si–N	Si–N–Si	950–830
	$Si-NH_2$	1250–1100
Si–O	R_3Si-OH	900–810
	Si–OC	1110–1000
		850–800
	Si–O–Si	1090–1030

6.4.16.3 Sulfur Compounds [6]

S–H	R–S–H	2590–2530
	R–CO–S–H	≈ 2550
S–C	$R-CH_2-SH$	730–570
	$R-CH_2-S-CH_2-R$	730–570
S–O	R–SO–OH	870–810
	R–O–SO–OR	740–720
		710–690
S=C	$R_2C=S$	1075–1030
	$(RS)_2C=S$	1080–1050

	(RO)$_2$C=S	1117–1075	
	RS(R)C=S	1210–1080	
S=O	R$_2$S=O	1060–1015	
	R–SO–OH	≈ 1100	
	(RO)$_2$S=O	1225–1195	
	R–SO$_2^{\ominus}$M$^{\oplus}$	1030 v_{as}	
		and 980 v_s	

		v_{as}	v_s
SO$_2$	R$_2$SO$_2$	1370–1290	1170–1110
	R–SO$_2$–OR	1375–1350	1185–1165
	(RO)$_2$SO$_2$	1415–1390	1200–1185
	R–SO$_2$–NR$_2$	1365–1315	1180–1150
	R–SO$_2$–Hal	1385–1375	1180–1170
	Aryl–SO$_2$–Hal	1410–1385	1205–1175
–SO$_3$	R–SO$_2$–OH	1355–1340	1165–1150
	R–SO$_3^-$M$^+$	1250–1140	1070–1030
	RO–SO$_3^{\ominus}$M$^{\oplus}$	1315–1220	1140–1050
	R–SO$_2$–SO$_2$–R	1360–1300	1150–1110

6.4.16.4 Phosphorus Compounds [27, 28]

P–H	phosphines	2320–2275
	–PH$_2$	2440–2275
		1090–1080
P–C	P–C$_{aliphat.}$	700–800
	P–C$_{aromat.}$	1130–1090
P–N	P–N	1130–930
	P–NC$_{aliphat.}$	1110–930
		770–680
	P–NC$_{aliphat.}$	≈ 930
P–O	P–OC$_{aliphat.}$	1050–970
		830–740
	P–OC$_{aromat.}$	1260–1160
	P^V–O	995–915
	P^{III}–O	875–855
	P–O–P	970–940
	R–P(=O)OH	1040–910
P=O	(C$_{aliphat.}$)$_3$P=O	≈ 1150
	(C$_{aromat.}$)$_3$P=O	≈ 1190
	P–P(OH)=O	1200–1100
P=S	P=S	700–625

6.4.17 Inorganic Compounds

The spectra of purely inorganic compounds [29] already differ so much in the total appearance from those of organic compounds, that they are immediately recognized. In particular, the CH-stretching and deformation bands present in almost all organic compounds are missing here. Only a very few but, in part, very wide bands occur. The simply designed anions of inorganic salts are theoretically well studied; one finds their position, intensity and assignment in the excellent compilations of Siebert [30] and Nakamoto [31]. The spectra of compounds with similar symmetry properties are much more similar to one another than spectra of chemically related substances with different symmetries: Na_3PO_4, Na_2HPO_4, NaH_2PO_4 have very different spectra; in contrast, they strongly resemble those of Na_2CO_3, $NaNO_3$, $NaClO_3$ and $NaBrO_3$. The cations have only a small effect on the spectral image in the mid-IR region. In the far-IR, however, they quite extensively determine the radiation absorption behavior as the result of lattice vibrations. Many isomorphic inorganic compounds thus differ from each other only below about 800 cm^{-1} (e.g. [Pt en]Cl_2 and [Pt en]Br_2 [32]). The absorption regions of important inorganic ions are listed in Table 6-29.

6.5   Causes of Band Shifts,
   Effects on the Spectrum

Up to now, spectra have been described from the viewpoint of the individual substance classes. Thereby, we have always met effects, which influence the spectral image. We will again briefly reiterate the causes of band shifts that were randomly mentioned as yet [33].

6.5.1   Mass Effect

Vibrating molecules react very sensitively to the mass changes of the involved atoms; the exchange of H by D has the greatest effect:

$H_2C=CH_2$ $\nu(C=C)$: 1623 cm^{-1} $D_2C=CD_2$ $\nu(C=C)$[1]: 1515 cm^{-1},
     $\nu(C-H)$: 3019 cm^{-1} $\nu(C-D)$: 2251 cm^{-1}

---

[1] respectively from the Raman spectrum (see Sect. 8.3)

Table 6-29. Absorption regions of important inorganic anions and cations (band positions in cm^{-1}).

Substance	Absorption Region			
AsO_4^{3-}	≈ 800			
BrO_3^-	810–790			
CO_3^{2-}	1450–1410	880–800		
HCO_3^-	1420–1400	1000–990	840–830	705–695
ClO_3^-	980–930			
ClO_4^-	1140–1060			
CrO_4^-	950–800			
$Cr_2O_4^{2-}$	950–900			
IO_3^-	800–700			
MnO_4^-	920–890	850–840		
NH_4^+	3335–3030	1485–1390		
N_3^-	2170–2080	1375–1175		
NO_2^-	1400–1300	1250–1230	840–800	
NO_3^-	1410–1340	860–800		
NO_2^+	1410–1370			
NO^+	2370–2230			
PO_4^{3-}	1100– 950			
$S_2O_3^{2-}$	1660–1620	1000–990		
SO_4^{2-}	1130–1080	680–610		
HSO_4^-	1180–1160	1080–1000	880–840	
SO_3^{2-}	≈ 1100			
SiF_6^{-2}	≈ 725			
SiO_4^{2-}	1100–900			

The [13]C- and ^{15}N-substitution of the normally found ^{12}C and ^{14}N-atoms inherently has a considerably lesser influence. However, [50]Cr/[53]Cr-isotopes still give measurable band shifts of the corresponding metal-ligand frequencies [34]. Because of its importance, the problem of vibrational coupling is again discussed using the case of the angular antisymmetric trimass model (according to Kohlrausch [35]). Example molecules of this are the ethyl derivatives, when we consider the methyl- or methylene group to be pure points of mass:

$$CH_3 \longrightarrow CH_2 \longrightarrow X$$

$$m_1 = 15 \quad m_2 = 14 \quad m_3$$

Figure 6-65 illustrates the fundamental vibrations of the model as a function of the mass m_3. As agreed upon and seen in Sect. 2.2.2.6, the frequencies are denoted with v_1, v_2... From the values of Fig. 6-65 and 6-66, one recognizes that v_2 is a deformation vibration for all masses, m_3. At least 88% of the vibrational energy are explainable via the corresponding deformation force constant. We would label the vibration with $\delta(C-C-X)$.

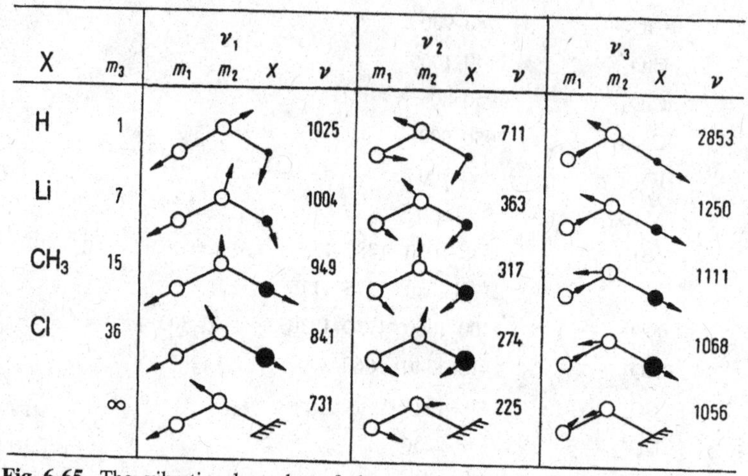

X	m_3	v_1				v_2				v_3			
		m_1	m_2	X	v	m_1	m_2	X	v	m_1	m_2	X	v
H	1				1025				711				2853
Li	7				1004				363				1250
CH$_3$	15				949				317				1111
Cl	36				841				274				1068
	∞				731				225				1056

Fig. 6-65. The vibrational modes of the angular model C–C–X as a function of the mass X (according to Kohlrausch [35]).

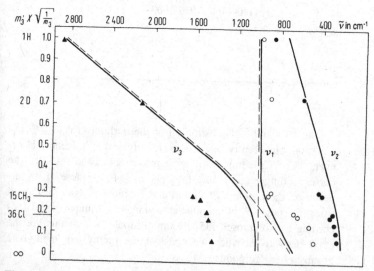

Fig. 6-66. Frequency course of the absorption bands of a non-linear trimass model $m_1-m_2-m_3$. Theoretical course (solid curves) and observed values ($m_3 = 1$ for hydrogen and m_3, corresponding to the varied substituents). See text for further explanation.

The conditions are different for v_1 and v_3. For heavy masses m_3, the vibration v_1 is a $v(C–X)$-motion yielding the greatest energy contribution.

For a smaller mass m_3, v_1 is a pure $(C–C)$-vibration. The case for v_3 is reversed. When the masses m_1 and m_3 are approximately equal, both oscillators C–C and C–X bind analogous amounts of energy. Thus, the characteristic properties of $v(C–C)$ and $v(C–X)$ get lost, and the term "group vibration" becomes senseless because of the coupling of the individual molecular motions.

Because for $m_1 = 15$ (CH$_3$) and $m_2 = 14$ (CH$_2$), $m_3 = 17$ (–OH) or $m_3 = 16$ (–NH$_2$) have similar masses, group frequencies for $v(C–O)$ or $v(C–N)$ cannot be set for ethanol and ethylamine (and their homologues). The actually observed vibrational frequencies of the ethyl derivatives in Fig. 6-66 follow the calculated frequency course only in the first approximation. The deviations are caused by model simplifications and other influences as the mass effect.

6.5.2 Steric Interactions

Mutual effects of this type can be presumed, when an expected frequency pattern does not follow a continuous path. Examples for this are the frequency positions of the NO$_2$-bands in *p*- and *o*-nitrochlorobenzene (in CHCl$_3$):

$$
\begin{array}{l}
NO_2 \\
CH_2 \\
CH_2 \\
CH_2 \\
CH_2 \\
CH_3
\end{array}
$$

$v_s(NO_2)$:	1553 cm^{-1}	1522 cm^{-1}	1540 cm^{-1}
$v_s(NO_2)$:	1345 cm^{-1}	1345 cm^{-1}	1358 cm^{-1}

Due to the voluminous Cl-substituent, the NO$_2$-group in *o*-nitrochlorobenzene is deflected out of the plane of the benzene ring, thereby suppressing the resonance effect of the phenyl ring. Similar effects are often found for 1,2-di-substituted compounds.

Another case is the carbonyl wavenumber in cyclic amino ketones [36]. The nucleophilic effect of the alkaline N-atom on the carbonyl group can only be realized for certain *n*-values. The value of *n* for the interaction shown on the side depends on geometric prerequisites (see Tab. 6-30).

$$(CH_2)_n \overset{\overset{\displaystyle R}{|}}{\underset{\underset{\displaystyle O}{\parallel}}{N}} (CH_2)_n$$

$$\left\{ \overset{\overset{\displaystyle H}{|}}{-N} \rightarrow \overset{|}{\underset{|}{C}} = O \longleftrightarrow \overset{\overset{\displaystyle H}{|}}{-N^{\oplus}} \overset{|}{\underset{|}{C}} - \overline{\underline{O}}^{\ominus} \right\}$$

Table 6-30. Effect of the ring size on the intramolecular interaction
$[\overset{|}{\underset{|}{>}}N| \rightarrow C=O]$ (*m*, number of ring atoms).

$R-N\overset{(CH_2)_n - C = O}{\underset{(CH_2)_n - CHOH}{<}}$			$v(C=O)$ in CCl$_4$	Structural suggestion for complete electron transfer
R	*n*	m		
C$_6$H$_5$	2	7	1701	
CH$_3$	3	9	1666	
C$_2$H$_5$	4	11	1705	
C$_2$H$_5$	5	13	1708	
C$_2$H$_5$	6	15	1709	

Ring-tension effects were described in Sect. 6.4.3, and the causes were attributed to coupling occurrences.

6.5.3 Resonance Effects

Interactions of the π-electrons can be extensively described by resonance structures. A discussion of such formulas according to the number and energy demand gives important information about the molecular geometry and binding influences in π-systems and, hence, about vibrational behavior. The individual examples presented during the course of the spectral interpretation ought to be summarized by a case from the aromatic compounds [37] (Tab. 6-31).

The $v(C=O)$-stretching vibrational frequencies follow the linear energy relationship according to Hammett [38] (see Sect. 6.4.12.3).

Table 6-31. Example for the effect of resonance on the band position of the carbonyl group.

	-X	-NH$_2$	-H	-Cl	-NO$_2$
	$v(C=O)$ (in CCl$_4$)	1667	1691	1692	1700
		→ electron affinity of X increases →			

6.5.4 Inductive Effects

Many bonds within a molecule have polar character. Heteroatoms affect the dipolar character of neighboring bonds and, thus, the frequency and intensity of IR absorption. This so-called "inductive effect" acts only along the chemical bonds depending on the electronegativity of the substituents or the substituent groups and on the molecular geometry. A simple case is represented by the substituted acetonitriles:

	$CH_3–CN$	$Cl–CH_2CN$	$F–CH_2–CN$
$\nu(C{\equiv}N)$ in cm^{-1}	2255	2259	2266

Inductive effects also affect the position of deformation vibrational bands due to their influence on the hybridization of neighboring bands. The case of the $\delta(CH_2)$-rocking frequency for halogenated methylene derivatives $X–CH_2–Y$ ought to illustrate this [39] (also see Tab. 6-32). Other examples depicting the influence of the inductive effect are presented in Tab. 6-33 [40].

Table 6-32. $\delta(CH_2)$-rocking wavenumbers of halogenated methylene derivatives $X–CH_2–Y$.

X	Y			
	F	Cl	Br	I
F	1176	1004	852	
Cl		940	810	801
Br				754
I				714

Table 6-33. Influence of the inductive effect on the carbonyl stretching vibrational frequency.

$\nu(C{=}O)$ in CHCl$_3$	1724	1747	1765	1770 cm^{-1}

Here, the incorporation of further nitrogen atoms in the ring increasingly draws the electrons. Thus, the negative charge on the oxygen atom is inductively shifted towards the carbon, thereby increasing the double-bond character of the C=O-bond and, as a result, shifting the absorption into the short-wave region.

6.5.5   Dipole-Dipole Interactions (Field Effect)

In Section 6.5.4, frequency shifts were attributed to inductive effects that act along the bond. Besides this, there are also electrical interactions that are transferred through space. Such effects are seen, for example, for *a*-chlorocarboxylic acid chlorides. They show two carbonyl bands in the spectrum, which can be attributed to the various conformational isomers:

I (*cis*)   II (*gauche*)   III

For the molecular structures I and II, the electrical effect of the C–C-dipole acting through space has a different effect on the C=O-group.

For structure II, one finds the position of the C=O-band relative to the methyl derivative II to be somewhat shifted towards shorter wavelengths by the inductive effect of the *gauche* *a*-Cl atom. With regard to structure I, the field effect of the *cis* *a*-Cl atom also comes into play (Tab. 6-34). Bellamy gives an explanation for this [41]. A positive charge is induced on the oxygen atom in the slightly polarizable C=O-bond due to the spatial proximity of the electronegative chlorine atom (Fig. 6-67). The same applies also conversely. Thus, the bonds lose a little of their polarity and the vibrational frequencies increase [37].

The C=O-absorption in acetophenone lies at 1691 cm^{-1} (Tab. 6-34). The first *a*-Cl atom shifts it to 1694 cm^{-1} (inductive effect). At the same time, a new high-frequency band appears at 1714 cm^{-1}. The introduction of a second Cl-atom has almost no effect on v(C=O), but does affect the intensity ratio of both bands. The higher-frequency becomes more intense (statistical effect). Only with the third Cl-atom does the inductively shifted C=O-band at 1693 cm^{-1} disappear, because a *cis*-(H–C–C=O)-conformation can no longer occur.

The intensity ratio of both C=O-bands depends on the temperature as well as on the solvent [43]. The studies were derived from the behavior of *a*-chloro-2-ketosteroids. Axially bound chlorine has no effect on the frequency of the v(C=O)-vibration. For *a*-chlorocyclohexanone, other authors [42, 44] base the frequency-shifting effect on Fermi resonance.

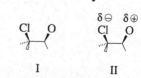

$\delta\ominus$ $\delta\oplus$

I   II

Fig. 6-67. Charge distribution for Structure I without any influence of the field effect. In Structure II, the field effect diminishes the polar character of the C=O and of the C–Cl bond, when the double-bond character of the C=O bond increases.

Table 6-34. Influence of the field effect on the position of the ν(C=O)-band for α-chlorinated acetophenones; data in cm^{-1}, solvent is CCl$_4$ [42].

	cis	gauche
⟨C₆H₅⟩–C(=O)–CH₃		1691
⟨C₆H₅⟩–C(=O)–CH₂–Cl	1714	1694*
⟨C₆H₅⟩–C(=O)–CHCl₂	1716*	1693
⟨C₆H₅⟩–C(=O)–CCl₃	1717	

* For these conformations, respectively only one mirror-symmetric structure exists relative to the CCO-plane.

6.5.6 Intermolecular Interactions

Intermolecular interactions do not let themselves be easily classified into defined categories, because there are continuous transitions from the gaseous to the solid state. Consequently, these effects are manifested throughout the spectrum [45]. Nevertheless, a certain classification could be derived. The regions, however, vary extensively.

Solvent Effects
Solvent-related band shifts are small, insofar as no specific molecular interactions occur. The best-known relationship between the solvent shift and the dielectric constant of the solvent originated from Kirkwood, Magat and Bauer. They give the following relationship between the frequencies in the gaseous state (ν_{gas}) and in solution (ν_{sol}) with the dielectric constant D (KMB-relationship [46], abbreviated after the authors' names):

$$\frac{\nu_{gas} - \nu_{sol}}{\nu_{gas}} = \frac{k(D - 1)}{(2D + 1)} \tag{6.1}$$

The term k is a proportionality factor. The equation applies only for isotropic and non-polar substances and mixtures. Already benzene doesn't apply. For solvent relationships that do not obey an equation, association phenomena play an essential role.

For substance class-specific solvent interactions, measuring the relative band shifts leads to more satisfactory relationships than using the KMB-method [47, 48]. Also for studying Fermi resonance splittings, one can apply the solvent dependency for the bands concerned: through the solvent-related shift of the absorption maximum of one of the bands involved in the Fermi resonance, one can eradicate the coincidence of the overtone with the fundamental vibration of such a resonance and thus clarify this cause for the band doubling [49].

The temperature dependency of solvent effects was investigated for substances with functional groups such as OH, NH_2, and C=O. Temperature changes hereby noticeably shift the equilibrium of rotational isomers [50]. Studies on $v(OH)$-bands are typically popular. Another study regarding solvent dependency of $v(OH)$-vibrations including the band shape analysis ought to be cited [51]. This research group measured the IR spectra of o-halogenated phenols and derivatives in which intramolecular hydrogen bonds occur. The band shifts could be correlated with the electro-optical properties of the solvent via the so-called Buckingham relationship. In general, studies on band shapes provide conclusions about the interaction of the molecules with their surroundings and about the time-related vibrational behavior of a molecule ensemble (also see [52]).

Complexes

Complexes may form when intermolecular interactions exist. Weak, but noticeable intermolecular attractive forces are the rule. Mostly, complexes exist with the solvent. The solubility increases through the complex formation. However, the solubility limit is rapidly attained upon attachment of the dissolved molecules onto each other as exemplified by the phthalocyanines.

Hydrogen Bonds

A special case of intermolecular interactions is the dimerization of organic carbonic acids. As we have seen in Section 6.4.9, this can go so far that such stable complexes have to be treated as a whole. However, the bonds directly involved on the H-bond formations are always affected so that one can recognize them from the shift of the O–H- and of the C–O-bands towards lower frequencies together with the increased half-width of the deformation bands.

Solids

Intermolecular effects continuously lead away from the ideal solution via complexes and H-bonds and finally result in the solid phase. Amides are a good example of this. In the solid state, their spectra are highly dependent on the pretreatment of the samples. In the crystalline state, the lattice vibrations in the long-wave region additionally have to be considered. Lattice vibrations are the motions that the molecules conduct as a whole against one another, namely, so-called external vibrations as opposed to internal vibrations having been only considered yet. For theoretically handling this, one must account for the spatial symmetry of the molecule in the elementary cell of the crystalline structure [53, 54].

Equilibria (Dynamic Effects)

When a substance exists in equilibrium with another substance, then the IR spectrum is a composite of the individual spectra. The vibrational processes are so fast that all molecular states have to be simultaneously analyzed. This especially applies to conformers. The intramolecular vibrations are much more frequent than the conversion of the various conformers into each other:

$$\nu(\text{C--Cl}): \quad 800 \text{ cm}^{-1} \quad \text{corresponding} \quad 0.8 \cdot 10^{13} \text{ s}^{-1}$$
$$\nu(\text{C--H}): \quad 3000 \text{ cm}^{-1} \quad \text{corresponding} \quad 3 \cdot 10^{13} \text{ s}^{-1}$$

This means that the C–Cl-dipole vibrates about 10^9-times before the *cis*-conformation converts to the *trans*-conformation. Because a molecule inherently has to vibrate at least once (otherwise it would be dead!), always all possible molecular conformations are observed simultaneously in the IR spectrum. In contrast, NMR spectroscopy is a slow method. With it, one can no longer register dynamic processes of over 10^3 s^{-1}; an "averaged" spectrum of the components results. Only in one case is it possible to interpret line widening in the IR region analogous to NMR spectroscopy: the transfer of protons between various H-acceptors. Hence, a part of the continuous IR background absorption is explainable.

An interesting study on the lifespan of conformations was presented via a profile analysis of the OH-stretching vibrational bands of phenyl-substituted alcohols such as 2-phenylethanol [56]. By using hydrocarbons as solvent, one can observe the stretching vibration of the OH-group in the *gauche*-position besides the *trans*-conformation as a separate band. The various structures are illustrated in Fig. 6-68. The examination of the overtone spectra in the near-IR region produces an even better separation of these bands, with which a lifespan of at least 3 ps

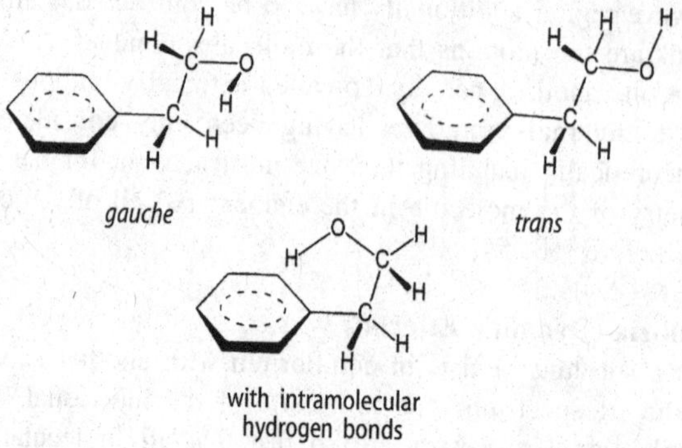

Fig. 6-68. Various molecular conformations of 2-phenylethanol.

can be estimated for the cited structures. One should note, however, that in the fundamental vibrational spectrum, the ν(OH)-vibration of the conformation with intramolecular hydrogen bonds to the aromatic π-electron system is clearly shifted by ca. 30 cm^{-1} to lower wavenumbers.

6.6 Spectral Interpretation as a Multidimensional Task

In Section 6.4, we were acquainted with spectra from the viewpoint of the chemical design of the molecule, i.e. we knew the substance and assigned the bands to certain groups in the molecule. Reading the spectrum in reverse, namely to figure out the unknown substance via the spectrum, is no longer so easy to do. Nevertheless, by cleverly applying all interpretation aids, one can make considerable headway when the problem is attacked from different sides.

Every substance has its own history: synthesis, isolation route, origin, application area, cleaning operations and many more aspects give initial tips about it. Similarly, every substance has its own elementary composition and physical properties. One should always start interpreting the spectrum considering these things. The more preliminary information the analyst has available, the clearer are his/her conclusions.

6.6.1   Band-Structure Correlations

For all chemical compound classes for which characteristic bands were found, in part very comprehensive tables of their absorption regions have been compiled [7, 8, 11, 57–62]. They give an initial indication about where bands for certain compounds will occur in the spectrum, or inversely, which functional groups come into question for a certain band in the spectrum. Such tables, however, are mostly complicated to handle, the intensity ratios are only roughly cited. Consequently, it is obvious that their interpretation is varied and, thus, their conclusion vague.

Not only are the visible bands in the spectrum important, but also the statement that an absorption is missing is just as important, especially when it concerns control bands. For such questions, one can refer to tables as was already mentioned. Detailed information, particularly in special cases, are found in special works such as those by Colthup-Daly-Wiberley [2], Szymansky [3] and Bellamy [4], in special literature articles or through spectra of similar compounds (see Chap. 9 regarding this).

After one has narrowed down the possibilities to particular structures, one then closely examines the expected spectra or band positions. Tables no longer suffice for this stage. The aforementioned special works are quite helpful but particularly suitable are the books edited by Hediger: the band tables [6] and the IRSCOT data cards [15, 63].

Since a few years, there are computer programs available for interpreting IR spectra. A commercially available program package is, for example, the IR MENTOR of the Sadtler/Bio-Rad company [64]. It contains an online database of characteristic group frequency regions and approximate band intensities, mainly taken from the works [2, 4]. The ca. 500 band assignments correspond to ca. 170 functional groups, which can be classified into 24 general chemical classes. Correlation and survey functions allow a fast assignment of importable digitized spectra.

Another IR spectroscopy training program for spectrum-structure correlation is IRTRAINS that can be used on PCs [65]. It lends knowledge for interpreting spectra based on selected FT-IR spectra, which are representative for the fields of organic and inorganic chemistry. It can be used for structure elucidation founded on the databases and search algorithms incorporated in the program. The new possibilities, particularly with expert systems including coupling with other types of spectroscopy, will still be handled further in Chapter 9.

6.6.2   Spectral Comparison

If only a few structures are under consideration, then one can search in special collections for reference spectra for these or similar compounds. The Sadtler catalogue and the DMS-records as well as a few other collections are suitable for this (see Chap. 9). To enable an easier review for spectral interpretation, a few publishing houses have published much abbreviated compilations of the large collections [66–68]. This is an attempt to prepare an overview about common characteristics of spectra within one substance class by using a limited number of spectra.

Because of application of spectrometer computers and the fact that today's spectra are available in digitized form, the spectral library search and the computerized comparison of spectra almost belong to the basic equipment of an IR spectrometer workplace. However, a cost factor is indeed the commercial spectral libraries that will still be discussed.

If no final indications for structure elucidation have been found, then the chemical conversion into known derivatives, the spectra of which exist in collections, is often a further help. Among these derivatives are especially decomposition products, since their spectra are easier to interpret and are often known.

It is seldom the case that structure elucidation is performable only with IR spectroscopy. The conclusions drawn from other methods are important [69]. They actually belong to preliminary information. For this reason, IR spectroscopy is applied mostly at the end of structure elucidation [70]. But also here, computer-aided systems exist which also incorporate, for example, additional information besides IR spectra such as MS- and NMR spectra, thereby helping to validate the structure elucidation (for further details, also see Chap. 9).

6.6.3   Examples for Spectral Interpretation

Two examples are presented to show how spectra can be interpreted:

Example 1: Figure 6-69 shows the IR spectrum of a by-product isolated from technical-grade adipic acid. In the meantime, we have gotten to know the spectra well enough that we can see, on hand of the C–H-bands at $\approx 2900 \text{ cm}^{-1}$ and 1460 cm^{-1}, that the unknown substance deals with an organic compound. Because we know from the previous history that the substance is a by-product of adipic acid production, we make sure that the main substance is no longer present as a disturbing component

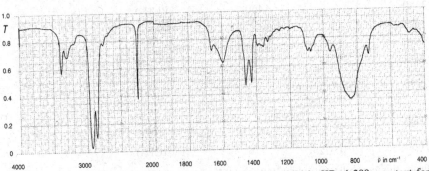

Fig. 6-69. IR spectrum of an unknown solid in KBr 1:300; see text for discussion.

(acid: 1705/3000/900 cm^{-1}; salt: 1500/1400 cm^{-1}; acid derivative: ν(C=O) 1800–1650 cm^{-1}). We try to roughly group the substance into the aliphatic-olefin-aromatic (or combination thereof) by finding the corresponding control bands. However, because the spectral region below 1000 cm^{-1} is free of intense bands with a narrow half-width, only the aliphatic basic type comes into consideration for our substance.

In the next step, we have to find out the functional groups. At first, we notice the sharp absorption at 2245 cm^{-1}. By using published tables, we can assign this band to a di-substituted acetylene, an alkyl nitrile and an aminonitrile. The band is too intense for it to be a dialkyl alkyne, so that we must presume either an aliphatic bound nitrile group or a dialkyl aminonitrile. A final decision about which assignment is correct cannot be made in this stage yet.

The further interpretation is based on the fact that a few very wide, little structured bands occur in the spectrum. One supposes that they are caused by strongly polar (preferably inorganic) compounds; polymeric substances, too, can give such spectra or systems with pronounced hydrogen bonds. We have already excluded inorganic compounds; the compound also cannot be polymeric because of its behavior during sample preparation (volatility etc.). Therefore, we have to narrow it down to alcohols and amines, since this substance class forms the strongest carbonyl-free H-bond systems.

The high-frequency double band at 3380 cm^{-1} and 3300 cm^{-1} definitely indicates a primary amine. We validate this assignment through the corresponding NH$_2$-deformation bands (1600/ 840 cm^{-1}).

In order to still find other functional groups, we have to assign the yet diagnosed partial structures to the IR bands for control purposes:

3380	$\left.\right\}\nu_{as,s}(NH_2)$	1460	$\delta(CH_2)$
3300		1428	(?)
2930	$\left.\right\}\nu_{as,s}(CH_2)$	$\left.\begin{array}{c}1190\\1175\end{array}\right\}$	$\nu(C-N)(?)$
2860			
2245	$\nu(-C\equiv N)$ or		
	$\nu(N-C\equiv N)$	960	(?)
1662	2×840 (?)	840	$\delta(NH_2)$
1660	$\delta(NH_2)$	728	$\delta(CH_2)$

From this compilation, which is the result of knowledge, supposition, trial, and assurance (it is only a work base), we see, for instance, that no CH_3-group appears which we simply did not consider upon the first rough classification. Furthermore, we observe on the occurrence of the CH_2-rocking band ($728\ cm^{-1}$) that a chain $(-CH_2)_{\geq4}$ must be present. The assignments of $1662\ cm^{-1}$, $1190/1175\ cm^{-1}$ and $960\ cm^{-1}$ are uncertain. Because they concern low-intensity bands, however, we may disregard them for the search for further functional groups. We cannot do anything at first with the moderately strong absorption at $1428\ cm^{-1}$, because it cannot be linked with a partial structure. However, it will important to us later during the spectral comparison. Because we could clarify most bands, now we have to connect the individual findings for building a structural formula. This inevitably leads to:

$$H_2N-(CH_2)_{\geq4}-C\equiv N \quad \text{and} \quad \begin{array}{c}H_2N-(CH_2)_n\\[4pt]\searrow N-C\equiv N\\[4pt]H_2N-(CH_2)_m\nearrow\end{array}$$

$$\text{I} \qquad\qquad\qquad \begin{array}{c}n=0,1\dots\\m\geqslant4\end{array} \quad \text{II}$$

Because structure II is the less probable formula due to the origin of the sample, we concentrate on the suggested structure I and apply the reference spectra of the corresponding ω-aminonitrile $H_2N-(CH_2)_n-CN$ with $n=4,5,6\dots$. In this way, and of course via the relative molecular mass, CH-analysis, NMR, MS, retention index etc., one finds that 6-aminocapronitrile is present.

Example 2: Figure 6-70 shows the IR spectrum of a solid from an illegibly labeled bottle. We suppose that it must be an organic compound, because bands occur in the C–H-stretching vibrational region at $3000\ cm^{-1}$. At the same time, bands are recognized "above" and "below" the $3000\ cm^{-1}$ mark, which belong to vinyl (=C–H) and aliphatic (–C–H) hydrogen.

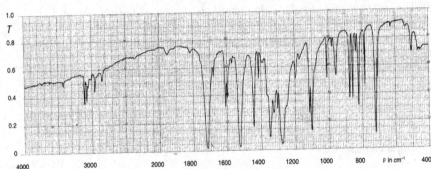

Fig. 6-70. IR spectrum of an unknown solid in KBr 1:200; see text for discussion.

In order to further narrow down the design of the C-skeleton, we try to find out whether the unsaturated structural part is olefinic or aromatic, or whether both groups are present simultaneously. The decision is made immediately in favor of the aromatic, because several bands lie at 1600 cm^{-1}, 1440 cm^{-1}, and between 910 and 660 cm^{-1} that may originate from an aryl nucleus. An olefinic group is not present, although the weak band at 1680 cm^{-1} could suggest this: the corresponding δ(CH)-o.o.p.-vibration would lead to a corresponding band below 1000 cm^{-1}. The 958 cm^{-1} band is not intense enough for a *trans*-olefin, and the intense 720 cm^{-1} band is too sharp to be a *cis*-olefin. Other olefins do not come into question because of the band position. Furthermore, we conclude from the intensity of the alkyl-C–H-stretching vibrational bands at 2955 and 2850 cm^{-1}, and relative to the aryl-C–H-bands between 3108 and 3010 cm^{-1}, that the aromatic structural fraction is the essential building block of the CH-skeleton and not the alkyl part. Moreover, we find only two bands below 3000 cm^{-1}, so that only $-CH_2$- or only $-CH_3$-groups can be present. A decision about this is still impossible due to the position (2955 and 2850 cm^{-1}), since we have at least one shoulder at about 1460 cm^{-1}.

After being sure about the nature of the C-skeleton, i.e. it deals with an aromatic compound having only few saturated C-atoms, we turn to the functional groups. By all means are they shown to be intense bands in the spectrum.

At first, we notice a carbonyl band at 1720 cm^{-1}, the nature of which, whether it be a ketone, aldehyde, acid, acid chloride, ester, anhydride or amide, must be decided from the control bands. According to the presentations in Sect. 6.4.11, we narrow down the strong band at 1274 cm^{-1} to possibly be an ester. Nonetheless, we still have to account for the possibility of it being a combination of aliphatic ketone/diaryl ether. Phenols are excluded because of the missing OH-absorption.

The next intense band at $1523\ cm^{-1}$ is so characteristic through its position, intensity and shape that we assign it, possibly using tables, to the $-NO_2$-group and verify this supposition by the similarly intense $1344\ cm^{-1}$ band. The next intense band at $720\ cm^{-1}$ is presumed to belong to an aryl-CH-o.o.p.-band, because it cannot stem from an aliphatic chain already for intensity reasons relative to $v(C–H)$. At first, no explanation dawns on us for the band at $1103\ cm^{-1}$, so that we attempt to design a molecule from the knowledge at hand. We have the following structural elements:

$$
\text{Aryl ring} \quad -\overset{\overset{\displaystyle O}{\|}}{C}-O- \quad \text{or} \quad -CH_2-\overset{\overset{\displaystyle O}{\|}}{C}-CH_2- \quad \text{and} \quad Ar-O-Ar
$$

$$
-CH_2- \quad \text{or} \quad -CH_3 \quad \text{and} \quad -NO_2
$$

Based on the band position, the NO_2- must sit directly on the aryl nucleus. The same applies for the carbonyl group when it belongs to an ester. The individual findings have solidified to the following possibilities:

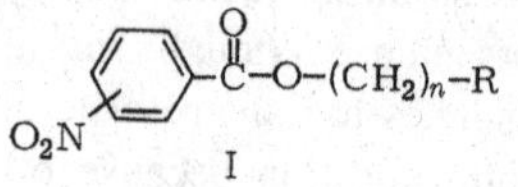

with R = H for $n=1$
$\neq$ H for $n >1$, because
otherwise a $-CH_3$-group
would appear.

R = $-O–C(O)–C_6H_4–NO_2$ for $n \geq 1$; the $-NO_2$-group would have to be on the same position as in structure I, because otherwise two $v(C=O)$-bands would be present.

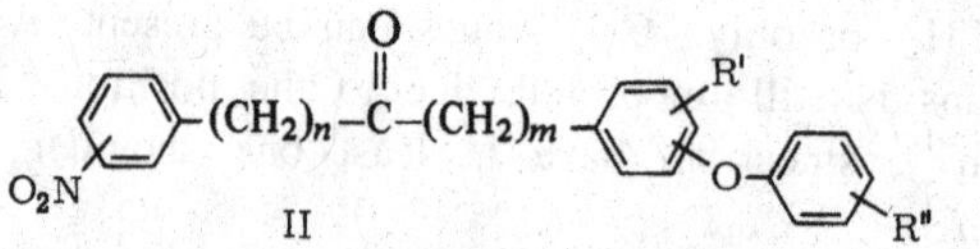

R′ = H, $-NO_2$ in any position, (II) because there is no effect anymore on $v(C=O)$; $n+m \geq 2$ (low).
R″ = H or one of the groups being discussed.

It makes sense now to first establish the alkyl group, whether it be $-CH_2-$ or $-CH_3$. By using the literature, we search for band positions (data in cm^{-1}) for the corresponding partial structures and find:

$$\overset{O}{\overset{\|}{-C}}-O-CH_2-$$

2980
1475

$$\overset{O}{\overset{\|}{-C}}-O-CH_3$$

2960
1440

$$\overset{\diagdown}{\diagup}C-CH_2-$$

2940–2915
2865–2800
1455–1435

$$-CH_2-\overset{O}{\overset{\|}{C}}-CH_2-$$

3000–2900
1440–1405

The 2960 and 1440 cm^{-1} combination is certain enough in order to preliminarily set aside the suggested structure II and to more closely examine methyl nitrobenzoate. We find from the literature that benzoates strongly absorb at frequencies between 1300 and 1250 cm^{-1} and that the absorption maximum often lies near 1270 cm^{-1}. Furthermore, benzoates and phthalates have two strong bands between 1150 and 1100 cm^{-1}. This information helps us in assigning the yet non-interpreted bands in this region, making structure II quite unlikely. Thus, we are dealing with a methyl nitrobenzoate. Still unresolved, however, is the type of substitution. Because the nitro group as well as the ester group disturb the substitution pattern bands of the CH-o.o.p.-bands, we cannot support our assumptions on the long-wave absorptions. We also cannot fall back upon the weak bands at 1952 and 1818 cm^{-1}, because their position also depends on the o.o.p.-bands. Moreover, a large partial region is masked by the carbonyl group.

It is now worthwhile to again make a band assignment with the supposed structure, analogous to the previous example. Here, a possibly extensive assignment should be attempted in the region from 4000–1400 cm^{-1} and from 1000 cm^{-1}. In the fingerprint region, the most intense bands should find an explanation. In our case, the following preliminary assignment results:

3420

2×1720

1720 $\nu(C=O)$

$\left.\begin{array}{l}1322\\ \text{or}\\ 1302\end{array}\right\}$ $\nu(C-N)$

$\left.\begin{array}{l}3108\\3075\\3050\\3020\\3010\end{array}\right\}$

$\underset{\text{(aromatic)}}{(=\overset{|}{C}-H)}$

$\left.\begin{array}{l}1609\\1598\end{array}\right\}$ $\nu(C=C)$ of the aryl ring, doubled by π-resonance

1274 $\overset{O}{\overset{\|}{\nu(C}}-O)$

		1523	$\nu_{as}(NO_2)_{aryl}$		
2955	$\nu_{as}(CH)$			1117 ⎫	
2850	$\nu_{as}(CH)$	1442	$\delta(CH_3)$	1103 ⎭	$\nu(\overset{\displaystyle O}{\overset{\displaystyle \|}{C-O-C}})$
		1412 ⎫		1112	$\delta(=CH)_{i.p.}$
1952 ⎫	combination and	1390 ⎭ (?)			
1818 ⎭	overtones of the aryl nucleus				
		1344	$\nu_s(NO_2)_{aryl}$	876 ⎫	
				858	
				838	
				821 ⎬	$\delta(=CH)_{o.o.p.}$
				784	
				720	
				710 ⎭	

The fine details become apparent in this work: the 1442 cm^{-1} band was first assigned to the $\delta(CH_3)$ vibrational mode; however, both 1500 cm^{-1} aryl bands are missing. Then again, the 1442 cm^{-1} band absorption is relatively intense, so that one may assume a random coincidence of $\delta(CH_3)$ and $\nu(aryl-C=C)$. The second component of the 1500 cm^{-1} aryl duplet would then be the shoulder on the lower-frequency side of the 1523 cm^{-1} nitro band. The $\delta(=CH)$-o.o.p. bands cannot be further decoded without reference spectra, because we already know about the disturbance by polar substituents. Nevertheless, we have reduced the number of possible compounds to the extent that we have expendable time to search for the three o-, m- and p-methyl nitrobenzoate spectra in the literature. Finally, we will find the reference spectra by one of the ways detailed in Chapter 9 and thereby verify the assignment for 4-methyl-nitrobenzoate.

6.7   Particularities and Artifacts

We have gotten to know the IR spectra closer in the previous sections. Thereby, we could always determine that the individual substance classes have very different absorption spectra. Nevertheless, how is the specificity of the spectra for very similar compounds?

Actually, the n-C_{18} and n-C_{16}-carbonic acids have very similar absorption patterns. Only with an exact comparison of both spectra in the fingerprint region at larger pathlengths are differences apparent. Such difference become increasingly smaller the longer the CH$_2$-chain is; i.e. for homologous series for long C-chains, we get typical spectral images, which differ from one another only in particularities. The same is true for polymers. This applies only so long the skeleton frequencies of the molecule remain unaffected by additional substituents.

Typical spectra, however, are more the exception than the rule, which exactly constitutes the specificity of the IR spectra. This is why a partial-structure analysis is so dangerous in contrast to UV spectroscopy. The bands of functional groups act additively, insofar as they are not mutually affected by any of the effects described in Sect. 6.5. This means that, for example, an equimolar mixture of 1-nitrohexane and 1-hexanoic acid can be distinguished from ω-nitrododecanoic acid primarily only by the missing CH_3-group absorptions. With an unfavorable composition, one cannot recognize mixtures (also see Sect. 9.4). The IR interpretation of functional groups for unknown samples, thus, does not yet need to state anything about the molecular size and the substance purity.

A large help for spectral comparisons is when one can recognize often-occurring disturbing bands of foreign substances. Section 10.1 summarizes the most important absorptions. These may include bands of solvent traces that adhere to the substance during re-crystallization. Hygroscopic substances attract water (KBr). If doubt arises about the correct assignment, one can mix the substance in question in with the sample. The corresponding bands will then appear stronger (compare with gas chromatography, peak identification by spiking). Interference bands for films reveal themselves by their regularity.

Shoulders on the flank of a band are often ignored. Despite their unobtrusiveness, they may lead to important conclusions, for instance, about vinyl-bound hydrogen, about rotational isomers, contaminants or changes in the aggregate state. On the other hand, one must check if not the Christiansen effect is simulating a shoulder. Regarding this, the selection of an unsuitable apodization function, which may produce band side maxima, is to be mentioned here. Moreover, poor spectral resolution causes shoulders. Bands that appear in the spectrum of the liquid sample as individual absorption bands sometimes split in solids. One must recall that the sample measurement techniques (measurement in transmission, with ATR or diffuse reflection) likewise affect the shape, intensity and position of bands.

In general, comparisons of the band intensity with reference spectra are helpful. One can often only differentiate between similarly designed substances of a substance class through such comparison; primarily, one can limit individual molecular properties: monool-diol-polyol; the length of the CH_2-chain in relationship to CH_3-groups is estimable; the intensity of the 1250 cm^{-1} C–O-ester band has about the same order of magnitude as the 1740 cm^{-1} ν(C=O)-ester band.

There are cases, for example in quantitative research or in assignment problems, in which the bands of the OH-group act disturbing. Through conversion into the –OD-group, analogous to NMR spectroscopy, the OH-dependent bands shift into the long-wave region. This is a simple example of derivation. Moreover, other chemical conversions can serve for verifying a spectral interpretation (hydration products, salt formation for amines, dehydrogenation to aromatic derivatives etc.). On the other hand, the spectrum can also lose specificity: the 2,4-dinitrophenylhydrazones of the isomeric C_7-ketones can only be differentiated through particularities in the spectrum, because the 2,4-dinitrophenylhydrazone-group dominates too much in the spectrum.

An important criterion is also the reproducibility of a spectrum. Spectral acquisitions of chlorides or bromides e.g. in CsI pellets are not reproducible because of halogen exchange (see Sect. 4.2.1.7). Likewise, one must recall that the type and manner of sample preparation may influence the spectra. Liquid-phase and solid-phase spectra of the same substance can so differ from each other that one occasionally would not like to assign them to the same substance (see Fig. 4-99. With respect to this, the abscissa imprecision should be mentioned here: depending on the instrument type and especially for older dispersive spectrometers, position differences of 1–5 cm^{-1} may, by all means, significantly differ from one another in the ν(C=O)-region. As was already noted, slight band shifts can be explained by different measurement techniques applied.

6.8   Spectral Calculation

Our entire interpretation work was based on empirical rules and extensive material gained from experience. For our purposes, namely structural elucidation by IR spectroscopy, there is no other way of interpreting a spectrum, unless it occurs by the relatively complicated route of normal coordinate analysis. We will only shortly graze the fundamental theory here (also see [2, 71–74]).

One considers the molecule as a system of harmonically vibrating masses, which are kept in equilibrium through spring forces. Each point of mass (atom) vibrates dependently on the motion of all other masses. Its deflection from equilibrium can be described relative to any given oriented Cartesian coordinate system (external coordinates). The calculation becomes simpler, when coordinates are defined in such a way that possibly many vibrational deflections occur parallel to the new coordinate axes.

Translation and rotation of the molecule do not apply hereby (inner coordinates are e.g. in the direction of bond lengths or angular changes). One obtains simpler calculation possibilities based on the so-called symmetry coordinates through linear combination of the internal coordinates by accounting for certain symmetry conditions. Via the classical mechanical principle about the constant sum of potential and kinetic energy, a system of linear equations results describing the relationship between known mass (atomic weight), known vibrational frequency (spectrum) and unknown force constants. The contribution of the force constants is adapted according to the method of the smallest square of deviation, so that preliminary calculation and the spectrum optimally agree with one another regarding the vibrational frequencies. The magnitude of the force constant can often be estimated from normal coordinate analyses of similar compounds. Useful here is the transferability of these effective force constants among corresponding molecular fractions. The general practical importance of this procedure for describing and pre-calculating fundamental vibrations is to be accentuated.

The atomic motions are thus dependent on one another. The atoms move in a valence force field that depends on the molecular relationships. The mathematical solution of this problem for calculating force constants is not exactly possible. In the normal coordinate analysis, one must fall back upon approximation techniques, because the number of unknown force constants is generally greater than the number of available data such as fundamental vibrational frequencies, centrifugal expansion parameters etc. This problem can be reduced in that one examines isotoped compounds, which have the same force field but other mass dependencies. Additional input is obtained with knowledge of the fundamental vibrations of these molecules. The complete determination of these vibrational frequencies can be difficult, because some fundamental vibrations are not IR-permissible for symmetry reasons or are only found in low intensity, so that the assignment of combination bands and the Raman spectrum can yield further information. An example of this procedure is found in the normal coordinate analysis for propionitrile [75].

Another point of discussion is that mostly for the wavenumbers of the fundamental vibrations, the contributions of anharmonicities or Fermi resonance effects were not taken into account so that the calculation of a harmonic force field necessarily leads to small errors. In the literature, different authors additionally apply normal frequencies of the liquid or solid phase that can show considerable shifts compared to the gaseous phase frequencies due to intermolecular interactions. Thus, for example, Fermi resonance effects in the various aggregate states act differently, thereby allowing a determination of resonance parameters in multiatomic molecules with which the undisturbed vibrational frequencies may be calculated [76]. Often simplified molecular structures are prescribed as the basis for the force field analysis, so that at least the same structural parameters exist for chemically similar molecules. This prerequisite, however, is often only approximately fulfilled.

Various force field models are used for theoretically handling the molecular vibrations. An overview on this is found in Matsuura et al. [77]; a few practical examples are discussed by Duncan [78]. Only the historically most important models for multiatomic molecules are cited here:

General Valence Force Field (GVFF)
This stance accounts for all interactions, also those of atoms not directly bound to each other. The method is successful only for small, highly symmetrical molecules.

Internal (or Modified) Valence Force Field (IVFF)
Based on internal coordinates, one considers only a certain number of interaction force constants proven to be significant within the scope of the force field analysis. This very widespread model is often called **Simplified General Valence Force Field (SGVFF)**. Redundancies in the internal coordinates and their force constants can be eliminated by using symmetry coordinates.

Local Symmetry Force Field (LSFF)
This force field model was also named **Group Coordinate Force Field (GCFF)**. The base coordinates are mainly defined here via the intramolecular groups and their local symmetry properties. Because the force field within the respective organic substructural units works within narrow limits and neighboring group effects are relatively low, the transferability between chemically similar molecules functions well. In addition, interaction constants between various groups are defined in order to obtain a fitting to the experimental data.

Modified Urey-Bradley Force Field (MUBFF)
In this model, force constants for valence and bonding angle changes, torsions and interactions also between non-bound atoms are considered. This model is named after the authors Urey and Bradley. The original model is suited less for molecules with non-localized electrons such as in π-bonds or with hydrogen groups. Thus, additional interaction constants were necessary. A typical example is the Kekulé interaction constant for benzene. Further details and extensive literature data are found in [77].

Standard computer programs for the normal coordinate analysis were described differently [79–81]. Important is that the various models, partially having very different values for the force constants, lead to comparable results for the vibrational frequencies. An important conclusion of the theoretical calculations is the potential energy distribution (PED). With this, a quantification of the potential energy of a fundamental vibration occurs on the different e.g. internal coordinates. For a "pure" C=O-stretching vibration, the potential energy contribution of the vibration would be completely ascribed to the C=O-coordinate. On the other hand, for cyclohexanone, only 70% of the potential energy of the band at 1715 cm^{-1} is bound with v(C=O), and at least four other internal coordinates change themselves additionally with this motion. This is also a reason why vibrating molecules must be considered as a whole. A critical treatment of the fundamental vibrations of cyclohexanone shows a few aspects of the theory [82].

By mean of quantum chemistry, an independent calculation of the force constants is possible with the help of an efficient computer. The molecular energy is calculated here as a function of the atomic coordinates. Via the search of the energy minimum, the equilibrium configuration of the molecule can also be calculated in this manner. This aspect is important for conformation analysis. The force constants can be determined by the two-fold differentiation of the energy according to the coordinates. Besides *ab initio* methods, which are especially applicable for small molecules, semi-empirical methods such as CNDO/2 [83] or PM3 [84] are applied. They necessitate scaling in the results because of systematic errors in the quantum chemical models; a comparison of the results for pyrazine is found in [85]. Refinements of the respective models may result in better concurrence between calculated and measured vibrational frequencies [86–88].

While the form of the vibrational spectrum is defined by band intensities, the changes in the dipole moment finally determine the intensity of the vibrational transfer. Besides quantum chemical methods, classical so-called parametric methods are available in order to ascertain changes of the molecular electrical dipole moment during the molecular vibration. Various stances are discussed in the literature, e.g. the presentation of the molecular dipole moment as the vector sum of bond moments or via atom-polar tensors (see e.g. [89]). Besides these, molecular-dynamic methods are applied, in which vibrational processes in a molecule are described via classical mechanics whereby the structure and force field are already given. For a molecular ensemble, the time-dependent vectors are calculated for the dipole moment and are the starting point for further steps via statistical methods [90, 91]. In addition, intermolecular potential functions can be considered as to simulate interactions between molecules and their effects on the spectra. Besides the calculated vibrational band intensities, particular importance is ascribed to the band assignment as well as to the discrimination between various topological and conformational structures.

On hand of the fundamental vibrations of benzene and of the phenyl group, we will get to know the advantages of a normal coordinate analysis [92]. As we have seen in Sect. 6.3.2, it is common to describe bands of an infrared or Raman spectrum as so-called group frequencies or also as characteristic frequencies [5]. This deals with vibrations that are common to certain functional groups of different molecules. Accordingly, the corresponding bands will always occur in narrow regions of the infrared or Raman spectrum. With respect to this, the phenyl group also has its own group frequencies. Figure 6-71 shows the wavenumbers of all 30 vibrations of a phenyl group (IR and Raman). For 25 vibrations, the cited values apply with ± 30 cm^{-1},

Table 6-35. Fundamental vibrations and symmetry species of benzene and phenyl groups (for the latter, wavenumber regions were calculated via various normal coordinate systems; the Raman-active vibrations of benzene are denoted by *Ra* and the IR-active by IR; also see Sect. 2.2.2.6 regarding IR- and Raman activity of the phenyl vibrations).

D_{6h}	Herzberg ν_i		Wilson ν_i	C_6H_6	$C_6H_5X_i$	Wilson ν_i	C_{2v}
a_{1g}	Ra	1	2	3074 i.p.	3050	2	a_1
	Ra	2	1	993	850–650	1	a_1
a_{2g}		3	3	1350 i.p.	1320	3	b_2
a_{2u}	IR	4	11	674 o.o.p.	750	11	b_1
b_{1u}		5	13	3057 i.p.	3050	13	a_1
		6	12	1010	1000	12	a_1
b_{2g}		7	5	990 o.o.p.	985	5	b_1
		8	4	707	700	4	b_1
b_{2u}		9	14	1309 i.p.	1275	14	b_2
		10	15	1150	1150	15	b_2
e_{1g}	Ra	11	10	847 o.o.p.	835	10a	a_2
					250–150	10b	b_1
e_{1u}	IR	12	20	3064 i.p.	1300–1060	20a	a_1
					3050	20b	b_2
	IR	13	19	1484	1500	19a	a_1
					1450	19b	b_2
	IR	14	18	1038	1075	18b	a_1
					1025	18a	b_2
e_{2g}	Ra	15	7	3057 i.p.	3050	7a	a_1
					3050	7b	b_2
	Ra	16	8	1601	1605	8a	a_1
					1585	8b	b_2
	Ra	17	9	1178	1175	9a	a_1
					1155	9b	b_2
	Ra	18	6	608	540–260	6a	a_1
					610	6b	b_2
e_{2u}		19	17	967 o.o.p.	960	17a	a_2
					900	17b	b_1
		20	16	398	410	16a	a_2
					560–415	16b	b_1

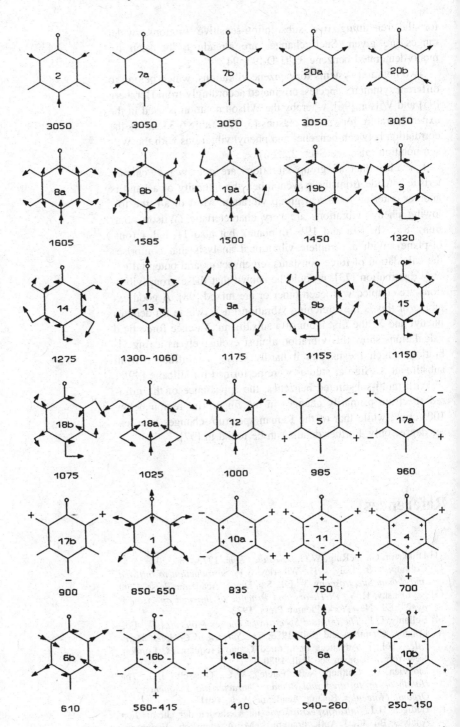

Fig. 6-71. Normal vibrations of the phenyl group (accord. to [92]); vibrational frequencies in cm^{-1}; also see Tab. 6-35 for further elucidation.

for the remaining five substitution-sensitive motions, wider ranges are given. Such changes are already to be seen for mono-deuterated benzene, C_6H_5D [93, 94].

Two notation systems for benzene vibrations, which belong to different symmetry species, originated accordingly from Herzberg [95] and Wilson [96], whereby the Wilson notation is used in the expert literature for various reasons [5, 97]. Table 6-35 shows the correlation between benzene- and phenyl vibrations with the Wilson notation (also see Sect. 2.2.2.6).

It is clear that all group vibrations are not always "characteristic". Aside from the wavenumber, the intensity of a band is another criterion for a real group vibration. Most of all, the following phenyl vibrations are very characteristic (Wilson notation): 8a, 8b, 19a and 19b (in-plane), but also 11 and 4 (out-of-plane). With a complete vibrational analysis that comprises the calculation of force constants, eigenvectors and potential energy distribution [73], it could be shown that these group vibrations are coupled with each other or are mixed [98]. A good example of this is the stretching vibration Ph-X, where Ph denotes phenyl and X, the first atom of a substituent. Valence force field calculations show this vibration almost exclusively as a ring vibration, which means that it hardly senses an influence from substituent X. Similar studies were performed for stilbene [99].

With multi-substituted benzenes, the appearance of the infrared spectrum primarily changes in the fingerprint region below 1000 cm^{-1}, while that of the Raman spectrum changes to a lesser degree only. Further details can be found in [97].

References

[1] Dijkkstra, G., DeRuig, W.G., *Z. Anal. Chem.* **1973**, 264, 204

[2] Colthup, N.B., Daly, L.H., Wiberley, S.E., *Introduction to Infrared and Raman Spectroscopy*, 3rd Ed., San Diego: Academic Press, **1990**

[3] Szymansky, H.A., *IR-Theory and Practice of Infrared Spectroscopy*, 2nd Ed., New York: Plenum Press, **1972**

[4] Bellamy, L.J., *The Infrared Spectra of Complex Molecules*, 3rd Ed., London: Chapman and Hall, **1975**; *The Spectra of Complex Molecules*, Vol. 2, Advances in Infrared Group Frequencies, London, New York: Chapman and Hall, **1980**

[5] Lin-Vien, D., Colthup, N.B., Fateley, W.G., Grasselli, J.G., *The Handbook of Infrared and Raman Characteristic Frequencies of Organic Molecules*, Boston: Academic Press, **1991**

[6] Hediger, H.J., *Infrarotspektroskopie* in: Methoden der chemischen Analyse, Bd. 11, 1. Aufl., Frankfurt a.M.: Akademische Verlagsgesellschaft, **1971**

[7] Cross, A.D., *Introduction to Practical Infrared Spectroscopy*, London: Butterworths Scientific Publ., **1960**

[8] Roeges, N.P.G., *A Guide to the Complete Interpretation of Infrared Spectra of Organic Structures*, Chichester: J. Wiley and Sons, **1994**

[9] Nakanishi, K., *Infrared Absorption Spectroscopy*, 2nd Ed., San Francisco: Holden Day, **1964**

[10] Hill, R.R., Rendell, D.A.G., *The Interpretation of Infrared Spectra – a Programmed Introduction*, London: Heyden and Son, **1975**

[11] Socrates, G.: *"Infrared Characteristic Group-Frequencies"*, 2nd Ed., J. Wiley and Sons, Chichester, **1994**

[12] Dempster, A.B., *J. Mol. Spectrosc.* **1970**, 35, 18

[13] Dempster, A.B., Uslu, H., *J. Mol. Struct.* **1974**, 21, 197

[14] Wexler, A.S., *Spectrochim. Acta* **1965**, 21, 1732

[15] Miller, R.G.J., Willis, H.A., Hediger, H.J. (eds.): *"IRSCOT-Infrared Structural Correlation Tables and Data Cards"*, London, Heyden and Son, **1969**

[16] van der Maas, J.H., Lutz, E.T.G.: *Spectrochim. Acta* **1974**, 30A, 2005; van der Maas, J.H., in: Heise, H.M., Korte, E.H., Siesler, H.W. (eds.): Proceedings of 8th International Conference on Fourier Transform Spectroscopy, *Proc. Soc. Photo-Opt. Instrum. Eng.* **1992**, 1575, 117

[17] Tichy, M., in: Raphael, D.A., Taylor, E.C., Wynberg, H. (eds.): *"Advances in Organic Chemistry"*, Vol. 5, Interscience Publ. **1965**, p 115

[18] Pimentel, G.C., McClellan, A.L.: *"The Hydrogen Bond"*, Chap. 3, Freeman, W.H. and Comp., San Francisco and London **1960**, p 67–141

[19] Geiseler, G., Seidel, H.: *"Die Wasserstoffbrückenbindung"*, 1. Aufl., Vieweg u. Sohn, Braunschweig, und Akademie-Verlag, Berlin **1977**, p 159

[20] Bellamy, L.J., Connely, B.R., Philpotts, A.R., Williams, R.L.: *Z. Elektrochem. Ber. Bunsenges. Physik. Chem.* **1960**, 64, 563

[21] Yamaguchi, M. in: Yukawa, Y. (ed.): *Handbook of Organic Structural Analysis*, Benjamin, W.A., New York, Amsterdam **1965**, p 399–410

[22] Braude, E.A., Nachod, F.C.: *"Determination of Organic Structures by Physical Methods"*, Academic Press, Vol. 2, New York **1955**, p 223

[23] Katritzky, A.R., Ambler, A.P. in: Katritzky, A.R. (ed.): *Physical Methods in Heterocyclic Chemistry*, Vol. 2, Academic Press, New York, London, **1963**, p 165

[24] Katritzky, A.R., Taylor, P.J. in: Katritzsky, A.R. (ed.): *Physical Methods in Hetrocyclic Chemistry*, Vol. 4, Academic Press, New York, London, **1971**, p 265

[25] Gerrard, W.: *"Organic Chemistry of Boron"*, Academic Press, London, New York, **1961**, p 223

[26] Meller, A.: *Organomet. Rev.* **1967**, 2, 1

[27] Dorbridge, D.E.C. in: Grayson M., Griffith, E.J. (eds.): *Topics in Phosphorous Chemistry*, Vol. 6: "The Infrared Spectra of Phosphorous Compounds", Intersci. Publ., John Wiley and Sons, New York, Sydney, Toronto, **1969**, p 235–365

[28] Thomas, L.C.: *"Interpretation of the Infrared Spectra of Organophosphorous Compounds"*, Heyden and Son, London, **1974**

[29] Nyquist, R.A., Kagel, R.O., Putzig, C.L., Leugers, M.A.: *"Handbook of Infrared and Raman Spectra of Inorganic Compounds and Organic Salts"*, Academic Press, London, **1996**

[30] Siebert, H.: *"Anwendungen der Schwingungsspektroskopie in der anorganischen Chemie"*, Springer-Verlag, Berlin, Heidelberg, New York, **1966**

[31] Nakamoto, K.: *"Infrared and Raman Spectra of Inorganic and Coordination Compounds"*, 4th Ed., J. Wiley and Sons, New York **1986**

[32] Berg, R.W., Rasmussen, K.: *Spectrochim. Acta* **1973**, 29A, 319

[33] Bellamy, L.J.: *Appl. Spectrosc.* **1979**, 33, 439

[34] Nakamoto, K.: *Angew. Chem.* **1972**, 84, 755

[35] Kohlrausch, K.W.F.: *"Ramanspektren"*, *Hand- und Jahrbuch der chemischen Physik*, Bd. 9, Abschn. VI, Akademische Verlagsgesellschaft 1943, p 227. Nachdruck. Heyden u. Son, **1973**

[36] Leonard, N.J., Oki, M.: *J. Amer. Chem. Soc.* **1954**, 76, 5708

[37] Gianturco, M. in: Freeman, S.K. (ed.): *"Interpretive Spectroscopy"*, 1st Ed., Reinhold Publ., New York, **1965**, p 95

[38] Jones, R.N., Forbes, W.F., Mueller, W.A., *Can. J. Chem.* **1957**, 35, 504

[39] Jones, R.G., Orville-Thomas, W.J., *Spectrochim. Acta* **1964**, 20, 291

[40] Lit. [37], p. 88

[41] Bellamy, L.J. et al., *J. Chem. Soc.* **1956**, 3704; **1957**, 4292

[42] Petrissans, J., Gromb, S., Deschamps, J.: *Bull. Soc. Chim. France* **1967**, 4381

[43] Jones, R.N., Spinner, E.: *Can. J. Chem.* **1958**, 36, 1020

[44] Reisse, J., Peters, R.A., Ottinger, R., Bervelt, J.P., Chiurdoglu, G.: *Tetrahedron Letters* **1966**, 23, 2511

[45] Young, R.P., Jones, R.N.: *Chem. Rev.* **1971**, 71, 219

[46] Caldow, G.L., Thompson, H.W.: *Proc. Roy. Soc.* **1960**, 254, 1

[47] Bellamy, L.J., Williams, R.L.: *Trans. Faraday Soc.* **1958**, 54, 1120; **1959**, 55, 14

[48] Bellamy, L.J., Hallam, H.E.: *Trans. Faraday Soc.* **1959**, 55, 220

[49] Ballester, L., Cario, C., Bertran, J.E., *Spectrochim. Acta* **1972**, 28A, 2103

[50] Nyquist, R.A.: *Appl. Spectrosc.* **1986**, 40, 79

[51] Broda, M.A., Hawranek, P.: *J. Mol. Struct.* **1988**, 177, 351

[52] Steele, D., Yarwood, J. (eds.): *"Spectroscopy and Relaxation of Molecular Liquids"*, Elsevier, Amsterdam **1991**

[53] Carter, R.L.: *J. Chem. Educ.* **1971**, 48, 297

[54] Schütte, C.H.J.: *Fortschr. chem. Forsch.* **1972**, 36, 57

[55] Husar, J., Kreevoy, M.M.: *J. Amer. Chem. Soc.* **1972**, 94, 2902

[56] Kirchner, H.-H., Richter, W.: *Ber. Bunsenges. Phys. Chem.* **1977**, 81, 1250; ibid. **1979**, 83, 192

[57] Ottin, W.: *"Spektrale Zuordnungstafel der Infrarot-Absorptionsbanden"*, 1. Aufl., Springer Verlag, Berlin, Heidelberg, New York, **1963**

[58] Weitkamp, H. in: Korte, F. (ed.): *"Methodicum Chimicum"*, Bd. 1, Teil 1: *"Analytik"*, 1. Aufl., Georg Thieme Verlag, Stuttgart u. Academic Press, New York, London **1973**, p 274

[59] Grasselli, J.G. (ed.): *"Atlas of Spectral Data and Physical Constants for Organic Compounds"*, The Chemical Rubber Co., Cleveland, Ohio **1973**

[60] Schrader, B.: *"Infrarot- und Raman-Spektroskopie"*, in: *Ullmanns Enzyklopädie der technischen Chemie*, Band 5, 4. Aufl., Verlag Chemie, Weinheim, **1980**, p 303

[61] Weitkamp, H., Barth, R.: *"Infrarot Strukturanalyse"*, Georg Thieme Verlag, Stuttgart, **1972**

[62] Dolphin, D., Wick, A.E.: *"Tabulation of Infrared Spectral Data"*, J. Wiley and Sons, London, Sidney, Toronto, **1977**

[63] Hedinger, H.J.: Kirba-Kartei: *"Kommentare zu Infrarot-Banden"*, Hedinger, H.J., Eigenacker 692, CH-8193 Eglisau ZH (vgl. [6], 151)

[64] Sadtler, I.R.: MENTOR, Fa. Bio-Rad Sadtler Division, 3316 Spring Garden Street, Philadelphia, PA 19104 (USA)

[65] Ebert, J.: *"IRTRAINS – Ein infrarotspektroskopisches Trainingsprogramm zur Spektrum-Struktur-Korrelation"*, Jürgen Ebert Verlag, Leverkusen, **1992**

[66] Simons, W.W. (ed.): *"The Sadtler Handbook of IR-Spectra"*, Heyden and Son, London, **1978**

[67] Craver, C.D. (ed.): *"The Desk Book of Infrared Spectra"*, The Coblentz Society, **1977**

[68] Pouchert, C.J. (ed.): *"The Aldrich Library of Infrared Spectra"*, Aldrich Chemical, Milwaukee, 3rd Ed., **1981**; vom gleichen Verlag und Herausgeber: *"The Aldrich Library of FT-IR-Spectra"*, **1985**; *"The Aldrich Library of FT-IR-Spectra: Vapor Phase"*, Vol. 3, **1989**

[69] Pretsch, E, Clerc, J.T., Seibl, J., Simon, W.: *"Tables of Spectral Data for Structure Elucidation of Organic Compounds"*, 2nd Ed., Springer Verlag, Berlin-Heidelberg-New York, **1989**

[70] Silverstein, R.M., Morrill, T.C., Bassler, G.C.: *"Spectrometric Identification of Organic Compounds"*, 5th Ed., J. Wiley and Sons, New York, **1991**

[71] Shimanouchi, T.: "The Molecular Force Field" in: Eyring, H., Henderson, D., Jost, W. (eds.): *"Physical Chemistry"*, Vol. VI, Chap. 6, Academic Press, New York, London, **1970**, p 233

[72] Barrow, G.M.: *"Introduction to Molecular Spectroscopy"*, McGraw-Hill, New York, **1962**

[73] Wilson, E.B. Jr., Decius, J.C., Cross, P.C.: *"Molecular Vibrations"*, McGraw-Hill, New York, **1955**

[74] Gribov, L.A., Orville-Thomas, W.J.: *"Theory and Methods of Calculation of Molecular Spectra"*, J. Wiley and Sons, Chichester, **1988**

[75] Heise, H.M., Winther, F., Lutz, H.: *J. Mol. Spectrosc.* **1981**, 90, 531

[76] Winther, F.: *Z. Naturforsch.* **1970**, 25a, 1912

[77] Matsuura, H., Tasumi, M.: "Force Fields for Large Molecules" in: Durig, J.R. (ed.): *Vibrational Spectra and Structure*, Vol. 12, Elsevier, Amsterdam, **1983**, p 69

[78] Duncan, J.L. in: *"Molecular Spectroscopy"*, Vol. 3, Chap. 2, The Chemical Society, London, **1975**

[79] Schachtschneider, J.H.: Shell Develop. Co., Technical Rpt. **1966**, 57–65; Schachtschneider, J.H., Snyder, R.G., Spectrochim. Acta **1963**, 19, 117

[80] Shimanouchi, T., *Computer Programs for Normal Coordinate Treatment of Polyatomic Molecules*, Universität Tokio **1968**

[81] Fischer, P., Bougeard, D., Schrader, B.: *"Simulation der Schwingungsspektren (SPSIM)"*, in: Gauglitz, G. (ed.): *Software-Entwicklung in der Chemie*, Bd. 3, Springer-Verlag, Berlin, **1989**, p 341

[82] Fuhrer, H., Karta, V.B., Krueger, P.J., Mantsch, H.H., Jones, R.N.: *Chem. Rev.* **1972**, 72, 439

[83] Sadley, J.: *"Semi-Empirical Methods of Quantum Chemistry CNDO, INDO, NDDO"*, Ellis Horwood, Chichester, **1985**

[84] Stewart, J.J.P.: *J. Comput. Chem.* **1989**, 10, 221

[85] Billes, F.: *Acta Chim. Hung.* **1992**, 129, 3

[86] Berces, A.R., Szalay, P.G., Magdo, I., Fogarasy, G., Pongor, G.: *J. Phys. Chem.* **1993**, 97, 1356

[87] Florián, J., Johnson, B.G.: *J. Phys. Chem.* **1994**, 98, 3681

[88] Billes, F., Mikosch, H.: *J. Mol. Struct.* **1995**, 349, 409

[89] Bougeard, D.: "Calculation of Frequencies and Intensities of Vibrations of Molecules and Crystals", in: Schrader, B. (ed.), *Infrared and Raman Spectroscopy*, VCH, Weinheim, **1995**, p 445

[90] Berens, P.H., Wilson, K.R.: *J. Chem. Phys.* **1981**, 74, 4872

[91] Smirnov, K.S., Le Maire, M., Brémard, C., Bougeard, D.: *Chem. Phys.* **1994**, 179, 445

[92] Baranović, G., Meic, Z., Mink, J., Keresztury, G.: *J. Phys. Chem.* **1990**, 94, 2833

[93] Goodman, L., Ozkabak, A.G., Thakur, S.N.: *J. Phys. Chem.* **1991**, 95, 9044

[94] Thakur, S.N., Goodman, L., Ozkabak, A.G.: *J. Chem. Phys.* **1986**, 84, 6642

[95] Herzberg, G.: *"Infrared and Raman Spectra of Polyatomic Molecules"*, Van Nostrand-Reinhold, New York, **1945**

[96] Wilson, E.B. Jr.: *Phys. Rev.* **1934**, 45, 706; *J. Chem. Phys.* **1993**, 7, 1047

[97] Varsányi, G.: *"Vibrational Spectra of Benzene Derivatives"*, Academic Press, New York, **1969**

[98] Meić, Z., Baranović, G., Suste, T.: *J. Mol. Struct.* **1993**, 296, 163

[99] Meić, Z., Suste, T., Baranović, G., Smrečki, V., Holly, S., Keresztury, G.: *J. Mol. Struct.* **1995**, 348, 229

7   Quantitative Spectral Assertions

Every substance property that clearly depends on the concentration of a substance can be made the basis of a quantitative analysis. With measurements in IR spectroscopy, the amount of absorbed electromagnetic radiation is determined upon its interaction with the sample. The measurement parameter is the ratio of the radiation intensity before and after passing through the sample at a certain wavelength. The Lambert-Beer law or commonly Beer's law describes the relationship of the experimentally determinable parameters with substance concentration.

7.1   Fundamentals

7.1.1   Beer's Law

Already in 1760, Lambert found that the decrease of radiation intensity with its path through material is proportional to the respective intensity (Fig. 7-1). If we divide the absorbing layer into differential lamellae of thickness dx, then one can express this concept by Eq. (7.1):

$$-\frac{dI}{dx} = aI \tag{7.1}$$

where I denotes the radiation intensity and the proportionality factor a, the so-called *absorption coefficient*.

The intensity I on any given site $x=l$ (see Fig. 7-1) is obtained by integration of Eq. (7.1) within the boundaries $x=0$ to $x=l$, whereby I_0 signifies the intensity of the incident radiation on the sample:

$$\int_{I=I_0}^{I} \frac{dI}{I} = - \int_{x=0}^{x=l} a\, dx \tag{7.2}$$

This results already in the basic form of Lambert's law:

$$\ln I - \ln I_0 = -a \cdot l \tag{7.3}$$

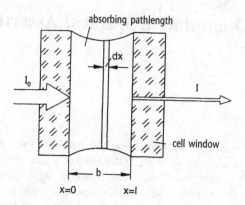

This is typically expressed as:

$$I = I_0\, e^{-al} \tag{7.4}$$

In 1852, Beer recognized the uniform influence of the concentration c on the radiation attenuation as that of the cell thickness b. Accordingly, a 1.0 molar solution of a substance in a 1.0 cm thick layer has the same absorbing effect as a 0.5 molar solution in a 2.0 cm thick layer. Consequently, one obtains Beer's law:

$$\ln I - \ln I_0 = -a \cdot b \cdot c \tag{7.5a}$$

where a' is the absorptivity. After conversion for the natural to the logarithm base 10, it follows that:

$$\log I - \log I_0 = -a \cdot b \cdot c \tag{7.5b}$$

with $a = 0.434\, a'$ (whereby $\log e = 0.434$). The typical expression, analogous to Eq. (7.4), is the exponential form:

$$I = I_0 \cdot 10^{-abc} \tag{7.6}$$

The parameter a is called the absorptivity to the base 10. Whereas the absorption coefficient a in the equations (7.1) thru (7.4) is dependent on the concentration c, a' and a are independent from this parameter and are considered to be constants specific for the substance at a particular wavelength. For easier handling Eq. (7.6), the absorbance A is introduced:

$$A \equiv -\log(I/I_0) = a \cdot b \cdot c \tag{7.7}$$

This allows calculation with additive parameters, because the logarithmic relationship between the measurement parameters I and I_0 with the concentration c are made linear with Eq. (7.7).

Hence, the absorbance A is a measure that is directly proportional to the concentration. Therefore, absorbance must be used for quantitative spectral analysis, while transmittance T often is taken for qualitative interpretation of spectra. The reason for this is that taking the logarithm of a number results in a small number. Therefore, bands with low intensity, but with significance for qualitative spectral interpretation might not be seen clearly in spectra presented as absorbance.

Eq. (7.7) describes the relationship of three variables (A, b, c) and a substance constant a. For quantitative measurements, one generally wants to determine the concentration c of a substance in a multi-component mixture; therefore, one must know the other parameters. To note hereby is that there exist measurement techniques other than the measurement in transmission. For instance, various reflection techniques were introduced in Sect. 5.1. Here, linear concentration dependencies likewise exist (often only in certain intervals) following signal transformation (logarithm formation or Kubelka-Munk transformation of the degree of reflectance). However, we will mainly limit ourselves to the absorbance and the determination of the parameters b and a for liquid samples.

7.1.2 Determination of Pathlength

If the spectrum of an empty, perfect liquid cell with small pathlength is recorded, one will obtain a sine wave pattern instead of a straight line.

This pattern is the result of interference phenomena within the cell (Fig. 7-2). In an air-filled cell, the difference of the refraction index between air and the window material is large enough to reflect a fraction of the radiation back and forth (also see Sect. 3.6.5. and 4.2.4.4). Depending on the path difference between the reflected and non-reflected beam, constructive or quenching interference occurs. The spectrum shows at which wavelengths quenching or intensity amplification happens. However, this depends on the distance b of the cell windows (see Fig. 7-3).

Fig. 7-2. The formation of the spectral interference pattern (only multiple reflections within the cell are shown). The conditions can be explained more easily with a slanted incident beam; they are identical to the perpendicular incident beam, whereby then $d=\overline{AB}$ applies: a part of beam 1 is reflected at A and, again, a fraction of this is reflected at B. Starting from the position B, this fraction interferes with beam 2 resulting in intensity amplification, provided that the wavelengths exist in an integer ratio of $2\overline{AB}$ The remaining wavelengths experience an intensity attenuation.

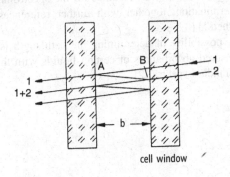

cell window

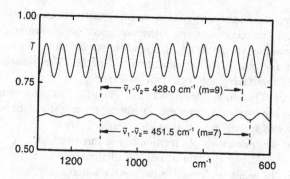

Fig. 7-3. Spectral interferences from two different cells with KBr-windows (shown above: pathlength 105.1 µm; shown shifted below: 77.5 µm with wedge-shape arranged windows); the pathlength is calculated according to Eq. (7.8).

Moreover important, however, is that the cell windows are parallel. Figure 7-3 illustrates the interference pattern of a good cell as well as that of a cell with somewhat "wedged" ordered windows. For the latter, one obtains a considerable reduction of the intensity differences between the minima and maxima, which makes the evaluation more difficult. The altered pathlength prescribed by the cell wedge can also lead to a non-linear signal dependency on the concentration (also see further below).

From the position of the maxima and minima in the spectrum, one can determine the pathlength b according to the following relationship:

$$b = \frac{m}{2} \frac{1}{\tilde{v}_1 - \tilde{v}_2} \ [\text{cm}] \tag{7.8}$$

The number of maxima (or minima) between the wavenumbers $\tilde{v}_2$ and $\tilde{v}$ is signified by m. But there are also more complicated methods than that presented, e.g. via the calculation of the transmittance spectrum of a cell and fitting of the cell parameters, whereby e.g. the non-parallel orientation of the windows and the convergent radiation in the sample compartment are taken into account. Alternatively, the spectral interference pattern can likewise be Fourier-transformed, with which a fast and precise method of the pathlength determination can be realized. A comparison of the various methods for the spectrometric pathlength determination, together with further references on this, was published [1].

Another possibility of determining the pathlength is by comparing measured absorbances of certain liquids with their corresponding standard spectra [2].

7.1.3 Determination of Absorbance

A measure for the radiation absorption of a substance is the intensity of its bands in the spectrum. The determination of the total area of a single band, the so-called "integral absorbance" requires somewhat more effort. Thus, one is satisfied mostly with measuring the intensity in the band maximum and thereby obtains the "maximum absorbance" A_{max} at the position $\tilde{v}_{max}$.

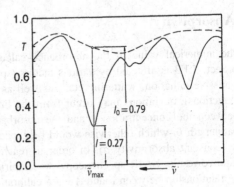

Fig. 7-4. For calculating the absorbance A_{max} (illustration with exaggeratedly curved, dashed baseline; for the example, one obtains $A_{max} = -\log (0.27/0.79) = 0.466$ (see text for further details).

The parameter A_{max} can be obtained from the spectrum in the following way (see Fig. 7-4): at first, the background is recorded. For this, the cell without sample, which ought to be present as a diluted solution, is put into the beam. Interference phenomena should be negligible for the cell only filled with pure solvent. This is generally the case when the difference of the refraction indices of the cell contents and the window material is not large. When finally the sample is put into the cell and its spectrum is recorded, the solid-line curve in Fig. 7-4 is obtained. It is easy to see that the band originated only from the sample. Now, its intensity needs only to be calculated.

To do this, a perpendicular line is drawn in the absorption maximum and its transmittance is obtained with its intersection with the baseline. The ordinate value of the baseline at the wavenumber $\tilde{v}_{max}$ can be used as the measure for the radiation fraction passing through the cell without the sample (corresponding to I_0). In the presence of the sample, only the radiation fraction, described by the spectral ordinate value at the position $\tilde{v}_{max}$, reaches the detector (corresponding to I). By definition, the absorbance A is obtained by first dividing the ordinate value of the sample spectrum at $\tilde{v}_{max}$, I by the corresponding ordinate value of the baseline, I_0 and then taking the negative logarithm of this quotient (see Eq. 7.7).

With today's computer-linked spectrometers, these values are obtained by scanning the spectrum with the cursor; wavenumber and transmittance values are indicated by the computer. The presentation in absorbance is also possible by simple conversion in the operating software. In this case, only the difference of both absorbance values must be taken or even the complete spectra are typically used. When the solutions are not very diluted, a scaled subtraction of the absorbance spectra must be performed in which the various solvent fractions in the cell are considered upon measurement of the solution and of the pure solvent.

7.1.4 Absorptivity

Because the numerical value of the absorbance calculated according to Sect. 7.1.3 is affected by various factors (spectral resolution, emissive radiation, scattering etc.[1], as well as intermolecular interactions), deviations may occur from the linear relationship between the concentration c and the absorbance A at constant pathlength b, which otherwise would be expected from Beer's law (constant absorptivity a). In order to reach dependable quantitative results by IR spectroscopy, we empirically determine the relationship between c and A via a calibration curve that should include the concentration range of future samples. Extrapolations to concentrations lying beyond the calibration interval might lead to larger systematic errors.

There are various sources of error, which are partially quite considerable. Some of them can be attributed to the spectrometer, whereby there are special types of errors implicit in the use of dispersive or FT spectrometers. Although a detailed discussion of the photometric precision would go beyond the scope of this book, a few important effects still ought to be cited here. For instance, one observes a drift resulting from e.g. an insufficient thermostatizing of the spectrometer or faulty radiation in grating spectrometers. Furthermore, detector non-linearity plays a role in errors as well (also see Sect. 3.4.3.3). Other sources of error are spectral resolution selected too low or also the beam divergence in the sample compartment. Uncertainties may also be related to the sample and cell: the variability of the pathlength, temperature deviations, non-parallel cell windows, reflection and interference effects, or also thermal emission have an effect. Contamination of the sample, e.g. by sample carryover in the cell can be prevented by careful work. A few selected publications, which contain references to further literature sources [3–5], depict the complexity of the spectral measurements when highest demands are placed.

---

[1] These parameters have to be reproducible in the practical analysis. For absolute measurements, they are extrapolated experimentally or mathematically to zero.

An important parameter of spectral acquisition is the spectral resolution that can considerably affect the measure band absorbances whenever they are selected too small compared to the band half-width. Already in the fifties, the effects of slit widths were investigated for spectrometers with monochromators. Regarding spectral resolution, the so-called Ramsay criterion [6] requires a resolution value of one-fifth of the band half-width when for the maximum band absorbance, only a deviation of less than 3% ought to result between the measured and true absorbance. These effects are attributed to the convolution of the transmittance spectrum with the corresponding spectrometer function (also see Sect. 3.6.2 and 3.6.4). Griffiths and Anderson [7] studied the effect of various apodization functions, as are used in FT spectroscopy. To note here is, that with insufficient resolution compared to the original true half-width of the absorption bands to be measured, larger non-linearities between the experimental maximum absorbance and the concentration result at absorbances as of about 1.0. This often plays a role for gas analysis by IR spectroscopy, because the band lines of the rotational-vibrational fine structure may show a half-width of less than 0.25 cm^{-1} (at atmospheric pressure). The next section will still present an example on this that clearly shows the influence of various apodization functions (and the different spectral resolution connected with this).

The determination of absolute absorptivities of pure liquids is very time-consuming, but corresponding techniques are known for this [8]. Such IR data exist for several substances so that these may be used as secondary standards for intensity measurements [9].

7.2   Calibration

7.2.1   Baseline Correction for Absorbance Determination

The calibration curve graphically shows the relationship between the absorbance A at a particular wavenumber and the concentration of a substance to be quantified. To establish the curve, mixtures of various concentrations must be produced which is relatively unproblematic when the components to be analyzed are present in pure form.

For determining A, one should avoid using the baseline as the basis for the evaluation – analogous to the curve illustrated in Fig. 7-4 as dashes. It brings uncertainties into the measurement and requires an additional spectral acquisition. Better is to select a line drawn through 2 points, i.e. the linear baseline (see (a) in Fig. 7-5). As the so-called base points, one takes two points in the spectrum, at each side of the measurement band, which lie at well reproducible positions of about the same transmittance. In general, these are absorbance minima. The further evaluation is then based on the corresponding procedure already mentioned in the previous section (Fig. 7-4).

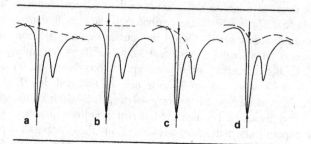

Fig. 7-5. The determination of the absorbance A according to the baseline technique: a) standard case: linear baseline through two defined points, b) horizontal base, c) curved base (approximated band splitting), d) background absorption as the baseline. The values between the arrows convert to the respective "maximum absorbance".

Evaluations with a strongly sloped baseline arise when the evaluation band lies in flanks of a different absorption. Such influences, however, are "calibrated in". Most absorbance determinations for quantitative evaluations are performed via the route described. The other possibilities for setting a baseline, like e.g. the vertical baseline thru a base point (Fig. 7-5b), a curved baseline (Fig. 7-5c) or also the background line, itself, (Fig. 7-5d; when a weak band of the main substance lies under the evaluation band) are exceptions, because their reproducibility is subordinate to the method first cited.

7.2.2 Determination of Calibration Data

The route for producing a calibration curve is elucidated by using the single determination of benzonitrile in carbon tetrachloride as an example. For this, different amounts of benzonitrile in CCl_4 are weighed in (see Tab. 7-1, columns 1-3).

Table 7-1. Experimental values for producing the calibration curve for benzonitrile in carbon tetrachloride.

Weighed Portion [g]		Mass Fraction w	
C_6H_5CN	CCl_4	C_6H_5CN in %	A
1.298	3.785	25.54	0.148
2.866	2.899	49.71	0.301
4.102	1.236	76.8	0.422
pure	0.0	100.0	0.479

From each sample, then the region of the nitrile band from 2500–2000 cm^{-1} is recorded and the absorbance in the band maximum according to the baseline method is determined (Tab. 7-1, column 4). Hereby, the pathlength b is selected so that the nitrile band maximum of the sample with the highest benzonitrile fraction lies at about $T=0.25$.

For producing the calibration curve, the absorbance values corresponding to the concentration c into a coordinate system with c as the abscissa and the absorbance A as the ordinate have to be recorded. The curve drawn through these points is subsequently used for quantitizing a sample with an unknown benzonitrile fraction (Fig. 7-6).

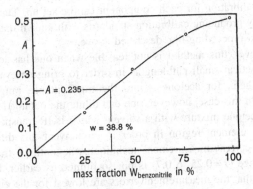

Fig. 7-6. Calibration curve for benzonitrile in CCl$_4$.

For measurement and evaluation, the same cell used for calibration is filled with the sample, the absorbance of the nitrile band is determined and from the calibration curve, the benzonitrile fraction corresponding to the calculated absorbance value is read. Upon assuming that the absorbance has been determined to be $A=0.235$, the carbon tetrachloride solution thus contains a mass fraction of 38.8% benzonitrile according to Fig. 7-6.

Transfer of calibration functions to other spectrometers is not possible without errors. In addition, literature values for absorptivities can be applied only under certain conditions, when e.g. the pathlength is exactly known and the spectrometer influences (limited resolution, apodization, etc.) can be ruled out. Nevertheless, there are multiple possibilities for errors (see Sect. 7.1.4). In such cases, a validation of the calibration function is absolute necessary.

For multivariate calibrations by using NIR spectra in which not only one absorbance value but a wide spectral interval is evaluated (also see Sect. 7.5), various tests were undertaken to obtain a spectrometer standardization by which a calibration transfer ought to be made possible. Bouveresse et al. examined the possibilities recommended in the literature and suggested improvements [10]. This is important for complicated calibration sets as are applied in the analysis of feed and foodstuffs.

7.2.3 Multi-Component Analysis with Separated Analytical Bands

In the previous section, we handled the determination of the concentration of one component in a solvent. One can proceed quite analogously if both substances are to be determined, for instance, benzonitrile in acetone. Because the evaluation bands at 2230 cm^{-1} and 1715 cm^{-1} do not disturb each other, both substances can be treated independently of each other and an individual calibration for each component can be set up. This is obtained by producing calibration standards with known mass fractions and proceeding like described above.

However, this method is not feasible when one has to resort to impractical small pathlengths in order to bring an evaluation band – here, for acetone – into the transmittance range over $T=0.2$. In this case, however, one can dilute the obtained benzonitrile-acetone mixture with a solvent which is IR-transparent in this measurement region in order to circumvent this difficulty. For the current measurement technique, the optimal transmittance range ($T=0.25$ to 0.7) is not so limited as earlier, but we will see that the measurement errors are lowest for the cited region.

In our example, we find that the optimal conditions are met at 10% of the binary mixture in CCl$_4$ with an acetone fraction $w_{acetone}$ of between 25% and 100%. For this range, corresponding calibration standards are placed at hand and are measured. The same route for measuring samples is followed: A solution with a mass fraction of the substance to be examined of about 10% in carbon tetrachloride is prepared and the respective maximum band absorbance is measured. If the degree of dilution of the sample slightly differs from that of the calibration standard, this can be taken into account by linear conversion. With the corrected value, the acetone fraction is read from the calibration curve.

To summarize: For every concentration range, there is an optimal calibration curve. In the above example, this exists for undiluted samples having a mass fraction of $w_{acetone}=25–100\%$ as well as of $w_{benzonitrile}=0–100\%$; for 10% solutions of $w_{acetone}=25–100\%$. In the case of benzonitrile, a dilution is not necessary, because the band intensity is not large. However, the quantification limit is not as favorable and the accuracy not as high as that of acetone.

7.2.4 Evaluation via Relative Absorbance Values

7.2.4.1 Quantitative Measurement of Liquids

The calibration curves $A = f(c)$ apply only if the pathlength of the sample does not change. Relative absorbance specifications offer a possibility to use calibrations regardless of larger changes in the pathlength. Here, not the absorbance A of the measurement band is labeled as the ordinate in the calibration curve but rather its ratio with the absorbance A_{St} of a band of a standard substance, namely A/A_{St}. For this, both measurements are performed in the same cell. Mathematically, this is equal to a division of the corresponding terms for Beer's law:

$$\frac{A}{A_{St}} = \frac{a \cdot c \cdot b}{a_{St} \cdot c_{St} \cdot b} \tag{7.9}$$

whereby a and a_{St} are substance constants and c ought to be determined or is known for calibration; the concentration c_{St} is likewise constant, since the pure standard substance is filled undiluted into the measurement cell. Thus, the pathlength b is eliminated, and one obtains the following equation for the calibration function:

$$\frac{A}{A_{St}} = k \cdot c \quad \text{with} \quad k = \frac{a}{a_{St} \cdot c_{St}} \tag{7.10}$$

Cyclohexane has proven itself useful as a standard substance, because three bands of various intensity appear in the spectrum, with which the substance is suitable for every pathlength range (Tab. 7-2).

Table 7-2. Evaluation bands of cyclohexane for various cell pathlengths.

Band position (cm^{-1})	Suitable for Pathlength b (mm)
906	up to 50
1013	50– 250
1109	250–2000

In practice, one measures the absorbance of the corresponding cyclohexane band before and after every measurement series. Pathlength changes, which occur during a measurement series, can be easily recognized and still be linearly interpolated.

7.2.4.2 Quantitative Measurement with Pellets

Pellets are poorly suited for quantitative studies (also see Sect. 4.2.1.8), the reasons for this mainly being the non-reproducible scattering of the IR radiation on the crystal surfaces, concentration inhomogeneity and irregular crystal size. Moreover, the prerequisite of an ideal solution of the substance in the matrix is rather the exception than the rule. The cited examples of the measurement of a few carbonates demand a careful preparation technique fitted to the particular problem [11]. The relative deviation lies at around 10%.

In some cases, one can obtain better than semi-quantitative results with relative measurements with reproducible preparation and by adding an internal standard. Proven useful for such evaluations are [12]:

K_2CO_3	875 cm^{-1}
DL-alanine	851 cm^{-1}
$K_3[Fe(CN)_6]$	2100 cm^{-1}
KSCN	2100 cm^{-1}

The conditions are different when measurement takes place in the long-wave IR region below 600 cm^{-1}. There, the scattering losses are no longer significant when the grinding time and pressing conditions are maintained in the calibration and preparation of the sample [13].

7.2.5 Quantitative Gas Analysis

Gases demand a somewhat modified evaluation method, because an additional complication, in contrast to the analysis in the condensed phase, is the dependency of the absorbance on the pressure and temperature. The basis again is of course Beer's law in the form:

$$P_c = \frac{A}{a \cdot b} \tag{7.11}$$

Often, only the partial pressure P_c stands for the concentration c. Inherently, pathlength changes are negligibly small for the gas cells used.

Creating a calibration curve is done analogously to that for solutions: at first, the pure gas is filled into the cell. In order to get clear conditions regarding the pressure widening of the bands, the cell mostly is filled with nitrogen to an always-constant total pressure, P_G, e.g. atmospheric pressure. Slight deviations here may be ignored. If the evaluation band is too intense, then one dilutes to a certain partial pressure until the measurement band lies in a favorable measurement range. Subsequently, the measurement gas is successively diluted to lower partial pressures, respectively is again filled with nitrogen to the previously selected total pressure, and the spectrum in the range of the band of interest is recorded. The calibration curve results from the dependency of the absorbance, calculated by the baseline technique, on the partial pressure P_c. It applies for the temperature T_k at which the calibration data were determined.

For the analysis, the gas mixture to be analyzed stands, after sample preparation, under the known loading pressure P in the gas chamber. If the gas temperature at the measurement (T) deviates from that of the calibration (T_k), it must be corrected by the 2nd Gay-Lussac law.

One example would be the partial pressure of the analytical mixture $P = 317$ hPa at $T = 294.15$ K (21 °C). Subsequently, it is filled up with N_2 to atmospheric pressure (e.g. 1000 hPa). The calibration curve applies for $T_k = 300.15$ K (27 °C). The partial pressure P' that the mixture would have at the calibration temperature T_k is then:

$$P' = \frac{P T_k}{T} = \frac{317 \cdot 300.15}{294.14} \text{hPa} = 324 \text{hPa} .$$

If the loaded gas were pure, then P' would correspond to a volume fraction of 100%. From the calibration curve, we find e.g. for an absorbance of $A = 0.345$, a partial pressure P_c of 234 hPa. From the relationship $P' : 100 = P_c : x$, follows that 72.3% is the relative partial pressure of the component (corresponding to the same volume fraction in %).

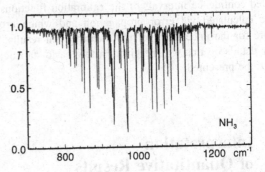

Fig. 7-7. Spectrum of the ν_2-band of ammonia (43 ppm (V/V) in N_2), acquired at an optical pathlength of 20 m (spectral resolution 1 cm^{-1}, boxcar apodization).

An example of this would be for analyzing room air. With a multi-reflection cell, test gases having various volume fractions of ammonia in nitrogen were produced via a dynamic dilution by means of a capillary dosing unit [14]. The test gases were analyzed by spectroscopy in a cell flow-through procedure after a short time span to rule out losses of the polar components by adsorption to the inner cell surface. The spectrum of one of the test gases is shown in Fig. 7-7.

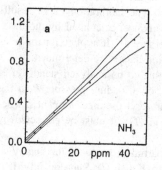 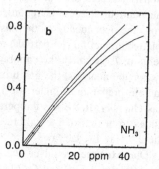

Fig. 7-8. Calibration curves with confidence intervals based on the same ammonia test-gas standard but with different calculation of the FT spectra (evaluation of the maximum absorbance at 967 cm^{-1}, nominal resolution 1.0 cm^{-1}; a) Boxcar apodization and b) triangular apodization.

Figure 7-8 illustrates two calibration curves for the same batch of calibration standards, but the spectra measured with one FT spectrometer at the same nominal resolution were calculated with different apodization functions. For the relatively wide Q-branch band at 967 cm^{-1} and, in contrast to the other rotational-vibrational lines, a linear dependency of the maximum absorbance is found as a function of the existing volume fraction range with boxcar apodization; the triangular apodization already results in a noticeable worsening of the spectral resolution, because the corresponding sinc^2-spectrometer functions show a greater half-width than the sinc function (boxcar apodization (see Sect. 3.4.4.1)). As theoretically predicted, this affects the experimentally measured absorbances to the extent that a non-linear calibration function results. Likewise shown are the so-called confidence intervals of the calibration functions which can be calculated via error propagation and which provide a measure for the uncertainty of the measurement results. Before further details on this will be given, a few basic statistical facts ought to be presented.

7.3   Interpretation
of Quantitative Results

The previous sections indicate how quantitative determinations can be obtained from IR spectroscopic measurements. However, one must always clearly understand the meaning of the results, whereby possibly much information ought to be extracted from the data. For evaluating and planning experiments, statistical and chemometric tools are used, whereby the latter actually represents mathematics with a special relationship to chemical analysis. Chemometrics often strives only in the last step of the experiments for determining the material composition, namely for the calculation and evaluation of analytical results. Here, the analyst should incorporate concepts into the planning stage, which will ultimately affect the quality of the quantitative results. A few examples ought to acquaint us with the simplest foundations. Please refer to special literature for more details, e.g. [15, 16].

7.3.1 Mean Value and Standard Deviation

In the acetone determination, a sample has been measured many times resulting in the absorbance results in column 2 of Tab. 7-3. From this, several acetone fractions from the calibration curve have been ascertained (column 3).

One can see from Tab. 7-3 that the values in columns 2 and 3 scatter. The arithmetic mean ($\bar{w} = 63.65\%$) is most considered to be the "best" value. However, had the first measurement in Tab. 7-3 not been listed, $\bar{w}$ would lie at 63.10%. We see that incomplete conclusions can be made from the measurement data alone: the single values fluctuate from $w_{acetone} = 61.7\%$ to 67.5%, and the mean depends on the number of measurements. Nonetheless, one can draw several conclusions with statistical methods.

Table 7-3 indicates that the individual measurement values concentrate around the mean value. For very many measurements, they can be grouped in classes (Fig. 7-9 a). A class comprises a narrow value range, e.g. 1.0 units. The number of measurement findings within one class is recorded on the ordinate.

Table 7-3. Multiple measurement of a sample for calculating the mean value w and the standard deviation s via Eq. (7.12).

Sample	A	w	$(w - \bar{w})$	$(w - \bar{w})^2$
1	0.5485	67.5	+3.85	14.82
2	0.5370	66.0	+2.35	5.52
3	0.5285	64.7	+1.05	1.10
4	0.5200	63.5	−0.15	0.02
5	0.5075	61.8	−1.85	3.42
6	0.5070	61.7	−1.95	3.80
7	0.5090	62.0	−1.65	2.72
8	0.5080	62.0	−1.65	2.72
		$\bar{w} = 63.65$		$\sum = 34.12$
$s = \pm(34.12/7)^{1/2}$			$= \pm 2.21$	($\pm 3.5\%$ rel.)

The graphic representation corresponds to a bar diagram as shown in Fig. 7-9 a. Upon transfer to an infinite number of measurements with differential class width, the draping of Fig. 7-9 a converts to a Gauss curve (Fig. 7-9 b). One then says that the measurement results follow a Gauss distribution or that the curve is distributed around the true mean μ with the true standard deviation σ (σ^2 is called the variance).

One of the most important parameters of a measurement series is the estimation of the standard deviation s. It is linked with the position of the turning point of the Gauss curve. For very many data, 68.3% of all single values lie within their borders ($\mu \pm \sigma$). Respectively 95.5% of all individual values fall within the interval $\mu \pm 2\sigma$ as well as 99.7% of all results at $\mu \pm 3\sigma$. The estimation s is a measure for the quality of the measurement results. The smaller s is, the narrower the result values lie next to each other and the more precise the measurement technique is. The numerical value of s can be calculated according to the following formula:

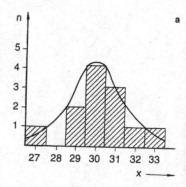

$$s = \pm \sqrt{\frac{\sum\limits_{i=1}^{n}(\hat{x} - x_i)^2}{n-1}} \qquad (7.12)$$

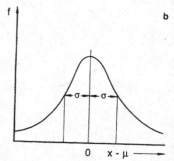

Fig. 7-9.
a) Bar diagram for a few measurement values ($n=12$),
b) standard distribution of the measurement values for an infinite number of measurements.

where $\bar{x}$ denotes the arithmetic mean of the individual values, x_i the value of the i-th single measurement and n the number of measurements. In analyses by IR spectroscopy, the relative standard deviation may generally lie better than $\pm 2\%$, e.g. provided that $n=10$ measurements in the range of $A=0.15$ to 0.70. In Tab. 7-3, an example is calculated for the acetone determination (see Sect. 7.3.3).

7.3.2 Scattering Range of Single Measurements and Confidence Interval of Mean Values

The following remarks apply for very many repeated measurements on the same sample ($n > \approx 30$). For eight measurements (as in the example of Tab. 7-3), one cannot yet base the conclusions on the ideal form of a Gauss distribution. Because of the too small number of measurement data, the distribution of the individual measurements must also be taken into account. This is done by the student factor t. The area in which the individual measurements scatter is given as $\bar{x} \pm s \cdot t$, whereby s is the standard deviation calculated from Eq. (7.12), and t accounts for the non-ideal measurement value distribution for few data. The factor t depends on two parameters: on the number of individual measurements and on the statistical certainty. The more results are available, the better the Gauss distribution is fulfilled and the smaller the factor t is (see Tab. 7-4).

Table 7-4. Student factors t as a function of the number of measurements and the statistical certainty (statistical tables contain the number of the degrees of freedom $n-1$ instead of n).

Number n of repeated measurements	2	3	4	5	6	7	8	9	10	∞
95%	12.71	4.30	3.18	2.78	2.57	2.45	2.37	2.31	2.26	1.96
99%	63.66	9.92	5.84	4.60	4.03	3.71	3.50	3.36	3.25	2.58

The statistical certainty is linked with the width of the confidence interval, whereby the specification with the factor $t_{95\%}$ means that 95% of the individual measurements lie within the boundaries $\bar{x} \pm t_{95\%} \cdot s$. In our example, this means that the individual measurements $w_{i,\,\text{acetone}}$ lie in the range of 58.4% to 68.9%.

Actually only the mean value of the measurements are of interest as the result and the decisive basis. Therefore, the uncertainty of the mean value $\bar{x}$ is often a more important parameter than the scattering of the single results. The confidence interval of the mean is $\bar{x}$ simply calculated from the number n and the standard deviation s of the individual measurements with $\pm t_{95\%} \cdot s\sqrt{n}$. The value of 95% certainty here means the probability of the found mean value lying within the boundaries $\bar{x} \pm t_{95\%} \cdot s\sqrt{n}$. In other words, of 20 results, one may lie outside of the cited confidence interval. For risky decisions, one will strive towards 99% or 99.7% certainty. For our example, 95% certainty gives: the mass fraction w_{acetone} of the sample lies somewhere between 61.8% and 65.5%. A narrowing of the range can only be obtained by additional measurements on the same sample or by measurements that are more precise.

7.3.3 Random and Systematic Errors

All above derived statements refer to the irregular deviations of the individual measurement results around a mean value. This is attributed to the random errors due to apparative and manual incongruities (e.g. detector and amplifier noise, deviations with weighing in). They are present for every physical measurement and can only be reduced by increased effort.

The effects of unsystematic errors for IR spectroscopic measurements lie only to a small part in the weighing in. A particular effect involves the absorbance. Near $A=0$ ($T=1$), uncertainty is given by baseline noise; for greater values of A ($T \to 0$), an enormous relative uncertainty likewise results because of the logarithmic dependency of the transmittance. The relative error σ_A/A of the absorbance measurement is calculated according to Eq. (7.13) with $A = -\log T$ as follows:

$$\frac{\sigma_E}{A} = \frac{\lg e \cdot 10^A \cdot \sigma_\tau}{A} \qquad (7.13)$$

whereby $\log e = 0.4343$, and σ_T represents the standard deviation in the baseline in transmittance units (e.g. estimated over a 100%-line). The function σ_A/A is illustrated in Fig. 7-10 for a baseline noise level with $\sigma_T = 0.001$. The curve minimum at 36.7% transmittance infers that the measurement precision is greatest there. With newer instruments, one can measure by all means in the range of $A = 0.03$ to 1.6.

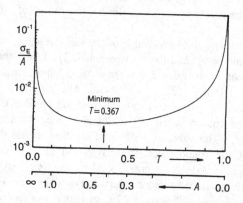

Fig. 7-10. Error curve of IR-spectroscopic measurements for which Beer's law applies ($\sigma_\tau = 0.001$).

A well-supported and reproducible analytical result nevertheless may be false. A maladjusted scale or a false factor e.g. changes the result one-sidedly and still reproducibly, which indicates a systematic error. Systematic errors are often difficult to recognize, and one can trace them only by applying principally different methods. Only by the concurrence of the results from e.g. IR, UV, GC within the standard deviation relatively assures certainty that no systematic error exists.

The data of Tab. 7-3 show a clear systematic error: the measurement values for the absorbance, and thus for the mass fraction, show a continuous decline towards smaller numerical values until sample 5. This trend is caused by the heating-up of the cell in the beam: the solvent volume increases until temperature equilibrium for approximately the same cell pathlength and the concentration decreases. From this, we can conclude that reproducible measurements can be obtained first after a heating-up period of the cell (about 10 min). Taking this basic rule into account for quantitative IR measurements, then the standard deviation decreases (with 4 measurements) to $s = \pm 0.15$ and the mean value lies with 95% probability at $\bar{w} = 61.87 \pm 0.24$ (corresponding to a relative error of $\pm 0.4\%$).

Values that do not fit into a series (so-called mavericks) are eliminated after which the statistical parameters are calculated anew. Mavericks may be treated as such only based on corresponding statistical tests [17]. Measurement value series that are not normally distributed (logarithmic distributions nearby the 0%- or the 100%-limit, binomial distributions on the basis of yes-no decisions etc.) must be evaluated according to other techniques.

7.4 Calibration Functions and Confidence Intervals

7.4.1 Linear Regression

Up to now, calibration curves have been presented only graphically. Of course, also calibration functions can be calculated here based on the data points that are prescribed by the calibration standard. In the calculations, it is assumed that the concentration data (independent variable x) are errorless opposed to the values of the dependent variable y. The simplest case is given when linear dependencies exist, so that a linear equation $y = a + bx$ can be the foundation, whereby the ordinate section a and the slope b are to be calculated. The optimal linear equation runs so that the sum of the deviation squares of the respective y_i-values are minimized to the corresponding ordinate values of the lines for the same abscissa x_i (compensation calculation).

Both constants of such a linear equation can be calculated as follows:

$$a = \bar{y} - b\bar{x} \tag{7.14}$$

and

$$b = \frac{\sum\limits_{i=1}^{n}(x_i - \bar{x})(y_i - \bar{y})}{\sum\limits_{i=1}^{n}(x_1 - \bar{x})^2} \tag{7.15}$$

where $\bar{x}$ and $\bar{y}$ represent the mean values based on the calibration value pairs. Another prerequisite is that within the working area of the calibration, the errors for the y_i-values are normally distributed and equally large. If the latter assumption is not fulfilled (so-called variance inhomogeneity), a weighted regression must be performed (see e.g. [18, 19]). As was already seen in Eq. (7.13), this applies for spectrometry, because the absorbance error is exponentially dependent on the absorbance values.

An important criterion for the quality of the line fitting is its standard deviation that is defined by the following equation, whereby $\hat{y}_i$ are the functional values of the calibration lines:

$$S_{y/x} = \pm\sqrt{\frac{\sum_{i=1}^{n}(y_i - \hat{y}_i)^2}{n-2}} \tag{7.16}$$

Another parameter that is frequently used for estimating a regression is the correlation coefficient r or r^2, the so-called measure of certainty:

$$r = \frac{\sum_{i=1}^{n}(x_i - \bar{x})(y_i - \bar{y})}{\sqrt{\sum_{i=1}^{n}(x_i - \bar{x})^2 \sum_{i=1}^{n}(y_i - \bar{y})^2}} \tag{7.17}$$

The correlation coefficient is often misused for statements on the linearity or on deviations, for which this is unsuitable but which a graphic representation of the x,y-value pairs apparently can achieve [20].

Confidence intervals of the calibration lines (see Fig. 7-8) can also be defined via the error calculation. For the analyst, in particular the uncertainties in the concentration data are important, whereby the statistical certainty can be guaranteed not only by the calibration, but rather also by the number of repeat measurements in the analysis [21].

More complicated is the calculation of non-linear calibration functions that may be approximated via polynomials. In addition, extensive literature exists including the calculation of confidence intervals, e.g. [22, 23].

7.4.2 Special Method Parameters

An important question in quantitative determinations arises with regard to methodical boundaries. The scattering of the measurement results via multi-measurements plays a large role, so that

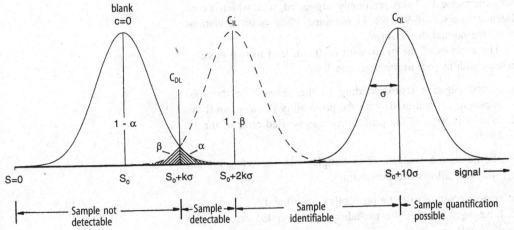

Fig. 7-11. Definition of various methodical parameters such as detection, identification and quantification limit (see text for further details).

statistical criteria must be drawn upon in order to be able to assess such important parameters such as the detectability of a method. Often, the sensitivity is also cited as an important parameter and is confused with the latter, whereby the sensitivity is clearly defined as the slope of the calibration function and represents a constant in the linear working range.

The detection-, identification- and quantification limit all have an extremely important place in analysis. The detection limit marks the possibility, e.g. of recognizing the smallest quantities of a substance, whereby this amount nevertheless does not suffice for a definite determination. The parameters can be defined via measurement of blanks that do not contain the component to be determined but also via the calibration function, which is typically used for the basis of a quantitative determination. Although its definition was established by a DIN norm [24], various terms and ranges are found in the literature for the same facts. A review on the variety and historical development is found in [25].

Figure 7.11 shows the normal distribution curves around three different signals including the value S_0, which can be assigned to the middle of the blanks. The total area of the Gauss curves is respectively normalized to 1, corresponding to the probability of 100%. The arbitrarily drawn limit $S_0+k\sigma$ (so-called critical value of the parameter) is used for characterizing the detection limit, whereby various factors k determine various probabilities for the measurement results beyond the limit. The value $k=2$ leads to a one-sided confidence interval $(1-a)$ of 97.7%, whereby this is reduced to only 75% in the absence of a normal distribution for the measurement errors. This is a reason why the factor $k=3$ was preferably suggested, with which a confidence interval of 99.86% is obtained (89% upon deviations from the normal distribution).

The analysis of the blank with $c=0$ can lead to two possible results with respect to the detection limit:

1. correct-negative (corresponding to the absence of the substance to be measured) with the probability $(1-\alpha)$ as well as
2. false-positive with the probability α (so-called error of the 1st kind).

Furthermore, the analysis of a sample with the concentration c_{EG} likewise allows two possibilities:

1. correct-positive with the probability $(1-\beta)$ and
2. false-negative with the probability β (so-called error of the 2nd kind).

For establishing the identification limit (other terms are used synonymously), the same error probability for α and β are given. The identification limit is the smallest value e.g. of a concentration that produces signals with a particular probability which are larger than that at the detection limit. It can be made clear that for concentrations above c_{EG}, the relative standard deviation (also called variation coefficient) is quite high, so that the quantification limit was introduced as a further parameter for which a relative minimum standard deviation can be found. The factor for σ is typically selected such that a value of 10% results for the variation coefficient.

Similarly, the detection-, identification- and quantification limit may be defined via the calibration function; also see [24] regarding this. Figure 7-12 exhibits both possibilities recom-

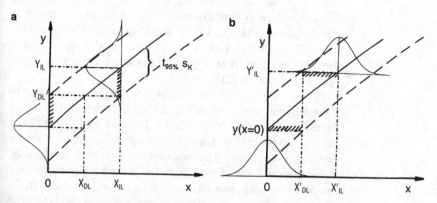

Fig. 7-12. Definition of the detection- and identification limit via the confidence intervals of the calibration in the signal (a) or concentration domain (b) ($t_{95\%}$ student factor (two-sided hypothesis), s_K standard deviation with variance fractions of the signal measurement and calibration).

mended in the literature: firstly, via the confidence intervals of the signal measurement including the calibration variance; secondly, via the corresponding confidence intervals around the *x*-values. However, also for non-linear calibrations with inconstant variance for the *y*-values, the respectively obtained limit values differ only unsubstantially from one another [25]. A comprehensive presentation on error propagation and confidence intervals was given by Ebel [26].

7.5 Multi-Component Analysis with Multivariate Evaluation

7.5.1 Classical Model Formation

The great number of multivariate methods can be explained by the fact that more comprehensive and improved conclusions are possible than with the consideration of single variables. Moreover, under certain conditions it is necessary to draw further information for evaluation. This especially applies for the case that no suitable undisturbed analytical bands are available for evaluation, because overlaps of bands of various components are present. The required strategy can be viewed on hand of an example.

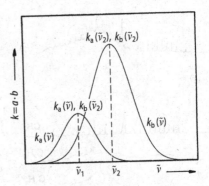

Fig. 7-13. Cross-sensitivities of two components during band overlapping.

Figure 7-13 illustrates two bands with their intensities resulting from the product of $\varepsilon_{a,b}(\tilde{v}) \cdot d$ which can be ssigned to one component each. With the selection of the wavenumbers $\tilde{v}_1$ and $\tilde{v}_2$ for the quantitative determination, respectively one particular cross-sensitivity is highlighted compared to the other component. Two linear equations can be set up which can be broken down according to the concentrations c_a and c_b:

$$A(\tilde{\nu}_1) = k_a(\tilde{\nu}_1) \cdot c_a + k_b(\tilde{\nu}_1) \cdot c_b$$
$$A(\tilde{\nu}_2) = k_a(\tilde{\nu}_2) \cdot c_a + k_b(\tilde{\nu}_2) \cdot c_b$$

(7.18 a)

Written in matrix form, this can be presented as follows, whereby the solution occurs via the matrix inversion:

$$\begin{bmatrix} A_1 \\ A_2 \end{bmatrix} = \begin{bmatrix} k_{1,a} & k_{1,b} \\ k_{2,a} & k_{2,b} \end{bmatrix} \begin{bmatrix} c_a \\ c_b \end{bmatrix}$$

or

$$\mathbf{A} = \mathbf{K} \cdot \mathbf{C}$$

(7.18 b)

More favorable for quantification, further absorbance values at additional wavenumber variables may be considered, whereby the vector **A** obtains a higher dimension and the matrix **K** is no longer square but rather rectangular (over-defined system). For solving such a linear equation system, one inserts the least-squares [15]. To note is that the number of spectral data points must be greater or equal to the number of components to be determined in order to obtain a sensible result. Figure 7.14 illustrates the foundation of the quantitative determination via the fitting of component spectra.

In principle, for multi-component mixtures with approximately ideal behavior of the components, their pure-substance spectra can be applied. However, it is recommended that, with

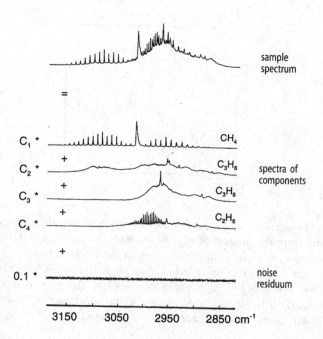

Fig. 7-14. Multi-component analysis via spectral fitting with spectra of pure components (reference spectra), the weights of which are estimated by means of least-squares determination (compensation calculation; the noise spectrum was enlarged by a factor of 10 for better presentation).

interactions with the matrix (pressure spreading in gas analysis, association, shifting of the bands due to solvent effects etc.), corresponding mixture ratios be given for the standards, similar to those expected during the analysis. Via different calibration standards with known composition, the respective "K-spectra" of the components can be calculated by means of determination over the least-squares.

Corresponding tests or model quality tests can be performed, to the extent that non-linear effects e.g. due to insufficient spectral resolution or falsely selected apodization occur. The confidence intervals of the concentration values can also be calculated via the linear regression [27]. Still to be mentioned is that, for example, a spectral baseline or non-linearity in the absorbance can be considered in the spectral fitting [28]. Model errors, such as the ignoring of components that contribute towards the mixed spectrum in the evaluated spectral interval, lead to systematic errors in the concentration calculation. Here, it is recommendable to look at the spectral residues after the spectral fitting in order to see if larger deviations from a pure noise spectrum exist (see Fig. 7-14, also).

7.5.2 Statistical Model Formation

Another way to obtain the chemical composition of a mixture via the measured spectrum is based on the inverse formulation of Beer's law (inverse least-squares technique) with which certain disadvantages of the classical technique through spectral fitting may be avoided. With this, it is possible, for instance, to analyze only one component without having to know the entire composition of the calibration samples. The calibration with three samples at two wavelengths is based then on the following equations:

$$c_1 = P_1 \cdot A_{1,1} + P_2 \cdot A_{1,2}$$
$$c_2 = P_1 \cdot A_{2,1} + P_2 \cdot A_{2,2}$$
$$c_3 = P_1 \cdot A_{3,1} + P_2 \cdot A_{3,2}$$

or

$$\mathbf{C} = \mathbf{A} \cdot \mathbf{P} \tag{7.19}$$

whereby the vector $\mathbf{C}$ incorporates the respective concentration values of the single component to be determined, and the rows of $\mathbf{A}$ contain the coordinates of the corresponding standard spectrum. The vector $\mathbf{P}$ with the "proportionality factors" can be calculated by least-squares fitting. Of course, a formulation for several components is also possible. The general prerequisite is that the number of standard samples is greater than the number of wavenumber data. The calibration samples, the number of which is mostly considerably larger than for the classical calibration batch, should not only include the concentration range of interest but should also include all other expected variance factors.

In general, the occurring linear systems of equations are partially poorly conditioned due to great similarity between the calibration spectra. Consequently, insufficient results are obtained if e.g. a regressor selection of absorbances at wavenumbers was not undertaken which is especially significant for the regression, whereby various strategies for multiple linear regression exist [29].

Another possibility for calculating optimal calibration models is the use of factor-analysis methods. The stances by means of Partial Least-Squares (PLS) and Principal Component Regression (PCR) clearly predominate during spectroscopic applications. Hereby, the calibration matrix is suitably broken down and, based on a certain significant number of factors, a matrix-inversion can be undertaken for estimating suitable regression vectors. In the PCR method, the matrix of the calibration spectra is applied alone for matrix factorizing, while for PLS, the correlation of spectral characteristics is considered with the sample concentrations in the definition of the first factor spectrum [30]. With PLS, a degree of freedom in the calculation of the optimal calibration is the number of successive factors, while differing for PCR, a selection is additionally undertaken according to statistical significance. Generally, the last factors include noise fractions of the spectra, whereby by disregarding these, an elimination of spectral noise is possible.

For poorly conditioned systems of equations, the standard deviation of the fitting is not a suitable criterion in statistical model formation in order to estimate the number of suitable factors. Instead, a validation of the predictability of the calibration model as a function of the number of factors must be performed with calibration-independent samples (also see [31, 32] about this).

An example for this is presented in Fig. 7-15. Of great interest for clinical chemistry is the reagent-free and fast multi-component analysis of human blood by means of IR spectrometry. The IR spectra of blood-plasma samples from a hospital were measured with the ATR method and evaluated mainly in the fingerprint region of mid-infrared [33, 34]. Besides studies on various blood substrates, including glucose, the predictability of calibration models for urea were determined. In Fig. 7-15 a, the independent Standard Error of Prediction (SEP) of the calibration model is recorded as a function of the number of PLS factors considered. An acceptable error of prediction results with a calibration model based on 20 factors, whereby in the selection, a possibly small number of factors also is to be taken into account because of the attempted robustness of the calibration model. Upon using all factors, one obtains, by the way, the least-squares model, the average error of prediction of which is generally almost twice that obtained with the optimal PLS model. Figure 7-15b represents the predicted concentrations by IR spectroscopy as a function of the reference values of the clinical-chemical laboratory.

The range of application of these calibration methods is enor-

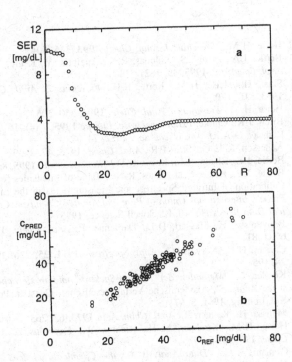

Fig. 7-15. PLS calibration results for urea in blood plasma: a) standard error of prediction (SEP) as a function of the number of PLS factors used for the calibration model, b) validation of the concentrations predicted by IR spectroscopy with an optimal PLS model in comparison to the clinical-chemical laboratory values.

mous by using IR spectra. In particular, many applications have been established in investigating textiles and polymers, foodstuffs and other agricultural products as well as in the area of petrochemistry and medicine, to name just a few examples and illustrate the range. The interested reader may find further information on this in the monographs [35, 36].

To note here is that not only substance concentrations can be recorded against spectra but also other properties of sample systems that are somehow manifested in the spectrum. For instance, the salt content of seawater, the temperature or other physical-chemical properties of water were determined, among others, via the spectrum measured in the near-infrared region (see e.g. [37]). Other so-called "exotic" applications in the literature include the determination of the taste of peas or the digestibility of animal feeds. In all, many applications are found in the area of online processing analysis or sample input control, whereby the special advantage of non-invasive IR spectroscopic methods is its reliability, speed and range of application possibilities.

References

[1] Heise, H.M., *Fresenius J. Anal. Chem.* **1993**, 346, 604
[2] Bertie, J.A., Zhang, S.L., Jones, R.N., Apelblat, Y., Keefe, C.D., *Appl. Spectrosc.* **1995**, 49, 1821
[3] Staat, H., Heise, H.M., Korte, E.H., *Fresenius Z. Anal. Chem.* **1983**, 316, 170
[4] Heise, H.M., *Fresenius J. Anal. Chem.* **1994**, 350, 505
[5] Birch, J.R., Clarke, F.J.J., *Spectrosc. Europe* **1995**, 7 (4), 16
[6] Ramsay, D.A., *J. Am. Chem. Soc.* **1952**, 74, 72
[7] Anderson, R.J., Griffiths, P.R., *Anal. Chem.* **1975**, 47, 2339
[8] Bertie, J.E., Zhang, S.L., Keefe, C.D., *Vibr. Spectrosc.* **1995**, 8, 215
[9] Bertie, J.E., Keefe, C.D., Jones, R.N.: "Table of Intensities for the Calibration of Infrared Spectroscopic Measurements in the Liquid Phase", *International Union of Pure and Applied Chemistry Chemical Data Series* No. 40, Blackwell Science, **1995**
[10] Bouveresse, E., Massart, D.L., Dardenne, P., *Anal. Chem.* **1995**, 67, 1381
[11] Chasan, D.E., Norwitz, G., *Appl. Spectrosc.* **1971**, 25, 226; **1974**, 28, 195
[12] Kössler, I., *"Methoden der Infrarot-Spektroskopie in der chemischen Analyse"*, Akademische Verlagsgesellschaft Geest u. Protig KG, Leipzig, **1961**, S. 97
[13] Malissa, H., Kellner, R., *Anal. Chim. Acta* **1973**, 63, 263
[14] Heise, H.M., Kirchner, H.-H., Richter, W., *Fresenius Z. Anal. Chem.* **1985**, 322, 397
[15] Graham, R.C., *"Data Analysis for the Chemical Sciences – A Guide to Statistical Techniques"*, VCH Publ. New York, **1993**
[16] Adams, M.J., *"Chemometrics in Analytical Spectroscopy"*, Royal Society of Chemistry, Cambridge, **1995**
[17] Rechenberg, W., *Fresenius Z. Anal. Chem.* **1982**, 311, 590
[18] Bubert, H., Klockenkämper, R., *Fresenius Z. Anal. Chem.* **1983**, 316, 186
[19] Massart, D.L., Vandeginste, B.G.M., Deming, S.N., Michotte, Y., Kaufmann, L., *"Chemometrics: A Textbook"*, Elsevier, Amsterdam, **1988**
[20] Ripley, B.D., *Analyst* **1988**, 113, 1469
[21] Ebel, S., *Comp. Anw. Lab.* **1983**, 1, 55
[22] Ebel, S., Alert, D., Schaefer, *Comp. Anw. Lab.* **1983**, 2, 172
[23] Schwartz, L.M., *Anal. Chem.* **1979**, 51, 723
[24] DIN 32645 *"Nachweis-, Erfassungs- und Bestimmungsgrenze"*, DIN Deutsches Institut für Normung, Beuth Verlag, Berlin, Mai **1994**
[25] Heise, H.M., *Fresenius Z. Anal. Chem.* **1986**, 323, 368
[26] Ebel, S., "Fehler und Vertrauensbereiche analytischer Ergebnisse", in: *Analytiker-Taschenbuch*, Bd. 11, Günzler, H., Borsdorf, R., Danzer, K., Fresenius, W., Huber, W., Lüderwald, I., Tölg, G., Wisser, H. (Hrsg)., Springer-Verlag, Berlin, **1993**

[27] Saarinen, P., Kauppinen, J., *Appl. Spectrosc.* **1991**, 45, 953

[28] Haaland, D.M., Easterling, R.G., Vopicka, D.A., *Appl. Spectrosc.* **1985**, 39, 73

[29] Draper, N.R., Smith, H., *"Applied Regression Analysis"* 2nd Ed., Wiley & Sons, New York, **1981**

[30] Marbach, R., Heise, H.M., *Chemom. Intell. Lab. Syst.* **1990**, 9, 45

[31] Marbach, R., Heise, H.M., *Trends Anal. Chem.* **1992**, 11, 270

[32] Martens, H., Næs: *"Multivariate Calibration"*, Wiley & Sons, Chichester, **1989**

[33] Heise, H.M., Marbach, R., Koschinsky, Th., Gries, F.A., *Appl. Spectrosc.* **1994**, 48, 85

[34] Heise, H.M., Bittner A., *J. Mol. Struct.* **1995**, 348, 127

[35] Burns, D.A., Ciurzak: *"Handbook of Near-Infrared Analysis"*, Marcel Dekker, New York, **1992**

[36] Murray, I., Cow, I.A.: *"Making Light Work: Advances in Near Infrared Spectroscopy"*, VCH, Weinheim, **1992**

[37] Lin, J., Brown, C.W., *Appl. Spectrosc.* **1993**, 47, 1720

8 Spectroscopy in Near- and Far-Infrared as well as Related Methods

8.1 Spectral Regions Beyond Mid-Infrared

For the classical uses of IR spectroscopy, one limited oneself mainly to the mid-IR region where the fundamental vibrations are found. Tab. 8-1 presents an overview of the various regions and the basic phenomena.

Table 8.1. Overview of the various IR spectral regions.

	Near-IR		Mid-IR		Far-IR
Wavelength:	0.78	– 2.5	–	25	– 1000 µm
Wavenumber:	12800	– 4000	–	400	– 10 cm^{-1}
Phenomena:	overtones and combination vibrations mainly involving molecular groups with C-H, O-H and N-H long-wave electron transitions		fundamental and combination vibrations involving light and heavy atoms		fundamental and combination vibrations involving heavy atoms furthermore: molecular skeleton vibrations, molecular torsions, crystal lattice vibrations, rotational spectra of small molecules

8.1.1 Near-Infrared Spectroscopy

8.1.1.1 Comparison of Mid-Infrared and Near-Infrared Spectroscopy

Near-infrared spectroscopy as a valuable tool for qualitative and quantitative analysis [1, 2] has experienced a revival through the development of chemometric methods and the introduction of fiber optics during the 1980s, and high-quality near-infrared spectra are provided by improved conventional and by Fourier transform instruments. Various aspects differentiate a mid-infrared spectrum from the corresponding near-infrared spectrum (Tab. 8-1). In the mid-infrared region, fundamental molecular vibrations always oc-

cur between 4000 and 400 cm^{-1}, while in the near-infrared region
overtones and combination bands of these fundamentals are observed between 12 800 and 4000 cm^{-1}. Near-infrared bands are
due primarily to hydrogenic stretches of C–H, N–H, and O–H
bonds whose fundamental vibrations give rise to strong bands between 4000 and 2000 cm^{-1}, so that their overtones and combinations occur above 4000 cm^{-1} as the most intense absorption bands
in the near-infrared spectrum. Since the near-infrared absorptions
of polyatomic molecules thus mainly reflect vibrational contributions from very few functional groups, near-infrared spectroscopy
is less suitable for detailed qualitative analysis than mid-infrared
spectroscopy, which shows all (active) fundamentals and the overtones and combinations of low-energy vibrations. The near-infrared overtone and combination bands depend more on their environment than does the fundamental of the same vibration; slight
perturbation in the bonding scheme causes only small changes
in the fundamental but significant frequency shifts and amplitude
changes in the near-infrared. In going from the fundamental to the
first overtone the intensity of an absorption band decreases by a
factor of about 10–100, so that the sensitivity of near-infrared
spectroscopy is reduced in comparison with the mid-infrared.
This is a disadvantage when gases are to be measured, but for liquids, it is a considerable advantage as regards sample handling
because cells with convenient path lengths between 1 mm and
10 cm can be used. Glass or quartz is used as window material.
Near-infrared bands of liquids and solids have relatively large
bandwidths between 30 and 60 cm^{-1}, so that they strongly overlap and direct assignment of bands is generally not possible for
larger, complex molecules. This is why, in the near-infrared, easy
quantitative analysis using isolated bands, as in the mid-infrared,
is not possible. Chemometrics [3–5], however, by the application
of mathematical and statistical procedures, generates correlations
between experimental data and the chemical composition or physical properties of the sample, and can be used in a general manner
to solve quantitative and qualitative analytical problems.

8.1.1.2 Applications of Near-Infrared Spectroscopy

Applications of near-infrared spectroscopy include chemistry [6,
7], the oil industry [8], clinical analysis [9], biological and med-
ical analysis [10], and the pharmaceutical industry [11, 12].
Since the intensities of characteristic near-infrared bands are independent of or only slightly dependent on the state of the system, near-infrared spectroscopy is widely applicable for the
quantitative study of liquid and compressed gaseous systems, including fluid systems, up to high pressures and temperatures.
The application of near-infrared data to routine quantitative

analysis was initiated by Norris [13], who used diffuse reflectance measurements to quantitatively determine major components, such as moisture, protein and fat, in agricultural commodities. In comparison with mid-infrared, near-infrared diffuse reflectance [14] is able to measure powdered samples with minimal sample preparation, and lends itself extremely well to quantitative analysis because the smaller absorptivities and larger scattering coefficients at shorter wavelengths result in a larger scattering/absorption ratio. Although near-infrared reflectance analysis was initially developed for agricultural and food applications [15], in combination with chemometrics it is now applied to many other areas, e.g., polymers [16], pharmaceuticals [17], organic substances [18], geological samples [19], and thin-layer chromatography [20, 21]. The excellent transmittance of quartz in the near-infrared range has led to a further enhancement of the potential of near-infrared spectroscopy by the introduction of fiber optics. Fiber-optic waveguides are used to transfer light from the spectrometer to the sample and, after transmission or reflection, back to the spectrometer. Most fiber optic cables consist of three concentric components: the inner part, through which the light propagates, is called the core, the middle layer is the cladding, and the outer protective layer is the jacket. Crucial to the light throughput of the fiber optic cable are the optical properties of the core and cladding. The difference in refractive index of the two materials defines the highest angle at which the core/cladding interface exhibits total internal reflection, and hence allows the light to propagate through the cable. This angle is related to the numerical aperture (N.A.) of the fiber by the following equation:

$$\sin a = \text{N.A.} = (n_{\text{core}}^2 - n_{\text{cladding}}^2)^{1/2} \qquad (8.1)$$

where is the half-angle of acceptance and n is the refractive index of the material. The greater the difference in refractive index the greater the half-angle of acceptance. Any light entering at an angle greater than the half-angle of acceptance is lost to the cladding through absorption. The core and cladding are usually of quartz (of different refractive index), while the outer protective layer is a polymer; other fiber optic materials are zirconium fluoride (ZrF) and chalcogenide (AsGeSe) [22]. The obvious advantage of the optical fiber technique is that the location of measurement can be separated from the spectrometer by between two and several hundred meters, depending on the length of the waveguide. Since almost all practically relevant substances show characteristic near-infrared absorption bands, quantitative analysis by near-infrared spectroscopy is generally applicable to on-line concentration measurements in connection with chemical reactions, chemical equilibria, and phase equilib-

ria. Various probes can be integrated into different reactors or
bypasses and offer numerous remote-sensing and on-line process control applications [23]. In the analysis of hazardous or
toxic materials, this has proved of outstanding importance. In
the refining, petrochemical, polymer, plastics, and chemicals industries, typical applications of near-infrared remote spectroscopy are the determination of the octane number and oxygenates
in gasoline, aromatics in fuels, the composition of solvent mixtures [24, 25], the hydroxyl number of polyols, low contents of
water in solvents, and for polymer analysis.

8.1.2 Far-Infrared Spectroscopy

Towards the long-wave side, the spectrum is mostly limited by
the window material of the cell. By suitable selection, e.g. polyethylene, wavenumbers down to $10 \, \text{cm}^{-1}$ can be easily reached.
Much more aggravating in this range is the low available energy
that originates from thermal radiators. Dispersive instruments
could be realized by wide slits and dry-air purging or evacua-
tion of the instrument because of the considerable steam absorption [26]. Their further development was interrupted by FT
spectroscopy (see Sect. 3.4.2). The first FT spectrometers were
only intended for the long-wave IR region due to technical reasons. The advantages of Fourier instruments have already been
mentioned in Sect. 3.4.3.3, whereby this has introduced a decisive improvement for spectroscopy in far-IR because of the limited spectral radiation power.

In far-IR, we find the stretching vibrations of molecules having heavy atoms (also see Tab. 8-1). In inorganic compounds
and metalorganic complexes [27], primarily the central atom-ligand vibrations and the crystal lattice vibrations show absorption bands below $400 \, \text{cm}^{-1}$. Moreover, molecular deformation
vibrations have their frequencies in the far-IR. Thus, new studies about ring puckering, torsion and skeletal vibrations have
been generally possible, see e.g. [28]. For determining thermodynamic data, spectroscopy in this region is extremely useful. In
addition, rotational spectra of gas molecules with two or three
atoms as well as optical properties of semiconductors and di-
electrics can be measured here [29]. Besides, spectroscopy of
oxidative superconductors is an interesting area that experienced
great interest for a longer time [30]. Moreover, Raman spectroscopy can also principally deliver the same or partially complementary statements (see Sect. 8.3).

8.2 Infrared Laser Spectroscopy

According to the classical scheme of spectrometer design with a radiation source, monochromator and detector, one could do without the monochromator part when a possibly monochromatically tunable radiation source is available. This path can be followed since the use of lasers has come into being. These radiation sources are characterized by the acronym for "**L**ight **A**mplification by **S**timulated **E**mission of **R**adiation". In the laser medium, coherent, i.e. equal phase amplification occurs by stimulated emission, for which the prerequisite is a greater occupational density in the upper energy level than in the lower one, which can be obtained by external energy addition, so-called "pumping" [31].

In the IR lasers, one can differentiate between gas and solid lasers. The CO_2 laser is the best-known gas laser; besides, this is also used for the analysis of atmospheric air, e.g. for LIDAR (**L**ight **D**etection **a**nd **R**anging). Among the tunable lasers, the diode laser has gained great importance. One often finds these in recreational electronic instruments (CD-players) or in communicational technology. The IR spectroscopy inherently profits from such developments. One can group the commercial diode lasers generally into two categories: one is based on III–V-semiconducting materials which allow diodes for the spectral range of between 0.63 and about 1.55 µm. The other class is produced form IV–VI-semiconducting materials that cover the range of between 3 and 30 µm. Unfortunately, their operation requires very low temperatures in contrast to the former type (see Fig. 8-1).

In general, the excess electrons in an n-conductor recombine with the holes of a p-conductor in the interface, and thereby give off heat. For the laser diodes, the result of this recombination process is the release of coherent, extreme monochromatic IR radiation with a high energy density, whereby a spectral resolution of 10^{-4} cm^{-1} is obtained with low effort (see Fig. 8-2). The tuning takes place via a change of the diode current or of the temperature in a range of 15 to 100 K at a temperature sta-

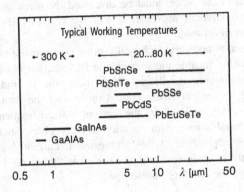

Fig. 8-1. Overview regarding the application range of various diode laser materials.

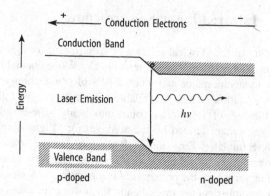

Fig. 8-2. Energy diagram of a diode laser.

bility of 3×10^{-4} K (tuning region of $4\,\mathrm{cm^{-1}/K}$). The tunability is understandable by the change of the refraction index of the semiconducting material and of the therein-connected change of the resonance frequency in the Fabry-Pérot resonator.

By coupling to a changeable external resonator, laser diodes can likewise be tuned. The advantage is a higher spectral resolution on the one hand, and, on the other hand, a better reproducibility and independence from aging processes in the laser material; nonetheless, additional effort is required for the resonator.

Despite the high spectral resolution, commercial laser spectrometers are still equipped with a monochromator for mode selection, if one cannot fall back upon expensive monomodal lasers. In general, laser spectroscopy allows the acquisition of highly resolved spectra of gases, which is advantageous for the determination of more exact molecular parameters. However, such extremely strong detection systems are also applied excellently for gas analysis, whereby multi-reflection cells with optical pathlengths of around 100 m are used under vacuum conditions. However, photoacoustic detection for extremely low sample-gas volumes can be considered to be just as successful [32]. NIR lasers, which are operated at room temperature, naturally have operative advantages over the lead salt lasers, but the lower absorption coefficients of the then existing overtone bands, compared to the latter, must be discussed. Nevertheless, strong detection analytical instruments, e.g. for gaseous HF or NH_3, have been developed.

Another, nonetheless technically more complicated possibility for producing tunable monochromatic IR radiation is offered by optical parametric oscillators. With them, the IR radiation is produced by the non-linear optical process of the formation of difference frequencies from the light of a fixed-frequency laser and of a tunable laser with emission in the visible or near-IR region. The advantages of this arrangement, currently used predominantly in basic research, are the wide tuning regions, high time-resolution and large pulse power.

8.3 Raman Spectroscopy

8.3.1 Physical Foundations

IR spectroscopy is very closely related to Raman spectroscopy [33–36], which is likewise a method for observing molecular vibrations. Both types of spectroscopy often give similar spectra, but also enough differences exist so that they can provide supplemental or complementary information. There are particular applications for which IR spectroscopy is superior to Raman spectroscopy and *vice versa* (also see Tab. 8-2).

Table 8-2. Overview of the particularities of Raman spectroscopy compared to IR.

- Other selection rules; i.e. partially complementary information,
- Low sensitivity, because of the weaker scattering effect,
- Well-applicable to aqueous solutions, because there are no significant bands causing interference in the solution spectrum (limitation: FT Raman spectroscopy with long-wave excitation),
- With visible excitation, the fluorescence of the sample or of contaminants may overlap the signal and prevent a measurement,
- Generally, no sample preparation is necessary, measurement through optically transparent materials (including packaging, glass windows, crystals),
- Bands below $400\ \text{cm}^{-1}$ down to a few cm^{-1} (e.g. crystal lattice vibrations) are measurable with the same apparatus,
- Good suitability for microanalysis.

The Raman effect is based on a different interaction between material and electromagnetic radiation than that for IR spectroscopy. If a sample is irradiated by a laser with monochromatic radiation, the following physical phenomena occur:

- *Absorption and fluorescence*: if the energy from the $h\nu_0$-quanta is sufficient to excite an electron jump, then the applied radiation is absorbed. The excitation energy of the electrons can partially or totally be converted into heat, but also be given up again as longer wave emission radiation, the so-called fluorescence.
- *Raleigh scattering*: a part ($\approx 10^{-4}\ \Phi_0$) is scattered with the original frequency ν_0 in all spatial directions. One can conceive this scattering fraction as arising from elastic collisions of the radiation quanta ($h\nu_0$) with the molecules of the sample.
- *Raman scattering*: a small fraction of the light ($\approx 10^{-8}\ \Phi_0$) is scattered non-elastically, e.g. with alteration of the frequency. The frequency distribution or the spectrum of the scattered light shows bands with frequency differences to the

excitation that correspond to the vibrational frequencies of the molecule (Fig. 8-3).

One can imagine the formation of a Raman spectrum as follows: a part of the incident monochromatic radiation is scattered elastically by the molecules (Raleigh scattering), but a few of the quanta release a part of their energy to the molecules and again exit the sample scattered with a lower frequency ν_R, as they are now energy-poorer by the amount of energy transferred to the molecules. The difference $\Delta E = h \cdot (\nu_0 - \nu_R)$ is consumed by the molecule for increasing its vibrational energy. We thus have a process here that is analogous to the Compton effect. If the

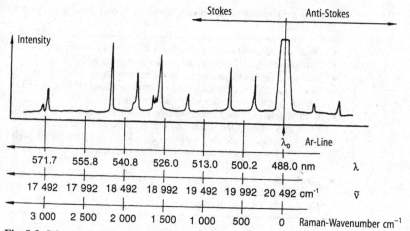

Fig. 8-3. Scheme of a Raman spectrum (λ_0 = wavelength of incident radiation, e.g. for Ar-laser 488.0 nm, corresponding to $\tilde{\nu}_0 = 20\,492$ cm^{-1}). Upper scale: wavelength of the scattered radiation; middle scale: wavenumbers of the scattered radiation; lower scale: difference between the wavenumbers of the scattered radiation and the excitation radiation $\tilde{\nu}_R - \tilde{\nu}_0$.

excitation energy $h\nu$ is sufficient for an electron jump, then the Raman effect can be totally masked by the fluorescence. These processes can be presented in a term scheme (see Fig. 8-4).

Figure 8-4c shows another possible case: when the excitation radiation hits the vibration-excited molecules, then they can transfer their vibrational energy to the incident radiation. Then, an energy-richer quantum is emitted from the sample. Hence, there are also Raman lines on the short-wave side of the Raleigh line; these are called "anti-Stokes lines" in contrast to the "Stokes-lines" in the long-wave region (see also Fig. 8-3).

With these basic principles, the essential prerequisites for Raman spectroscopy can be understood:

- The excitation should be done with monochromatic radiation. With several excitation frequencies, each would produce a

Raman spectrum for itself, thereby resulting in interlocking spectra.

- Small fractions of fluorescing foreign materials in the study sample may mask the Raman spectrum by their intense fluorescence.
- Due to the low yield of Raman scattered radiation, one needs very intense radiation sources.
- Because upon Raman scattering, the entire spectrum is formed simultaneously, a parallel measurement of all spectral elements is appropriate for optimal detection limits.

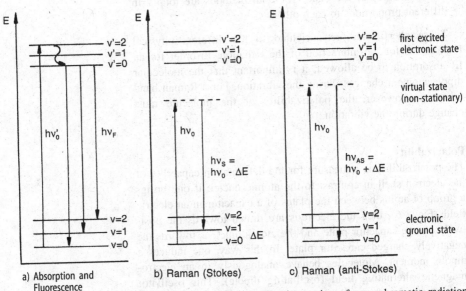

Fig. 8-4. Term scheme for the interaction of monochromatic radiation with molecules (v and v′ vibrational quantum numbers of vibrational states in the electronic ground state and excited state, respectively; v_0 and v'_0 excitation frequencies, v_F fluorescence frequencies, v_0–v_S Stokes-Raman line, v_0–v_{AS} anti-Stokes-Raman line.

These physical prerequisites principally affect the instrument design (see Sect. 8.3.2). One should note that the resonance-Raman effect occurs when the virtual state approaches an electronic excited state, thereby resulting in an observable extensive increase of the Raman scattering up to around a factor of 10^6.

For selecting the excitation wavelength, one finds that the emission power of the induced dipoles is proportional to $v^4{}_S$, with which advantages result from choosing a high excitation frequency v_0. On the other hand, disturbing fluorescence can be excited with large quantum energies $h v_0$, which is why a relatively low excitation frequency should be chosen. Compromises consist in either using argon-ion lasers (wavelengths at 514 nm (green) and 488 nm (blue-green) with intense Raman scattering, but occasionally disturbing fluorescence) or using Nd:YAG lasers at 1064 nm, with which practically no fluorescence occurs but an about 20-fold weaker scattering is measured compared to the Ar^+-laser.

We have found that Raman and IR spectroscopy can produce complementary information. Upon comparing a Raman spectrum with an IR spectrum (Fig. 8-5), clear differences are observed that cannot be explained only by the cited fundamentals:

- Especially for symmetrical molecules with an inversion center, bands appear in the Raman spectrum that are not found in the IR spectrum, whereby the reversed case can occur.
- The intensities of the bands for both methods are totally in different proportions to each other.

These phenomena are attributed to the different physical foundations for the interaction of the radiation. In order for an IR absorption to be allowed, it is important that the molecular dipole moment changes during the vibration. For a Raman band to occur, however, the polarizability of the molecule must change during the vibration.

Polarizability

The polarizability is a measure for the deformation capability of the electron shell in contrast to the atomic nuclei: if one brings a group of atoms between the plates of a capacitor in an electric field (Fig. 8-6), then the electrons are drawn towards the positively charge capacitor plate and the atomic nucleus towards the negatively charged capacitor plate. In this way, one induces a dipole moment. Molecules behave analogously in an electromagnetic alternating field (oscillating dipole). This oscillator emits radiation of its vibrational frequency corresponding to the Raleigh line and the Raman transitions (see Fig. 8-4).

Considering the polarizability of a CO_2 molecule, it depends on the type of vibration of the atomic nuclei against each other (Fig. 8-7). In the stretched state, the polarizability is smaller than in the crunched state. Hence, the polarizability changes with the symmetrical stretching vibration (form I and II). However, it is to be noticed that the dipole moment, as a whole, has not changed.

The opposite is true for the antisymmetric stretching vibration. Here, the polarizability for the transition from form III to form IV does not change. Thus, this vibration is forbidden in the Raman spectrum. Similar conditions are often found for such symmetrical molecules. The change of the polarizability is dependent on the molecular geometry. Figure 8-8 elucidates the conditions by using a tetragonal bi-pyramid as an example.

The equation scheme in Fig. 8-8 shows the dependency of the induced dipole moment $\mu_{x,y,z}$ on the components of the elec-

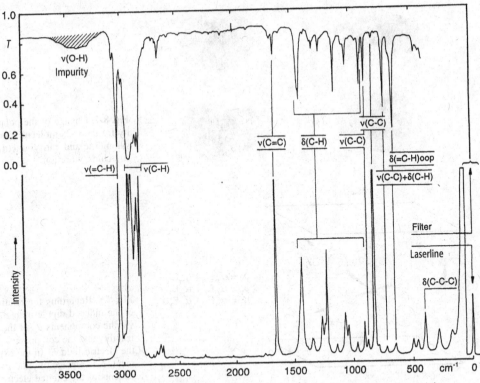

Fig. 8-5. IR- and Raman spectrum of cylcohexane for comparison.

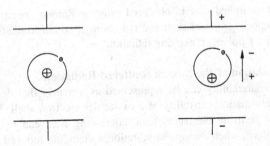

Fig. 8-6. Effects of an electric field on the charge-distribution in the atom.

tric field strength $E_{x,y,z}$. The proportionality factors a_{ij} are the polarizabilities, which represent the elements of the so-called polarizability tensor. The tensor, by the way, is symmetrical, which means $a_{ij} = a_{ji}$. It assigns fractions of the polarizability to the vectors of the electric field that comprise the respective dipole moment components, whereby outer-diagonal tensor elements such as a_{xy}, a_{yz} and a_{xz} describe the induction of the dipole moment fractions which do not concur with the electrical field direction. It is decisive which components of the polarizability change during the vibration of the nuclei. With this, an

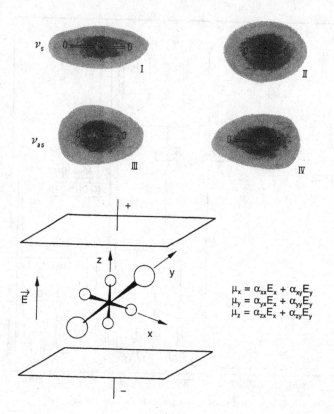

ν_s

I

II

ν_{as}

III

IV

Fig. 8-7. Change of the polarizability for the CO_2-molecule during the symmetric and antisymmetric stretching vibration.

$$\mu_x = \alpha_{xx}E_x + \alpha_{xy}E_y$$
$$\mu_y = \alpha_{yx}E_x + \alpha_{yy}E_y$$
$$\mu_z = \alpha_{zx}E_x + \alpha_{zy}E_y$$

Fig. 8-8. Regarding the relationship of the induced dipole moment $\mu_{x,y,z}$ via the components a_{ij} of the polarizability and the components $E_{x,y,z}$ of the electric field $\vec{E}$. In the experiment, $\vec{E}$ is replaced by the $\vec{E}$-components of the excited electromagnetic field.

effect is linked that is observed only in Raman spectroscopy: the depolarization of scattered radiation, although irradiation occurs with polarized exciting radiation.

Polarization of the Raman Scattered Radiation

The polarizability can be represented as a matrix that describes the deformation capability of a molecular electron shell. Energy uptake from an electromagnetic alternating field can only then take place, when during the vibration a changing induced dipole moment results parallel to the components of the electric vector. For gases and liquids, this is always the case for the entire sample, because the molecules have no fixed orientation. However, the polarization of the scattered radiation depends on the components of the induced dipole moment. Let us first take the total symmetric vibration of the CCl_4 molecule as an example [37]. An isotropic, i.e. ball-formed distribution of the vibration-dependent polarization change exists here.

The exciting field induces a dipole parallel to its vibrational orientation. This dipole emits secondary radiation perpendicular to its vibrational orientation (Fig. 8-9), but not parallel to it. Upon measuring the polarization components of the scattered ra-

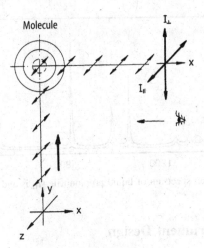

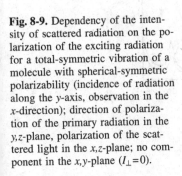

Fig. 8-9. Dependency of the intensity of scattered radiation on the polarization of the exciting radiation for a total-symmetric vibration of a molecule with spherical-symmetric polarizability (incidence of radiation along the y-axis, observation in the x-direction); direction of polarization of the primary radiation in the y,z-plane, polarization of the scattered light in the x,z-plane; no component in the x,y-plane ($I_\perp=0$).

diation once parallel to the z-axis and another time parallel to the y-axis, various values are found.

The ratio $I_\perp/I_{II}$ is called the degree of depolarization σ. Hereby, I_{II} is the intensity of the scattered beam measured parallel to the polarization direction of the exciting radiation, and $I_\perp$ is the perpendicular component of the Raman radiation (parallel to the direction of incidence of the exciting radiation). For the above example of the CCl_4 vibration, the degree of depolarization is $\rho=0$.

For all other molecular vibrations which are not totally symmetric, the Raman scattered radiation has components in both directions of y and z. The scattered radiation is "depolarized". Thus, it is possible via the degree of depolarization to draw a conclusion about the symmetry of the vibration. For non-polarized (e.g. natural) exciting radiation, totally symmetrical vibrations have a value for ρ_n of $0\leq\rho\leq\rho_n<6/7$, others have a degree of polarization of $\rho_n = 6/7$. Upon using linearly polarized radiation, as e.g. in laser light, totally symmetric vibrations show a degree of polarization ρ_L of $0\leq\rho_L<3/4$. This applies for substances with statistical molecular orientation such as gases and liquids.

In all, up to six different polarization components can be measured on single crystals by corresponding combination of sample orientation as well as incidence- and observed polarization. Because such studies are easy to perform, they have extensively displaced polarization measurements in IR and additional information can be obtained about the symmetry of vibrations. Important is that conclusions can be drawn about the molecular design and about band assignment, because with this, one can individually examine the polarizability tensor.

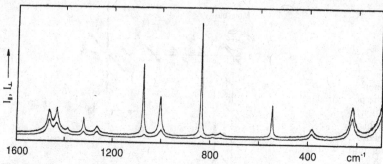

Fig. 8-10. Raman spectrum of liquid propionitrile (I_{II} is and $I_{\perp}$).

8.3.2 Instrument Design

8.3.2.1 Conventional Grating Spectrometer

The small fraction of frequency-shifted scattered radiation in the Raman effect requires an intense radiation source. For special applications with a weak signal (e.g. for gases), observation occurs perpendicular to the irradiation direction (see Fig. 8-11). The 180°-arrangement is simpler, because here the foci of the laser and of the collection optics automatically agree (Fig. 8-12).

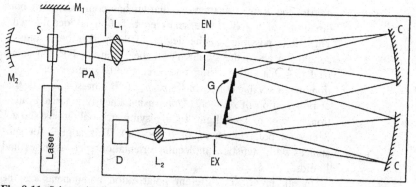

Fig. 8-11. Schematic diagram of a Raman spectrometer (S sample, PA polarization analyzer, M_1 and M_2 mirrors, $L_{1,2}$ lenses, EN and EX entry and exit slit, G grating, C collimator mirror, D detector).

The laser emits monochromatic radiation that penetrates the study substance S. The mirror M_1 reflects the excitation beam and doubles its effect. This, however, can be problematical because of the retro coupling in the laser. In this example, the scattered light is focused 90° to the direction of incident radiation through the lens L_1 onto the entry slit EN. The mirror M_2

doubles the intensity of the collected scattered radiation. It is reflected from the collimator mirror C onto the grating G, spectrally separated there, and it finally hits the detector D after passing through the exit slit EX. Double or triple monochromators are often applied to minimize undesired scattered light.

If one uses radiation in the visible spectral region, photomultipliers can be applied as detectors. Instead of PA, an analyzer can be applied for studying the polarization direction of the scattered radiation. Many problems of low signal intensity such as which occur in IR spectrometers are eliminated by the photomultiplier detection. One can easily register lower-frequency vibrations until the so-called "Raleigh line" masks these by its intensity. Without making particular provisions, one reaches Raman shifts down to about 100 cm^{-1}, and with more effort, down below 5 cm^{-1} (also see Fig. 8-10).

Nowadays, multi-channel detectors are mostly used with which so-called polychromators can be designed (see Fig. 8-12). For simultaneously detecting an entire spectral interval, a detector array is needed. Cooled CCDs (charge-coupled devices), developed on a silicon semiconductor basis, are extremely sensitive for the region of 300 to 1100 nm [38]. Because of the existing multiplex advantage, weaker lasers, e.g. diode lasers, can suffice for Raman measurement.

Until about 1960, large sample quantities and the low power of a mercury low-pressure lamp limited the applicability of this method. With introduction of the laser, the irradiation density rose by 10 orders of magnitude. Argon-ion lasers with radiation powers in the Watt range are often applied. A measurement volume of ca. $1 \text{ } \mu m^3$ (corresponding to about 10^{-12} g) can be realized with a con-focal Raman microscope [39], and the recording

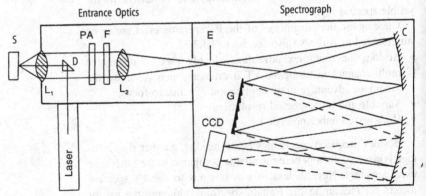

Fig. 8-12. Schematic diagram of a Raman spectrometer with multi-channel detection (S sample, PA polarization analyzer, D deflection prism, $L_{1,2}$ lenses, F filter for suppressing the Rayleigh line (on interference basis or holographically), E entry slit, G grating, C collimator mirror, CCD charge-coupled device.

times decreased from·hours to minutes with simultaneously definite improvement of the signal-to-noise ratio.

The line widths of the incident light no longer play any role
for highly resolved gas spectra: Hg-burners ($\lambda=435.8$ mm):
$\Delta\tilde{v}=0.3$ cm^{-1}; Ar-lasers ($\lambda=514$ nm): $\Delta\tilde{v}=0.15$ cm^{-1} with eta-
lon (single-mode operation): 0.001 cm^{-1} [40]. Such single-mode
lasers are used for measuring non-linear Raman effects (see e.g.
[41]), for which the experimental expense is considerably greater, however. Especially the coherent anti-Stokes Raman spectroscopy (CARS) is very interesting for studying highly resolved gas spectra, because herewith spectral resolutions better
than 0.001 cm^{-1} can be obtained. Nonetheless, please refer to
further literature for more details, e.g. [42].

8.3.2.2 FT-Raman Spectrometer

An important limitation of Raman spectroscopy was always the
presence of fluorescence from the sample, whether it is from
analytes or possible contaminants. Thus, colored substances
could hardly be analyzed by spectroscopy. A solution to this
problem was found with excitation at longer wavelengths. The
application of Nd:YAG lasers is very well suited for this
($\lambda_0=1064$ nm), because its radiation generally cannot cause any
electron excitation. However, one trades off the background
freedom for very small signal intensities due to the small cross-
sections of Raman scattering that, as was already mentioned in
Sect. 8.3.1, are proportional to the function v^4. An additional aspect is that the semiconducting detectors, available in the near-
IR region and mostly consisting of InGaAs or Ge, do not show
the same detectivity as e.g. photomultipliers or CCDs in the
visible spectral region.

Once again, the advantages of the FT spectrometers are to be
summarized as follows (also see Sect. 3.4.3.3):
- Besides the multiplex advantage already used in dispersive
  multi-channel instruments, FT instruments may also use the
  Jacquinot-advantage upon application of a macro-focus,
- Variable and high spectral resolution,
- High wavenumber precision.

Besides precision mechanics being needed, another disadvantage is worse detectors being available compared to the otherwise
used detectors. Nevertheless, it was obvious to use FT spectrometers for measuring the Raman spectrum, although the use of
such instruments was very controversial for a long time. Of para-
mount importance for such a system are extremely efficient filters
(e.g. holographic filters) in order to suppress the Rayleigh scattered radiation. Please refer to the literature for a further discus-

sion on the advantages and disadvantages of these instruments compared to dispersive Raman spectrometers [34, 43].

8.3.3 Applications

Many assertions of vibrational spectroscopy can be derived from the IR as well as the Raman spectrum. Because IR spectroscopy is easier to work with, IR spectra were mostly resorted to for acquiring information. However, the use of Raman spectroscopy has increased since the application of laser light sources and, recently, also through the FT Raman spectroscopy. While the IR spectroscopy is excellently suited for studying polar compounds, predominantly non-polar substances are analyzed by Raman spectroscopy. Various fields of application are based on this.

8.3.3.1 Assignment of Vibrational Bands

Each spectral interpretation lastly includes the assignment of all fundamental vibrations to particular bands, the determination of power constants, the study of vibrational interactions and the potential energy distributions, besides intensity specifications or assertions about thermodynamic parameters. For that purpose, knowledge of the Raman spectrum is absolutely a prerequisite. For highly symmetrical molecules, stereochemical problems cannot at all be answered alone from the IR spectrum. Hence, a large application area of Raman spectroscopy concerns interpretation- and assignment problems (normal coordinate analysis).

8.3.3.2 Aqueous Samples

Water is an unsuitable solvent in IR spectroscopy, with the exception that the ATR technique is applicable. Inversely, water has a line-deficient and low-intense Raman spectrum, with which aqueous solutions can be analyzed by Raman spectroscopy without any problem. This opens up wide possibilities for analyzing exclusively water-soluble substances.

Nevertheless, quantitative analyses e.g. for metabolites such as glucose, urea and lactic acid in aqueous solution are still better to be performed with IR than Raman spectroscopy [44, 45]. However, the characterization of various types of human body tissue, including breast cancer tissue, via Raman spectroscopy was recommended by various authors [46, 47].

8.3.3.3 Quantitative Analysis

The technique of sample preparation has been refined recently. In particular, for example, air-sensitive substances can be directly taken up in the glass ampoule, so that complicated transfer work with protective gas is no longer required. Quantitative measurements can be conducted via corresponding calibration curves, as in IR spectroscopy. Mainly inorganic substances, which often show only few bands in the IR spectrum due to high symmetry, can be measured quantitatively e.g. on Raman-permitted but IR-inactive bands. A prerequisite for this is a very stable sample holder, so that the cells can be inserted reproducibly. Moreover, multivariate calibrations were introduced by means of PLS e.g. for determining octane numbers of fuels [48].

An interesting application of Raman spectroscopy has been described for the analysis of automobile exhaust. Advantageous here is that also components can be analyzed such as O_2, N_2 and H_2 that cannot show any IR spectrum due to their symmetry (exception: "forbidden transitions" of O_2 in near-IR) [49, 50]. Because the Raman scattering cross-sections are relatively small, one uses another effect, i.e. *Surface Enhanced Raman Scattering* (*SERS*). Substances are hereby analyzed by spectroscopy, which are adsorbed to rough metallic surfaces or metal particles consisting of silver or gold [51]. The observed amplification factors lie at about 3 to 6 orders of magnitude, so that this measurement technique, combined with miniaturized Raman spectrometers, comes into question for the development of sensor systems (see e.g. [52, 53]).

8.3.3.4 Investigations of Solids

In the region below $30 \, \text{cm}^{-1}$, phenomena can be studied that cannot be reached by IR spectroscopy. Pure rotational spectra, phonone waves, lattice vibrations are mentioned here as examples. An overview on this is found in Brandmüller and Schrötter [54].

Otherwise, Raman spectroscopy is advantageous because the laser is extremely well suited for focusing onto minute surfaces and volumes. The multiple applications for non-invasive examination of micro-samples are founded on this. For this, a fitting is necessary of micro-arrangements and microscopes to the Raman spectrometer (see e.g. [55]).

References

[1] Workman, J., Jr., *J. Near Infrared Spectrosc.* **1993**, 1, 221
[2] Murray, I., Cowe, I.A. (eds.), *Making Light Work: Advances in Near Infrared Spectroscopy.* Weinheim: VCH Verlagsgesellschaft, **1992**
[3] Geladi, P., Dabbak, E., *J. Near Infrared Spectrosc.* **1995**, 3, 119
[4] Massart, D.L., Vandeginste, B.G.M., Buydens, L.M.C., de Jong, S., Lewi, P.J., Smeyers-Verbecke, J.: *Handbook of Chemometrics and Qualimetrics, Part A.* Amsterdam: Elsevier, **1997**
[5] Massart, D.L., Vandeginste, B.G.M., Buydens, L.M.C., de Jong, S., Lewi, P.J., Smeyers-Verbecke, J.: *Handbook of Chemometrics and Qualimetrics, Part B.* Amsterdam: Elsevier, **1998**
[6] Donahue, S.M., Brown, C.W., Caputo, B., Modell, M.D., *Anal. Chem.* **1988**, 60, 1873
[7] Lin, J., Brown, C.W., *Vibr. Spectrosc.* **1994**, 7, 117
[8] Westbrook, S.R., *SAE Tech. Paper Ser.* **1993**, 930734, 1
[9] Heise, H.M. in Gremlich, H.-U., Yan, B. (eds.): *Infrared and Raman Spectroscopy of Biological Materials.* New York: Marcel Dekker, **2001**, 259
[10] Dempsey, R.J., Davis, D.G., Buice, R.G., Jr., Lodder, R.A., *Appl. Spectrosc.* **1996**, 50, 18A
[11] Kirsch, J.D., Drennen, J.K., *Appl. Spectrosc. Rev.* **1995**, 30, 139
[12] Plugge, W., van der Vlies, C., *J. Pharm. Biomed. Anal.* **1996**, 14, 891
[13] Ben-Gera, I., Norris, K.H., *Israel J. Agric. Res.* **1968**, 18, 125
[14] Wetzel, D.L., *Anal. Chem.* **1983**, 55, 1165 A
[15] Williams, P., Norris, K. (eds.): *Near-Infrared Technology in the Agricultural and Food Industries.* St. Paul: AACC, **1987**
[16] Miller, C.E., *Appl. Spectrosc. Rev.* **1991**, 26, 277
[17] MacDonald, B.F., Prebble, K.A., *J. Pharm. Biomed. Anal.* **1993**, 11, 1077
[18] Weyer, L.G., *Appl. Spectrosc. Rev.* **1985**, 21, 1
[19] Honigs, D.E., Hirschfeld, T.B., Hieftje, G.M., *Appl. Spectrosc.* **1985**, 39, 1062
[20] Ciurczak, E.W., Cline-Love, L.J., Mustillo, D.M., *Spectrosc. Int.* **1991**, 3, no. 5, 39
[21] Ciurczak, E.W., Murphy, W.R., Mustillo, D.M., *Spectrosc. Int.* **1991**, 3, no. 7, 39
[22] Smith, M.J., May, T.E., *Int. Lab.* **1992**, 22, 18
[23] McDermott, L.P., *Ad. Instrum. Control.* **1990**, 45, 669
[24] Guided Wave, Inc., Application Note no. A3-987, El Dorado Hills, CA 95630
[25] Guided Wave, Inc., Application Note no. A4-188, El Dorado Hills, CA 95630
[26] Kimmit, M.F., *"Far-infrared Techniques"* Pion, London, **1970**
[27] Schnöckel, H., Willner, H., "Inorganic Substances", in: *Infrared and Raman Spectroscopy*, Schrader, B. (ed.), VCH, Weinheim, **1995**, p. 223
[28] Durig, J.R., Sullivan, J.F., "Chemical Utility of Low Frequency Spectral Data", in: *Chemical, Biological and Industrial Applications of Infrared Spectroscopy.* Durig, J.R. (ed.), J. Wiley & Sons, Chichester, **1985**, p. 335
[29] Chantry, G.W., *"Long-wave Optics"*, Vol. 2, Academic Press, London, **1984**
[30] Kuzmany, H., "Conducting Polymers, Semiconductors, Metals, and Superconductors", in: *Infrared and Raman Spectroscopy*, Schrader, B. (ed.), VCH, Weinheim, **1995**, p. 372

[31] Demtröder, W., *"Laser Spectroscopy"*, 2nd Ed., Springer, Berlin, **1996**

[32] Heise, H.M., "Infrarotspektrometrische Gasanalytik – Verfahren und Anwendungen", in: *Infrarotspektroskopie*. Günzler, H. (Hrsg.), Springer, Berlin, **1996**, S. 1

[33] Colthup, N.B., Daly, L.H., Wiberley, S.E., *"Introduction to Infrared and Raman Spectroscopy"*, 3rd Ed., Academic Press, San Diego, **1990**

[34] Hendra, P., Jones, C., Warnes, G., *"Fourier Transform Raman Spectroscopy: Instrumental and Chemical Applications"*, Prentice-Hall, Englewood Cliffs, **1991**

[35] Lin-Vien, D., Colthup, N.B., Fateley, W.G., Grasselli, J.G., *"The Handbook of Infrared and Raman Characteristic Frequences of Organic Molecules"*, Academic Press, Boston, **1991**

[36] Grasselli, J.G., Bulkin, B.J., *"Analytical Raman Spectroscopy"*, Chemical Analysis Series, Vol. 114, J. Wiley & Sons, New York, **1991**

[37] Tobias, R.S., *J. Chem. Educ.* **1967**, 44, 2

[38] *The Photonics Design & Application Handbook*, Laurin Publ., Pittsfield, MA, **1995**

[39] Markwort, B., Kip, B., Da Silva, E., Roussel, B., *Appl. Spectrosc.* **1995**, 49, 1411

[40] Heise, H.M., Schrötter, H.W., "Rotation-Vibration Spectra of Gases", in: *Infrared and Raman Spectroscopy*, Schrader, B. (ed.), VCH, Weinheim, **1995**, S. 253

[41] Schrötter, H.W., Berger, H., Boquillon, J.P., Lavrel, B., Millot, G., "High-resolution nonlinear Raman Spectroscopy of Rovibrational Bands in Gases", in: *Progress in Molecular Spectroscopy*, Vol. 20, Salzer, R., Kriegsmann, H., Werner, G, (eds.), Teubner, Leipzig, **1988**, p. 102

[42] Kiefer, W., *"Nonlinear Raman Spectroscopy"*, in: *Infrared and Raman Spectroscopy*, B. Schader (ed.), VCH, Weinheim, **1995**, p. 162

[43] Chase, B., *Appl. Spectrosc.* **1994**, 48, 14A

[44] Wicksted, J.P., Erckens, R.J., Motamedi, M., March, W.F., *Appl. Spectrosc.* **1995**, 49, 987

[45] Heise, H.M., Bittner, A., *J. Mol. Struct.* **1995**, 348, 21

[46] Frank, C.J., McCreery, R.L., Redd, D.C.B., *Anal. Chem.* **1995**, 67, 777

[47] Keller, S., Schrader, B., Hoffmann, A., Schrader, W., Metz, K., Rehlaender, A., Pahnke, J., Ruwe, M., Budach, W., *Raman Spectrosc.* **1994**, 25, 663

[48] Cooper, J.B., Wise, K.L., Groves, J., Welch, W.T., *Anal. Chem.* **1995**, 67, 4096

[49] Hirschberger, R., D'Orazio, M., *Intern. J. Environ. Anal. Chem.* **1992**, 48, 115

[50] Hirschberger, R., *MTZ Motortechn. Zeitschr.* **1996**, 57, 2

[51] Otto, A., Mrozek, I., Grabhorn, H., Akemann, W., *J. Phys. Condens. Matter* **1992**, 4, 1143

[52] Hill, W., Wheling, B., Fallourd, V., Klockow, D., *Spectrosc. Europe* **1995**, 7, 20

[53] Hill, W., Wehling, B., Gibbs, C.G., Gutsche, C.D., Klockwo, D., *Anal. Chem.* **1995**, 67, 3187

[54] Brandmüller, J., Schrötter, H.W., *Fortschr. Chem. Forsch.* **1972**, 36, 85

[55] Schrader, B., "Infrarot- und Raman-Mikrospektroskopie", in: *Analytiker-Taschenbuch*, Bd. 13. Günzler, H., Bahadir, A.M., Borsdorf, R., Danzer, K., Fresenius, W., Huber, W., Lüderwald, I., Schwedt, G., Tölg, G., Wisser, H. (Hrsg.), Springer-Verlag, Berlin, **1995**, S. 3

9 Reference Spectra and Expert Systems

The elucidation or confirmation of the structure of a chemical compound is mostly based on empirical knowledge derived from partial spectra and characteristic band combinations, which can be assigned to certain molecular substructures. The calculation of a chemical structure based on physical modeling of the spectrum fails mostly because of the complexity of the problem. The spectral interpretation is easiest when one starts out from a known or supposed structural formula and can assign the bands. Much more difficult is the reversed route, i.e. to draw conclusions about the molecular structure from the spectrum alone. Several tables and collective works were already cited in Chapter 6 that may be used in the spectra-structure correlation (see Sect. 6.3 and 6.4). This knowledge basis can be used for computer-supported interpretation systems.

On the other hand, the identity of a substance can be determined via a reference spectrum. Concerning this, spectral library search has been successfully applied for years, whereby the availability of computers plays an important role. Both ways, the interpretation of IR spectra and the search procedure, are handled in detail in the following sections.

9.1 Spectral Collections

9.1.1 Reference Collections

In the past, various attempts were made to set up comprehensive spectral collections. Thus, a collective work, the Documentation of Molecular Spectroscopy (DMS) completed in 1973, exists for the qualitative spectral interpretation. This work lists 23 560 spectra found each on a DIN A-5 binder index card together with further information on the spectrometer, preparation, origin, literature citation, structural formula, name, band positions etc. The visual-hole card index allows some presorting according to special characteristics. Other book-bound spectral collections are cited here, e.g. the Raman/IR-Atlas by Schrader [1] and the handbooks by Hummel/Scholl on polymer analysis that is an invaluable work for every spectroscopist involved with polymers [2]. Another ordered and published collection is the Aldrich collection [3].

Exclusively FT-IR spectra (over 3000), acquired mainly with the KBr pellet technique, are found in the Merck FT-IR Atlas [4]. The high-quality spectra are also available in digital form, together with searchable information such as molecular empirical formula, structure, CAS ID-number etc. Works that are more recent are likewise the four volumes of the Sprouse collection (polymers, tensides, solvents and other organic substances); likewise, digitized band tables exist [5].

Bio-Rad/Sadtler Division markets a comprehensive number of spectral handbooks: about 3000 reference spectra (IR, NMR, UV) exist in a three-volume handbook [6]. A collection of 1000 spectra exists for the near-IR [7]. Furthermore available is a two-volume work on the interpretation of gaseous-phase spectra; in addition, 500 spectra are presented which were recorded by means of the GC/IR-technique [8]. Spectra were illustrated for over 345 inorganic substances [9]. Moreover, handbooks are available about minerals, monomers and polymers, polymer additives, adhesives, components of the rubber industry, substances which are especially important as air impurities and other toxic substances, tensides, solvents, esters etc. Further detailed information on this is found in [10].

A particular collection of transmittance spectra exists for inorganic minerals [11]. For other inorganic compounds, there are several older publications and monographs, including that by Nyquist and Kagel [12] for less common, simple inorganic compounds (salts), that by Siebert [13] and Nakamoto [14] for inorganic and coordination compounds and that by Ferraro [15], who primarily reports about central atom-ligand studies.

A particular atlas of NIR spectra was compiled by Buback and Vögele [16] (the corresponding digital database is available via Chemical Concepts, Weinheim [17]).

A gas-spectra library with data for quantitative analysis, also in digital form, can be obtained by the Infrared Analysis company. Interesting is an atlas with absorption lines of atmospheric gases in the range of $17900\,cm^{-1}$ down into the millimeter wavelength range with specification of line intensities [19]; this presentation is based on the so-called HITRAN library (also see Sect. 9.1.2). A double-band work by Guelachvili and Rao is to be cited regarding highly resolved spectra [20].

Section 7.1.4 already cited quantitative standard spectra of various organic solvents that can be used as secondary standards for checking the photometric precision [21].

9.1.2 Digital Spectral Libraries

Spectroscopic databases are valuable aids for structure elucidation. However, only computer-supported IR spectrometers provided digitized spectra, which served as the basis for spectral search systems on computers. Besides extremely long computing times and insufficient storage capacity, insufficient user interfaces of the operating programs and high costs also limited the use of such systems in the past. A typical data reduction for the coding of the spectra by storing binary data (e.g. presence or absence of bands) inherently reduced the information content of an IR spectrum considerably. The limitations nowadays, however, lie in the algorithms and partially in the quality and usability of the reference data.

The establishment of high-quality spectral libraries is further a priority, whereby various suggestions were made on quality criteria for the acquisition of IR reference data, e.g. [22]. Several databases, by all means, contain substantial errors that may limit their usability. The effort, like that expended for making intensity standards available via the determination of optical constants of liquids, is naturally not applicable for the great numbers of spectra acquired. Furthermore, one should note that libraries of various producers are typically incompatible. In the meantime, however, a generally accepted format of data exchange exists with JCAMP-DX (Joint Committee on Atomic and Molecular Physical Data) [23], which meanwhile can also be applied for NMR- and MS spectra. A further development allows the exchange of structure information [24]. One monograph particularly deals with the reproduction of structures [25].

The most comprehensive IR spectral databases with more than 130 000 spectra in over 50 different packages are marketed by Bio-Rad/Sadtler Division already mentioned in the previous section [10]. The biggest spectral libraries contain e.g. almost 72 000 IR spectra of substances in condensed phase and over 9000 spectra of substance measured in the gaseous phase. The substance classes already presented in the spectral handbooks are found here as separate databases, besides numerous particular categories such as pharmaceutical active ingredients, fragrances, metalorganic compounds and many more. Some databases contain molecular structures that can be presented and used for the substructure search. It is important that the databases are also usable on PCs.

Moreover, various spectrometer manufacturers offer a few databases, e.g. extensive collections of substances in condensed and gaseous phase (together, ca. 21 000) from Nicolet/Aldrich. The Sigma company set up over 10 000 spectra of biochemicals. Besides the ca. 3000 FT-IR spectra of Bruker-Merck, recently the same number of FT-Raman spectra has likewise been added

to the collection. The structure elucidation system of BASF
(IDIOTS [26]) contains 17000 spectra, which was presented as
an example of a company in-house database. Please refer to literature for a further discussion of the databases of IR spectra
[27–30].

The databases for calculating spectra of gases, that are
mainly present in the atmosphere, have another quality. To cite
here are the HITRAN database (High Resolution Transmission)
of Philips Laboratory, Air Force Department (USA) [31], the
GEISA database (Gestion et Etude des Informations Spectrosco-
piques Atmosphériques) of the ARA group at the Laboratoire de
Météorologie Dynamique, CNRS (France) [32], as well as the
ATMOS database (Atmospheric Trace Molecule Spectroscopy)
of the Jet Propulsion Laboratory (USA) [33], the application
areas of which lie in quantitative atmospheric spectroscopy,
whereby the ATMOS-experiment includes stratospheric measurements during the SPACELAB 3-mission. There are currently
about 720000 entries for 40 molecules and 86 isotope species.
These contain various spectroscopic parameters such as line positions and intensities as well as pressure-widening coefficients
for calculating line half-widths. The spectral parameters have
been determined via many high-quality, highly resolved spectra.
Various line profiles (Gauss, Voight- or Lorentz functions) are
calculable as a function of the pressure conditions. A comparison of the various line-data catalogs was published a few years
ago [34].

The gas-spectra database with about 250 different substances
by the Infrared Analysis company, used for e.g. the quantitative
analysis of air, was already mentioned in the previous section.
The spectra were measured with multi-reflection cells and exist
mostly with a resolution of $0.5 \, \mathrm{cm}^{-1}$, ca. 150 master spectra of
which also show a resolution of $0.125 \, \mathrm{cm}^{-1}$. The Entropy library, obtainable by ENTROPY, Inc., is also available for this
application area, having about 400 spectra with a spectral resolution of $0.25 \, \mathrm{cm}^{-1}$ of 120 components [35]. The data are also
available on the internet [http://info.arnold.af.mil/epa].

9.2 Computer-Supported Research

9.2.1 Library Search

Two different criteria are applied for the spectral search: spectral identity or spectral similarity. When spectral similarity exists, the underlying hypothesis is that the molecular structures

are approximately equal to each other. This quite large range for interpreting is not given, when an almost identical spectrum is actually found in the reference library [36].

The spectral search can also be instituted by comparing band positions and intensities (so-called peak search), but more often, it occurs with completely stored spectra, of which also partial regions can be applied. In alternative techniques, the use of transformed spectral data, e.g. interferograms or otherwise reduced data sets was recommended [37, 38]. Some library search systems allow the incorporation of further information during the search such as e.g. the empirical formula and other searchable parameters.

Prior to the spectral library search, a spectral preparation is typically undertaken which also accounts for the fitting of the spectral resolution to the parameters on which the database is based (often $4 \, \text{cm}^{-1}$ or $8 \, \text{cm}^{-1}$). Furthermore, the spectra are normalized in such a manner that the maximum absorbance of the strongest band in the spectrum assumes e.g. a particular value, mostly 1.0.

For estimating the similarity, various similarity parameters come into question. Most often, the following four definitions are applied and can be used for setting up an evaluation parameter, the so-called *Hit Quality Index* (HQI) [39]:

$$
\begin{aligned}
M_{\text{AB}} &= \sum |S_i - R_i| \\
M_{\text{SQ}} &= \sum (S_i - R_i)^2 \\
M_{\text{AD}} &= \sum |dS_i - dR_i| \\
M_{\text{SD}} &= \sum (dS_i - dR_i)^2
\end{aligned}
\tag{9.1}
$$

In the first algorithm, absolute differences between the absorbances of the samples and of the reference spectrum are calculated point-wise and compared as a sum value, while in the second algorithm, the quadratic deviations are added, whereby greater deviations, of course, may experience considerable weighting. In both last algorithms, the first derivative of the spectra is used respectively. This is inherently advantageous in that linear baselines or simple baseline shifts no longer have any negative effects on the calculation of the similarity parameter. In these cases, the best agreement occurs with a HQI near zero. In contrast, the evaluation numbers in other definitions assume values below 100 or also 1000.

Another possibility is the calculation of the Euclidean distance [40, 41], whereby also another spectral standardization occurs, e.g. via the spectral vector length:

$$
M_{\text{EU}} = \left\{ \sum (S_i - R_i)^2 \right\}^{1/2} .
\tag{9.2}
$$

or via the application of the scalar vector product [42] or of the correlation coefficient [43]. Another approach uses the "fuzzy-set theory" for comparing indistinct sample- and reference spectra (see e.g. [44]). To still note here is that the scalar size, as it represents a similarity parameter, does not always say that similar coefficients also cause spectra having similar bands. Regardless of this, one should always confirm the conclusions by a visual comparison in order to obtain a corresponding certainty (see Fig. 9-1).

Figure 9-1 illustrates the result of a full spectral search for a FT-IR spectrum of acrolein as it is obtained by the gas chromatographic separation of a solvent mixture (also see Sect. 5.7.1, Fig. 5-20 and 5-21). The search takes place in the 4000 to 700 cm^{-1} range with an intensity-normalized and baseline-corrected spectrum based on the sum of the squares of the deviation compared to various library spectra. The spectra of the first three substances from the hit list are presented in comparison to the measured GC/FT-IR spectrum. In addition, the first five components of a hit list, which were obtained via a peak search, are shown.

Obvious is the respective leap in the HQI after the first located, pure substance spectrum which is almost identical to that obtained by GC, whereby the sum of the squares of deviation presents a better measure of similarity than the HQI based on peak agreement (identity at 100). What is also interesting is that the spectrum of benzaldehyde is present in two different libraries; the slight differences in the found HQI infer experimentally related differences in the respective acquisition of the benzaldehyde spectra.

An important point for the practitioner is to sensibly choose the spectral library, for its particularity is decisive for the search result. A second point is the quality of the spectrum acquired, whereby good spectroscopic practice is decisive (the spectrum should be free of water- and CO_2-bands; for liquids, spectral interferences in the baseline should be avoided). The effects of a large noise level on the result of the spectral search were investigated by Hallowell and Delaney [45]. In this respect, also interesting are the effects of band shifts or different band intensities which are caused by various measurement techniques such as ATR or diffuse reflectance [46].

The quality of the search algorithms of library search systems was studied in various manners, such as by Clerc et al. [36, 47]. Harrington and Isenhour worked on a similar question that dealt with the effects of spectral noise and the results in mixed spectra. Tanabe et al. suggested a totally different approach for rapidly identifying spectra in the library search based on a neuronal network [49].

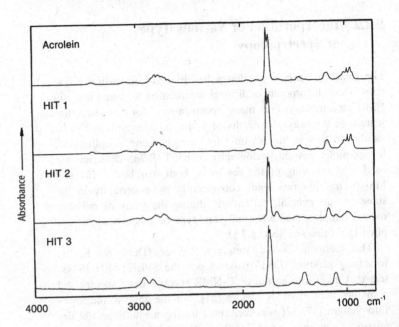

a) Hit list obtained via spectral search:

HIT	Library	ID No.	Name	HQI
1	EPA2	GC000696	ACROLEIN/VAPOR	0.08
2	GCPE2000	RK000141	CROTONALDEHYDE C4H6O	0.42
3	EPA2	GC000401	FORMAMIDE, N,N-DIMETHYL-	0.57
4	GCPE2000	RK000149	BENZALDEHYDE C7H6O	0.59
5	EPA2	GC000750	BENZALDEHYDE	0.61

b) Hit list obtained via peak search:

HIT	Library	ID No.	Name	HQI
1	EPA2	GC000696	ACROLEIN/VAPOR	89
2	EPA2	GC001681	HEXANETHIOL, 2-ETHYL-	50
3	GCPE2000	RK000141	CROTONALDEHYDE C4H6O	50
4	EPA2	GC000441	ISOVALERALDEHYDE	48
5	EPA2	GC000389	ETHANE, 1,2-BIS/2-CHLOROETHOXY-	46

Fig. 9-1. Spectrum of acrolein compared to the three most similar spectra obtained by spectral search, as well as hit-list results of various library searches (see text regarding standardization of spectra).

9.2.2 Incorporation of Various Types of Spectroscopy

The future of spectral databases lies in multi-dimensional systems which incorporate additional information acquired e.g. via NMR spectroscopy or mass spectrometry for the structure search. In this way, the results of a structure suggestion can be validated. This is the reason why corresponding couplings of, for example, gas chromatography with FT-IR/MS-detection are used. The coupling of the results of both searches in IR- and MS-spectral libraries lends considerably more certainty in the substance assignment, particularly during the study of environmental samples, than when only one type of spectroscopy is applied [50] (also see Sect. 5.7.1).

The German Cancer Research Center (Deutsche Krebsforschungszentrum (DKFZ)) developed the SPEKTREN II-system that takes into account ^{13}C-NMR spectra, mass spectra and a smaller number of IR spectra [51]. For the analyst, the Spec-Info-System [17, 52] was conceived having a multi-spectral database of currently ca. 135000 spectra and 110000 structures which can be downloaded online via STN International, c/o Fachinformationszentrum Karlsruhe, D-76012 Karlsruhe. Besides NMR- and MS-spectra that make up the principle part, about 20000 IR spectra are available. The retrieval language "Messenger" is capable of tracing spectra similar to the measured spectra, which is uploaded to the host computer. During the automatic structure elucidation, the production is essential of a structure suggestion based on the various spectra of a substance permitting the suggestions for the substructure. Here, one can start out from e.g. numerous plausible structures that belong to one empirical formula. The aim is the integration of a structure-generator, which possesses sufficient artificial intelligence for clearly predicting the structure by means of unanimous spectroscopic information.

9.3 Interpretative Systems

The derivation of a structure contains extremely complex steps making formalization difficult. Expert systems, the rules of which were derived via a knowledge base, attempt a general applicability but which has limitations. These limitations are partially attributed to symmetry particularities and vibrational couplings occurring over larger structural regions that may lead to problems in assigning partial structures with bands. Various developments were presented of which no algorithm has yet been shown to be exceptional, however.

9.3.1 Computer-Interpretation According to Empirical Rules

In an initial attempt to place the spectral interpretation on a somewhat stable base, one tries to exhaust the quantity of experiences with the least amount of effort. The limitations of manual interpretation strategies, however, are very quickly reached due to the multi-interpretation of the functional relationships. Often, one does not get beyond establishing a number of functional groups for drawing clear conclusions. Essentially, one can follow two possibilities: one can either begin with the band positions to which the functional groups are assigned, or one picks out various functional groups for which the presence or absence of bands is questioned.

Various computer programs are offered to the spectroscopist that ought to replace leafing through books of tables. One of these programs is the IR Mentor program of Bio-Rad/Sadtler, already mentioned in Sect. 6.6.1, which aids the spectral interpretation. Various chemical classes such as e.g. aromatics, esters and halogen compounds with almost 200 functional groups and corresponding band assignments are the current database. In the spectrum to be interpreted, all functional groups assignable to one selected band can be presented, whereby e.g. for each single group, also the spectral interval widths for the occurrence of the band and the respective relative band intensities can be requested.

A much more comprehensive program is SpecTool, developed by Pretsch and colleagues and marketed by Chemical Concepts (Weinheim) [53]. It is an electronic reference work (hypermedia application), which contains spectroscopic data for all important substance classes of organic chemistry, whereby characteristic features from IR-, ^{1}H-NMR-, ^{13}C-NMR-, MS- and UV-spectra can be assigned to each structure class. Reference data and spectra, as well as subprograms, among others for the different types of spectroscopy or for calculating the molecular formula from spectroscopic and elementary-analytical data [54], are linked through an efficient "navigation system" within the software.

Another computer-based interpretation system with possible user-interactive cooperation is the X-PERT-software of Bruker [55] which, besides the IR spectrum, also needs ^{1}H-NMR- and ^{13}C-NMR spectra as input data as well as the empirical formula of the substance to be investigated. Larger fragment sets, which are taken into account by a structure-generator, can be produced with the structural fragments recommended via the interpretation rules by considering the empirical formula. The structure elucidation is concluded by a verification of all structures produced by the generator via the spectroscopic data.

For the computer techniques, the spectral features are often given, whereby the substructures are to be assigned. A few publications exist that allow an automatic derivation of the spectrum-structure correlations from larger databases, which have various advantages over tabular works. Firstly, a clear definition is possible, e.g. based on statistical probabilities for the occurrence of correlations in contrast to possibly vague linguistically formulated relationships (also, yet unknown correlations can also be found). Secondly, a rapid update of library entries can be made or, by all means, also rules for larger substructures are derivable (see e.g. [56]).

Several computer programs for interpreting IR spectra were developed in the past, which are based on the logical linking of partial structures and corresponding characteristic spectral features. The programs are successfully applicable when the spectroscopic structure elucidation problem is falling within the presumed knowledge base. Reviews on this were given by Luinge [57] and Warr [58]. Regarding this in particular and in a wider respect, a monograph of the latter author [59] is recommended. To be explicitly cited are PAIRS [60], EXPEC [61], CHEMICS [62], STREC [63] and CASE [64], as well as EXPIRS [65]. With their use considered to complement existing programs for interpreting spectra, *Fuzzy* systems are formulated on the basis of neuronal networks as will be discussed in the next section.

9.3.2 Multivariate Methods for Spectral Interpretation

The use of neuronal networks in the area of IR spectroscopy ranges between quantitative and qualitative questions. For the latter, the interpretation of IR spectra is an already often-described task. Since the eighties, neuronal networks for "soft-modeling" approaches have been tested. These acquired their name by the comparison of their design from a massively parallel working system of identical units with models of neuronal networks. The network must be trained to solve a certain problem (calibration), the calculation of which partially proves to be very demanding, while the use for prediction only requires little mathematical effort. Empirical models can hereby be produced, the formulation of which otherwise is shown to be unknown or too difficult. Because the presentation of the basics as neuronal networks goes beyond the scope of this book, please refer to further literature [67, 68].

Complete spectra are typically used as input data for the prediction of substructure. Various publications exist for this field (see e.g. [69]). Interesting is that the correlations may also be used to predict spectra based on the given structures [70].

As an alternative to the neuronal networks, a classification was tested and compared with the results of the much more complicated technique by means of modified *Partial Least-Squares* (PLS) (also, see Sect. 7.5.2), which requires only little mathematical effort for the calibration. The prediction results for the projection of the spectra to the substructure space were comparable [71]. Griffiths and colleagues [72, 73] exhibited a rapid classification of GC/FT-IR spectra by means of principle-component analysis to certain substance classes such as alcohols, carbonic acids, ketones and aldehydes or also to larger structure units, as exemplified by barbiturates.

9.3.3 Pattern Recognition Methods

Another possibility for a substructure-assignment via the spectrum results from the statistical means of pattern recognition [74–78]. A feature pattern, which is not directly measurable, can be hereby assigned, by all means, to a certain class, the so-called cluster. This can take place via conventional projections, e.g. via a principle-component analysis, but also by means of a classifying neuronal Kohonen-network. Most methods are based on the already mentioned assumption that objects that resemble each other also lie near one another in the spectral space. The distance- and similarity parameters play an important role here. Before the pattern-recognition methods can be applied, a scaling of the variables might be advisable.

One differentiates between supervised and unsupervised learning. With the first approach, one can requiring a training principle for establishing the differentiating features – e.g. distances to the criterion for separating two classes, while in the second technique, absolutely no *a priori* knowledge is necessary for making a classification. When more than two classes must be separated, a series of binary classifiers is necessary, as e.g. in the KNN-method (K-nearest neighbor, K is a small integer), whereby for an object of unknown class assignment, its surroundings must be checked). The name is based on the idea that for K known objects nearby, their class assignment is determined and the object to be classified is assigned via the distance of the class of the majority of the nearest K reference objects. In the linear discriminatory analysis, a function is sought which separates the various classes from each other.

For the yet cited non-parametric methods, only distances between the data points, defined by multi-dimensional spectral vectors, are considered so far. However, each class can also be based on statistical models, for which then probability distributions can be defined.

The creation of so-called dendrograms, which comprise similar members of a population via hierarchic cluster, is based on algorithms of unsupervised learning for classifying similarities. Although there are many classification approaches, the certainly optimistic expectations of the pattern-recognition methods have not been fulfilled as yet [79].

9.4   Qualitative Mixture Analysis

Qualitative mixture analysis means the identification of components of a mixture, whereby a subsequent quantitative analysis can be certainly attempted. The identification can take place via a library search, whereby difficulties inherently arise. The reasons for this are that a mixed spectrum is composed of the sum of the individual component spectra with approximately ideal mixing behavior and that the database only contains spectra of pure substances. With a suitable search algorithm, possibly one of the components (in the best case, the principle component) should be at the top of the hit list. An example of this is the case of two non-separated GC-fractions (see Fig. 5-20). Figure 9-2 shows the mixed spectrum of propanol and toluene as well as their respective pure-substance spectra. A library search via spectral comparison yields a hit list, sections of which are presented. The first eight found spectra with the greatest similarity are exclusively alcohols, and an aromatic is found only at the 16th place. One can subtract the first suitable hit spectrum from the mixed spectrum by means of scaled subtraction and again undertake a library search for the spectral residue obtained, which then clearly produces, in this case, toluene as the substance with the best spectral agreement.

By all means can mixtures of two or three components still be identified reliably, whereby fractions present in small concentrations can only be clarified with difficulty.

Another constituted case is given when spectra of various samples of unknown composition, but with the respective components existing in variable concentration ratios such as in the characterization of overlapped HPLC-peaks, reactive substances or materials. An additional complication is the fact that the information (i.e. spectra) about the pure components is not

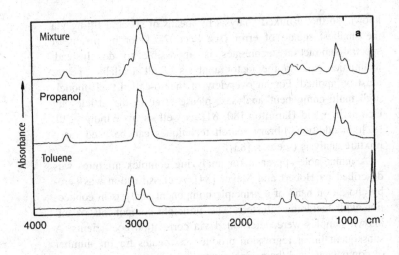

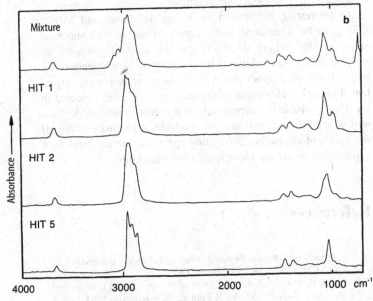

a) Hit list obtained via spectral search:

HIT	Library	ID No.	Name	HQI
1	EPA2	GC000681	PROPANOL	0.32
2	GCIRPE18	GC000016	1-BUTANOL	0.35
3	EPA1	GC000335	BUTANOL	0.36
4	GCIRPE18	GC000077	2-METHYLBUTANOL-1	0.36
5	EPA2	GC000206	PENTANOL,2-METHYL-	0.39
–	–	–	–	–
15	EPA2	GC000332	PENTANOL	0.42
16	EPA2	GC001038	BENZENE,HEXAMETHYL-	0.42

Fig. 9-2. Spectrum of a mixture of propanol and toluene, as well as their pure-substance spectra (a), and in comparison to this, three of the most similar spectra obtained via spectral search (b), as well as hit-list results of a library search by means of spectral comparison.

known, so that a mixed analysis by means of spectral fitting via the smallest square of error (see Sect. 7.5.1) or by means of spectral-subtraction techniques is impossible to do. Instead, techniques e.g. via the factor analysis of all available spectra must be applied. For an overview of the so-called "self-modeling" multi-component analyses, please refer to the articles by Gemperline and Hamilton [80, 81] as well as by Windig et al. [82]. In addition, library search techniques can be used for a mixture analysis (see e.g. [83]).

A remarkable approach for analyzing complex mixtures was described by Hobert and Meyer [84]. A classification was hereby chosen on hand of a principle-component analysis in connection with hierarchic clusters. The principle components of unknown samples were determined via correlation coefficients. A subsequent linear regression produces estimates for the number of components and their concentrations.

An interesting publication by Liang, Kvalheim and Manne [85] is to be mentioned with respect to the multi-component analysis. The authors discuss a classification of problems of mixture analysis and methods for quantitative determination, for which lastly the number and type of the components are important. Further developments of chemometric tools are expected in the future. However, experience shows that considerable time will pass until these will also be available in routine analysis, if the cycles of innovation concerning spectroscopic and analytical expert systems are not shortened in the near future.

References

[1] Schrader, B., "Raman/Infrared Atlas of Organic Compounds", 2nd Ed., VCH Verlagsges., Weinheim, **1989**

[2] Hummel, D.O., Scholl, F., "Atlas of Polymer and Plastics Analysis", Vols. 1–3, 3rd Ed., VCH Publishers, Weinheim, **1991**

[3] Pouchert, C.J. (ed.): "The Aldrich Library of Infrared Spectra" Aldrich Chemical, Milwaukee, 3rd Ed., **1981**; "The Aldrich Libraryof FT-IR Spectra", **1985**; "The Aldrich Library of FT-IR Spectra: Vapor Phase" Vol. 3, **1989**

[4] Merck, E., "Merck FT-IR Atlas", VCH-Verlagsges., Weinheim, **1988**

[5] Hansen, D.L., "The Sprouse Collection of Infrared Spectra, Vols. 1–4, Elsevier Science Publ., Amsterdam, **1987** and **1988**

[6] Simons, W.W., "The Sadtler Handbook of Infrared Spectra", Sadtler, Philadelphia, **1978**

[7] "The Atlas of Near Infrared Spectra", Sadtler, Philadelphia, **1981**

[8] Nyquist, R.A., "The Interpretation of Vapor-Phase Infrared Spectra Sadtler", Philadelphia, **1984**

[9] "The Infrared Spectra Handbook of Inorganic Compounds", Sadtler, Philadelphia, **1984**

[10] Bio-Rad/Sadtler Divison, 3316 Spring Garden Street, Philadelphia, PA 19104 (USA)

[11] Jones, G.C., Jackson, B., *"Infrared Transmission Spectra of Carbonate Minerals"*, Chapman & Hall, London, **1993**

[12] Nyquist, R.A., Kagel, R.O., *"Infrared Spectra of Inorganic Compounds* (3800 – 45 cm^{-1})", Academic Press, New York, **1971**

[13] Siebert, H., *"Anwendungen der Schwingungsspektroskopie in der anorganischen Chemie"* Springer Verlag, Berlin, **1966**

[14] Nakamoto, K., *Angew. Chem.* **1972**, 81, 755; Nakamoto, K., *"Infrared and Raman Spectra of Inorganic and Coordination Compounds"*, 4th Ed., J. Wiley & Sons, New York, **1986**

[15] Ferraro, J.R., *"Low Frequeny Vibrations of Inorganic and Coordination Compounds"*, Plenum Press, New York, **1971**

[16] Buback, M., Vögele, H.P., *"FT-NIR Atlas"*, VCH, Weinheim, **1993**

[17] Chemical Concepts, Boschstraße 12, D-69469 Weinheim

[18] Hanst, P.L., Hanst, S.T., Williams, G.M., *"Infrared Spectra for Quantitative Analysis of Gases"*, Infrared Analysis, Anaheim, CA, **1995**

[19] Park, J.H., Rothman, L.S., Rinsland, C.P., Pickett, H.M., Rickardson, D.J., Namkung, J.S., "Atlas of Absorption Lines from 0 to 17900 cm^{-1}", *NASA Reference Publication* 1188, Hampton, **1987**

[20] Guelachvili, G., Rao, K.N., *"Handbook of Infrared Standards: With Spectral Maps and Transition Assignments between 3 and 2600 μm"*, Academic Press, London, **1986**; *"Handbook of Infrared Standards II: With Special Coverage of 1.4 μm–4 μm and 6.2 μm–7.7 μm"*, Academic Press, London, **1993**

[21] Bertie, J.E., Keefe, C.D., Jones, R.N., Tables of Intensities for the Calibration of Infrared Spectroscopic Measurements in the Liquid Phase", *International Union of Pure and Applied Chemistry Chemical Data Series* No. 40, Blackwell Science, **1995**

[22] Griffiths, P.R., Wilkins, C.L., *Appl. Spectrosc.* **1988**, 42, 538

[23] McDonal, R.S., Wilks, P.A. Jr., *Appl. Spectrosc.* **1988**, 42, 151

[24] Gasteiger, J., Hendriks, B.M.P., Hoever, P., Jochum, C., Somberg, H., *Appl. Spectrosc.* **1991**, 45, 4

[25] Ash, J.E., Warr, W.A., Willett, P., *"Chemical Structure Systems"*, Ellis Horwood, Chichester, **1991**

[26] Passlack, M., Bremser, W., "IDIOTS – Structure-oriented Data Bank System for the Identification and Interpretation of Infrared Spectra", in: *Computer Supported Spectroscopic Databases*, Zupan, J. (ed.), Ellis Horwood Publ., Chichester, **1986**

[27] Davies, A.N., *Nachr. Chem. Techn. Lab.* **1989**, 37, 263

[28] Warr, W.A., *Chem. Intell. Lab. Sys.* **1991**, 10, 279

[29] Davies, A.N., McIntyre, P.S., "Spectroscopic Databases", in: *Computing Applications in Molecular Spectroscopy*, George, W.O., Steele, D. (eds.), Royal Soc. Chem. Cambridge, **1995**, p. 41

[30] Warr, W.A., *Anal. Chem.* **1993**, 65, 1045A

[31] Rinsland, C.P., Goldman, A., Flaud, J.M., Quant, J., *Spectrosc. Radiat. Transfer* **1992**, 48, 693; Rothman, L.S., Gamache, R.R., Tipping, R.H., Rinsland, C.P., et al., *J. Quant. Spectrosc. Radiat. Transfer* **1992**, 48, 469

[32] Husson, N., Bonnet, B., Chédin, A., Scott, N.A., Chursin, A.A., Golovko, V.F., Tyuterev, V.G., *J. Quant. Spectrosc. Radiat. Transfer* **1994**, 52, 425

[33] Brown, L.R., Farmer, C.B., Rinsland, C.P., Toth, R.A., *Appl. Optics* **1987**, 26, 5154

[34] Husson, N., Chédin, A., Bonnet, B., *NATO ASI Series*, Vol. I 9, "High Spectral Resolution Infrared Remote Sensing for Earth's Weather and Climate Studies", Chédin, A., Chahine, M.T., Scott, N.A. (eds.), Springer-Verlag, Berlin, **1993**, p. 443

[35] Entrop, Inc., P.O. Box 12291, Research Triangle Park, NC 27709-2291 (USA)

[36] Clerc, J.T., Pretsch, E., Zürcher, M., *Mikrochim. Acta* **1986**, II (1–6), 217

[37] De Haseth, J.A., Azarraga, L.V., *Anal. Chem.* **1981**, 53, 2292

[38] Owens, P.M., Isenhour, T.L., *Anal. Chem.* **1983**, 55, 1548

[39] Lowry, S.R., Huppler, D.A., Anderson, C.R., *J. Chem. Inform. Comp. Sci.* **1985**, 25, 235

[40] Hanna, A., Marshall, J.C., Isenhour, T.L., *J. Chrom. Sci.* **1979**, 17, 434

[41] Novi, M., Zupan, J., *Anal. Chim. Acta* **1983**, 151, 419

[42] McGrattan, B.J., Schiering, D.W., Hoult, R.A., *Spectrosc.* **1989**, 4, 39

[43] Powell, L.A., Hieftje, G.M., *Anal. Chim. Acta* **1978**, 100, 313

[44] Blaffert, T., *Anal. Chim. Acta* **1984**, 161, 135

[45] Hallowell, J.R. Jr., Delaney, M.F., *Anal. Chem.* **1987**, 59, 1544

[46] Rosenthal, R.J., Lowry, S.R., *Mikrochim. Acta* **1986**, II (1–6), 291

[47] Affolter, C., Clerc, J.T., *Fresenius J. Anal. Chem.* **1992**, 344, 136

[48] Harrington, P.B., Isenhour, T.L., *Anal. Chim. Acta* **1987**, 197, 105

[49] Tanabe, K., Tamura, T., Uesaka, H., *Appl. Spectrosc.* **1992**, 46, 807

[50] Cooper, J.R., Wilkins, C.L., *Anal. Chem.* **1989**, 61, 1571

[51] Förster, T., von der Lieth, C.W., Opferkuch, H.J., *GIT Fachz. Lab.* **1989**, 33 (4), 318

[52] Canzler, D., Hellenbrandt, M., *Fresenius J. Anal. Chem.* **1992**, 344, 167

[53] Cadisch, M., Pretsch, E., *Fresenius J. Anal. Chem.* **1992**, 344, 173

[54] Fürst, A., Clerc, J.T., Pretsch, E., *Chemom. Intell. Lab. Syst.* **1989**, 5, 329

[55] Bruker Analytische Meßtechnik GmbH, Wikingerstr. 13, D-76153 Karlsruhe

[56] Ehrentreich, F., Dietze, U., "Vergleich computergestützt abgeleiteter IR-Spektrum-Struktur-Korrelationen", in: *Software-Entwicklung in der Chemie 7*, Ziessow, D. (Hrsg.), Ges. Deutscher Chemiker, Frankfurt, **1993**, S. 53

[57] Luinge, H.J., *Vibr. Spectrosc.* **1990**, 1, 1

[58] Warr, W.A., *Anal. Chem.* **1993**, 65, 1087A

[59] Warr, W.A., Suhr, C., *"Chemical Information Management"*, VCH Verlagsges., Weinheim, **1992**

[60] Wyhoff, B., Hong-Kui, X., Levine, S.P., Tomellini, S.A., *J. Chem. Inf. Comp. Sci.* **1991**, 31, 392

[61] Luinge, H.J., *Trends Anal. Chem.* **1990**, 9, 66

[62] Funatsu, K., Miyabayaski, N., Sasaki, S., *J. Chem. Inf. Comp. Sci.* **1988**, 28, 18

[63] Elyashberg, M.E., Gribov, L.A., Serov, V.V., *"Molecular Spectral Analysis and the Computer"* (in russ.), Nauka Publ., Moskau, **1980**

[64] Munk, M.E., Farkas, M., Lipkis, A.H., Christie, B.D., *Mikrochim. Acta* **1986**, II, 189

[65] Andreev, G.N., Argirov, O.K., *J. Mol. Struct.* **1995**, 347, 439

[66] Otto, M., *Anal. Chim. Acta* **1993**, 283, 500

[67] Zupan, J., Gasteiger, J., *Anal. Chim. Acta* **1991**, 248, 1

[68] Zupan, J., Gasteiger, J., *Neural Networks in Chemistry and Drug Design*, Wiley-VCH, Weinheim, **1999**

[69] Smits, J.R.M., Schoenmakers, P., Stehmann, A., Sijstermans, F., Kateman, G., *Chemom. Intell. Lab. Syst.* **1993**, 18, 27

[70] Affolter, Ch., Clerc, J.T., *Chemom. Intell. Lab. Syst.* **1993**, 21, 151

[71] Luinge, H.J., van der Maas, J.H., Visser, T., *Chemom. Intell. Lab. Syst.* **1995**, 28, 129

[72] Perkins, J.H., Hasenoehrl, E.J., Griffiths, P.R., *Chemom. Intell. Lab. Syst.* **1992**, 15, 75

[73] Hasenoehrl, E.J., Griffiths, P.R., *Appl. Spectrosc.* **1993**, 47, 643

[74] Varmuza, K., Pattern Recognition, in: *Computer in der Chemie*, 2. Aufl., Ziegler, E. (Hrsg.), Springer, Berlin, **1985**

[75] Massart, D.L., Vandeginste, B.G.M., Deming, S.N., Michotte, Y., Kaufman, L., *"Chemometrics: a Textbook"*, Elsevier, Amsterdam, **1988**

[76] Luinge, H.J., "Multivariate Methods for Automated Spectrum Interpretation", in: *Computing Applications in Molecular Spectroscopy*, George, W.O., Steele, D. (eds.), Roy. Soc. Chemistry, Cambridge, **1995**, p 87

[77] Adams, M.J., *"Chemometrics in Analytical Spectroscopy"*, Royal Soc. Chem. Cambridge, **1995**

[78] Henrion, R., Henrion, G., *"Multivariate Datenanalyse"*, Springer-Verlag, Berlin, **1995**

[79] Woodruff, H.B., "Novel Advances in Pattern Recognition and Knowledge-Based Methods in Infrared Spectroscopy", in: *Computer-Enhanced Analytical Spectroscopy*, Meuzelaar, H.L.C., Isenhour, T.L. (eds.), Plenum Press, New York, **1987**

[80] Gemperline, P.J., *J. Chemom.* **1989**, 3, 549

[81] Hamilton, J.C., Gemperline, P.J., *J. Chemom.* **1990**, 4, 1

[82] Windig, W., Guilment, *Anal. Chem.* **1991**, 63, 1425

[83] Mauro, D.M., Delaney, M.F., *Anal. Chem.* **1986**, 58, 2622

[84] Hobert, H., Meer, K., *Fresenius J. Anal. Chem.* **1992**, 344, 178

[85] Liang, Y., Kvalheim, O.M., Manne, R., *Chemom. Intell. Lab. Syst.* **1993**, 18, 235

10 Appendix

10.1 Positions of the Most Frequent Disturbing Bands in the IR Spectrum

Wavenumber (cm^{-1})	Disturbed by:	Remarks
3450–3330	H_2O	water in the substance or in KBr
around 2345	CO_2	atmospheric non-compensation for samples and background spectrum
2325	CO_2	dissolved in liquids
2000–1280	H_2O	steam (e.g. from uncompensated moist spectrometer atmosphere)
1755–1695	$\text{>}C=O$	carbonyl compounds of various origins (softeners, acetone, oxidized products etc.)
1640	H_2O	liquid water
1610–1515	$-COO^{\ominus}$	organic salts, possibly from the reaction with the window material
1355	$NO_3^{\ominus}$	from KBr, H_2O residues etc.
1265	$Si–CH_3$	grinding fat
1100–1050	SiO_2 $Si–O–Si$	glass
973	CCl_4	solvent
837	$NO_3^{\ominus}$	from KBr, H_2O residues etc.
667	CO_2	atmospheric non-compensation for samples and background spectrum

10.2 Spectra of Frequently Used Solvents

The following spectra of frequently used solvents were acquired with a Perkin Elmer 2000 FT spectrometer equipped with Globar, KBr beamsplitter and DTGS detector. At a spectral resolution of 2.0 cm^{-1}, 20 interferograms have been co-added. A cell with KBr windows and a pathlength of 25 µm was used. A linear baseline correction was applied in order to compensate the effects of various reflection losses on the cell windows.

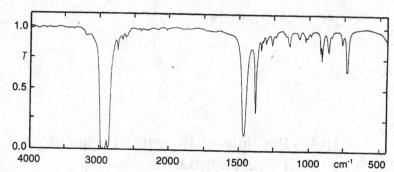

Fig. 10-1. *n*-Pentane (at least 99% pure (GC)).

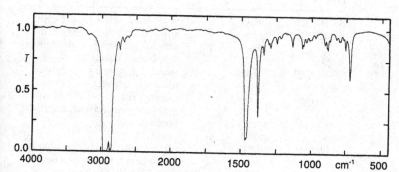

Fig. 10-2. *n*-Hexane (at least 98% pure).

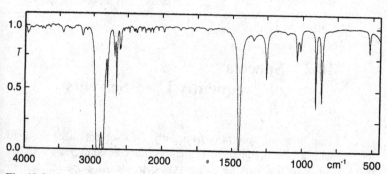

Fig. 10-3. Cyclohexane (at least 99.7% pure).

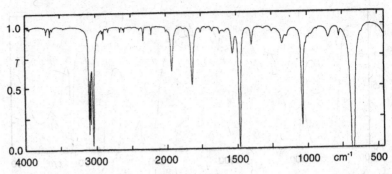

Fig. 10-4. Benzene (at least 99.7% pure (GC)).

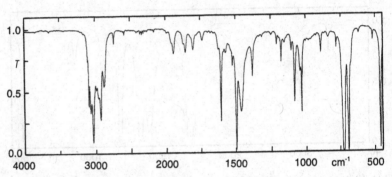

Fig. 10-5. Toluene (Uvasol, 99.9% pure).

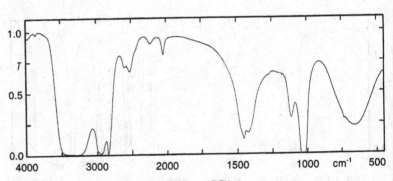

Fig. 10-6. Methanol (at least 99.8% pure (GC)).

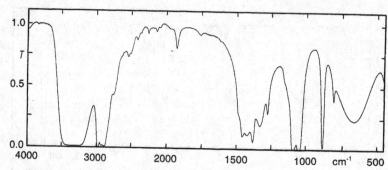

Fig. 10-7. Ethanol (Uvasol, at least 99.9% pure).

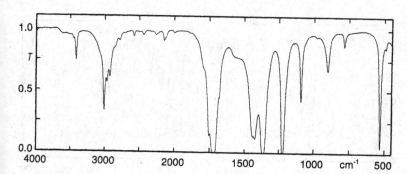

Fig. 10-8. Acetone (Uvasol, 99.9% pure).

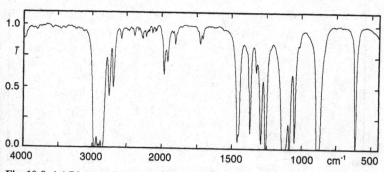

Fig. 10-9. 1,4-Dioxane (Uvasol, 99.5% pure (GC)).

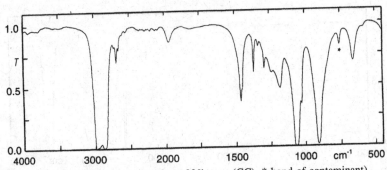

Fig. 10-10. Tetrahydrofuran (at least 99% pure (GC); * band of contaminant).

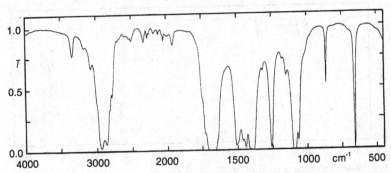

Fig. 10-11. *N,N*-Dimethylformamide (at least 99.5% pure (GC)).

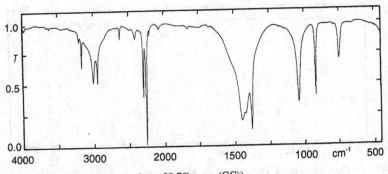

Fig. 10-12. Acetonitrile (at least 99.7% pure (GC)).

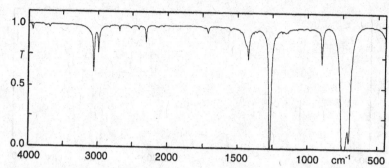

Fig. 10-13. Dichloromethane (99.9% pure).

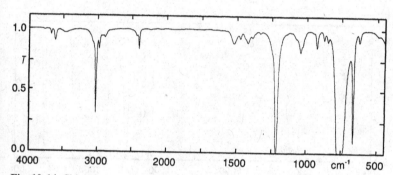

Fig. 10-14. Chloroform (Uvasol, 99% pure (GC)).

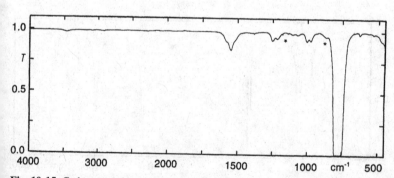

Fig. 10-15. Carbon tetrachloride (at least 99.7% pure;
* contaminant bands).

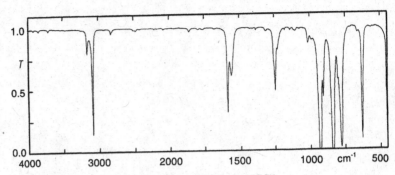

Fig. 10-16. Trichloroethylene (at least 99.5% pure (GC)).

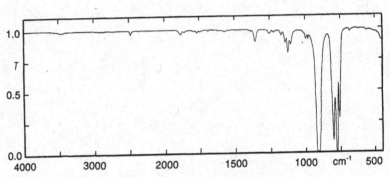

Fig. 10-17. Tetrachloroethylene (Uvasol, at least 99.7% pure).

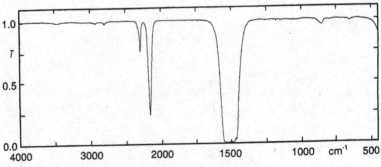

Fig. 10-18. Carbon disulfide (Uvasol, at least 99.9% pure).

Index